Travel Discount Coupon

This coupon entitles you to special discounts
when you book your trip through the

☀TRAVEL NETWORK®
RESERVATION SERVICE

Hotels ♦ Airlines ♦ Car Rentals ♦ Cruises
All Your Travel Needs

Here's what you get: *

♦ A discount of $50 USD on a booking of $1,000** or
more for two or more people!

♦ A discount of $25 USD on a booking of $500** or more
for one person!

♦ Free membership for three years, and 1,000 free miles
on enrollment in the unique Travel Network Miles-to-
Go® frequent-traveler program. Earn one mile for
every dollar spent through the program. Redeem
miles for free hotel stays starting at 5,000 miles. Earn
free roundtrip airline tickets starting at 25,000 miles.

♦ Personal help in planning your own, customized trip.

♦ Fast, confirmed reservations at any property
recommended in this guide, subject to availability.***

♦ Special discounts on bookings in the U.S. and around
the world.

♦ Low-cost visa and passport service.

♦ Reduced-rate cruise packages and special car rental
programs worldwide.

Visit our website at   nmer
or call us globally at   :all
toll-free at 1-888-94   ıda,
call at 1-905-707-72
60-3-7191044, or fax

D1456989

* To qualify for these travel discounts, at least a portion of your trip must
include destinations covered in this guide. No more than one coupon discount
may be used in any 12-month period, for destinations covered in this guide.
Cannot be combined with any other discount or promotion.

**These are U.S. dollars spent on commissionable bookings.

***A $10 USD fee, plus fax and/or phone charges, will be added to the cost of
bookings at each hotel not linked to the reservation service. Customers must
approve these fees in advance. If only hotels of this kind are booked, the traveler(s)
must also purchase roundtrip air tickets from Travel Network for the trip.

Valid until December 31, 1999. Terms and conditions of the Miles-to-
Go® program are available on request by calling 201-567-8500, ext 55.

TAO234

Frommer's® 98

Santa Fe, Taos & Albuquerque

by Lesley S. King

Macmillan • USA

ABOUT THE AUTHOR

Lesley S. King grew up on a ranch in Northern New Mexico, where she still returns on weekends to help work cattle. She's a freelance writer and photographer and a frequent contributor to *New Mexico Magazine*—as well as an avid kayaker and skier. Formerly the managing editor for *The Santa Fean,* she has written about food and restaurants for *The New York Times,* the Anasazi culture for United Airline's *Hemispheres* magazine, and ranches for *American Cowboy.* This is her first travel guidebook for Frommer's.

MACMILLAN TRAVEL

A Simon & Schuster Macmillan Company
1633 Broadway
New York, NY 10019

Find us online at **www.frommers.com**
or on America Online at Keyword: **Frommers**.

ISBN 0-02-861782-7
ISSN 0899-2789

Editor: Neil E. Schlecht
Production Editor: Whitney K. Ward
Design by Michele Laseau
Digital Cartography by Raffaele Degennaro, Roberta Stockwell, and Ortelius Design
Production Team: Eric Brinkman, David Faust, Heather Pope, and Karen Teo

SPECIAL SALES

Contents

List of Maps

AN INVITATION TO THE READER

In researching this book, we discovered many wonderful places—hotels, restaurants, shops, and more. We're sure you'll find others. Please tell us about them, so we can share the information with your fellow travelers in upcoming editions. If you were disappointed with a recommendation, we want to know that, too. Please write to:

Frommer's Santa Fe, Taos & Albuquerque '98
Macmillan Travel
1633 Broadway
New York, NY 10019

AN ADDITIONAL NOTE

Please be advised that travel information is subject to change at any time—and this is especially true of prices. We therefore suggest that you write or call ahead for confirmation when making your travel plans. The author, editors, and publisher cannot be held responsible for the experiences of readers while traveling. Your safety is important to us, however, so we encourage you to stay alert and be aware of your surroundings. Keep a close eye on cameras, purses, and wallets, all favorite targets of thieves and pickpockets.

WHAT THE SYMBOLS MEAN

✪ Frommer's Favorites

Our favorite places and experiences—outstanding for quality, value, or both.

The following abbreviations are used for credit cards:

AE	American Express	EU	Eurocard
CB	Carte Blanche	JCB	Japan Credit Bank
DC	Diners Club	MC	MasterCard
DISC	Discover	V	Visa

FIND FROMMER'S ONLINE

Arthur Frommer's Outspoken Encyclopedia of Travel (www.frommers.com) offers more than 6,000 pages of up-to-the-minute travel information—including the latest bargains and candid, personal articles updated daily by Arthur Frommer himself. No other Web site offers such comprehensive and timely coverage of the world of travel.

Introducing Northern New Mexico

This land, once the site of live volcanoes and cataclysmic ground shifts, has a tumultuous character that not only marked the past but continues to inform the present. Northern New Mexico witnessed the epic clash of Spanish, Native American, and Anglo cultures; today, disparate but overlapping identities continue to negotiate for space. But, just as geological mutations gave rise to a desert and mountain landscape that is by turns austere and lushly beautiful, cultural conflicts have produced immeasurable richness and hard scars. Today, it is a land of immense cultural diversity, creativity, appreciation for the outdoors, and a place where people very much pursue their own paths.

The center of the region is Santa Fe, a hip, artsy city that wears its 400-year-old mores on its sleeve. Nestled on the side of the **Sangre de Cristo Mountains,** it's an adobe showcase of century-old buildings that hug the earth. Many of these are artist studios and galleries set on narrow streets, ideal for desultory browsing. And then there's upstart Taos, the little arts town and ski center of just 5,000 people that lies wedged between the 13,000-foot Sangre de Cristo Mountains and the 700-foot-deep **Rio Grande Gorge.** Taos is a place of extreme temperatures and temperaments. Winter snows here bring light powder, excellent for skiing; spring's warmth fattens the rivers with runoff, allowing for terrific rafting and kayaking; and summer and fall are full of sun, great for a variety of outdoor activities. *Taoseños,* the locals, eschew regular work schedules—some businesses even shut down on good ski days. Albuquerque is the big city, New Mexico style, where people from all over the state come to trade. You'll see cowboys and Native Americans with pickups loaded with everything from saddles to swing sets to solar windmills, heading in all directions.

Not far from these three cities are the 19 settlements and numerous ruins of the Native American Pueblo culture, an incredible testament to the resilience of a proud people. And through it all weave the **Manzano, Sandia, Sangre de Cristo,** and **Jemez mountains,** multimillion-year-old reminders of man's recent arrival to this vast and unique landscape.

Northern New Mexico

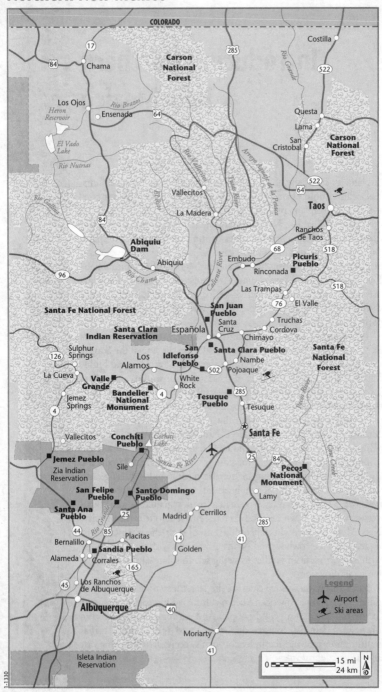

1 Frommer's Favorite Northern New Mexico Experiences

- **High Road to Taos.** This spectacular 80-mile route into the mountains between Santa Fe and Taos takes you through red painted desert, villages bordered by apple and peach orchards, and the foothills of 13,000-foot peaks. You can stop in Cordova, known for its wood-carvers, or Chimayo, known for its weavers. At the fabled Santuario de Chimayo, you can rub healing dust between your fingers. You'll also pass through Truchas, a village stronghold on a mountain mesa 1¹/₂ miles above sea level.

- **Pueblo Dances.** These native dances, related to the changing cycles of the earth, offer a unique chance to see how an indigenous culture worships and rejoices. The winter dances (around Christmastime) focus mainly on animals such as the deer and turtle. Taos is noted for the deer dance, during which men wearing antlers and hides come down from the hills at dawn. The Matachine dances, which depict the Spanish conquest of the Indians, are performed by both Hispanic and Pueblo communities and are also worth experiencing. San Juan Pueblo usually does this dance on Christmas morning. In late August or early September, San Ildefonso Pueblo performs the corn dance. Throughout the year other pueblos celebrate the feast days of their particular saints, so you're likely to catch a dance at any time of year—all in the mystical light of the Northern New Mexico sun.

- **Santa Fe Opera.** One of the finest opera companies in the United States—thought by some to rival New York City's Metropolitan—has called Santa Fe home for 40 years. Performances are held during the summer months in a hilltop, open-air amphitheater. You'll see well-known classical works such as Verde's *La Traviata,* as well as world premieres—Peter Lieberson's *Ashoka's Dream* was performed in 1997. If you're lucky, you may see lightening bolts strike in the distance during the performance.

- **Taos Ski Valley.** World renowned for its difficult runs (such as Al's) and the ridge where skiers hike for up to 2 hours to ski fresh powder, Taos has long been a pilgrimage site for extreme skiers. Today this southwestern play spot with a Bavarian accent is no longer a well-guarded secret. Over the years, the ski area has opened up new bowls to accommodate intermediate and beginners too (though many beginners I've spoken to seem daunted by the mountain's verticality).

- **Museum of International Folk Art.** Santa Fe's perpetually expanding collection of folk art is the largest in the world; it contains 130,000 objects from more than 100 countries. You'll find an amazing array of imaginative works, ranging from Hispanic folk art *santos* (carved saints) to Indonesian textiles and African sculptures.

- **Chaco Canyon.** Though it's a long drive west of Santa Fe, it's worth a visit to these spectacular ancient ruins set in a beautiful canyon—a site believed to be the center of Anasazi culture. An ancient people best known for its pottery making, the Anasazi lived in the area from around A.D. 850 to around 1200. Remains of 800 rooms exist there, many built on top of each other in structures that once towered four and five stories high. Most spectacular is Pueblo Bonito, with its giant *kivas* (underground chambers used by the men especially for ceremonies or councils); the structure was once a religious gathering place for Indians all over the area.

- **Sandia Peak Tramway.** The world's longest tramway ferries passengers 2.7 miles, from Albuquerque's city limits to the summit of the 10,378-foot Sandia Peak. On the way you'll likely see rare Rocky Mountain bighorn sheep and birds of prey. In the summer, you may see hang gliders taking off from the giant precipice to soar in the updrafts that sweep up the mountain. Go in the evening to watch the sun burn its way out of the western sky; then enjoy the glimmering city lights on your way down.

- **Kodak Albuquerque International Balloon Festival.** The world's largest balloon rally assembles more than 800 colorful balloons and includes races and contests. Highlights are the mass ascension at sunrise and the "special shapes rodeo," in which balloons in all sorts of whimsical forms, from liquor bottles to milk cows, rise into the sky. As befits the sponsor, the festival is a photographer's delight. During the "balloon glow," hundreds of balloons take flight after dark.

- **Bandelier National Monument.** Along with Puye Cliff Dwellings, these ruins provide a spectacular peek into the lives of the Anasazi Pueblo culture, which flourished in the area between A.D. 1100 and 1550. Less than 15 miles south of Los Alamos, the ruins spread across a peaceful canyon. You'll probably see deer and rabbits as you make your way through the canyon to the most dramatic site, a *kiva* and dwelling in a cave 140 feet above the canyon floor—reached by a climb up long pueblo-style ladders.

- **Northern New Mexican Enchilada.** The food here is unique, reflecting the rich cultural heritage of the region. And few things are more "New Mexican" than the enchilada. You can order red or green chile, or "Christmas"—half and half. Sauces are rich, seasoned with *ajo* (garlic) and oregano. Unlike Mexican food in other Pueblo locales, New Mexican cuisine isn't smothered in cheese and sour cream, so the flavors of the chiles, corn, and meats can really be savored. Enchiladas are often served with *frijoles* (beans), firmer than the refried bean paste you might be used to getting in Tex-Mex restaurants. They're also served with *posole* (hominy) and *sopaipillas* (fried bread), both culinary elements derived from local native cultures.

- **The Galleries Along Canyon Road.** Originally a Pueblo Indian route over the mountains and later an artists' community, Santa Fe's Canyon Road is now gallery central—and arts capital of the Southwest. The narrow one-way street is lined with more than 100 galleries, in addition to restaurants and private residences. You can visit the elaborate gardens of El Zaguán, where native plants mingle with roses and irises, or the quaint shops of Gypsy Alley, where buildings date back to 1753. Artwork on display (and for sale) ranges from the beautiful to bizarre. You can step into artists' simple studio-galleries as well as refined galleries showing world-renowned artists' works, such as paintings by Georgia O'Keeffe and sculpture by Frederic Remington. Be sure to stop for lunch at one of the streetside cafes.

- **Pecos National Historical Park.** It's hard to rank New Mexico's many ruins, but this one, sprawled on a plain about 25 miles east of Santa Fe, is one of the most impressive, resonating with the history of the Pueblo Revolt of 1680, the only successful revolt of indigenous people in the New World. You'll see evidence of where the Pecos people burned the mission church before joining in the attack on Santa Fe. You'll also see where the Spanish conquistadors later compromised, allowing sacred *kivas* to be built next to the reconstructed mission.

- **Rio Grande Gorge.** A hike into this dramatic gorge is unforgettable. You'll first see it as you come over a rise heading toward Taos, a colossal slice in the earth formed during the late Cretaceous period 130 million years ago and the early Tertiary period, about 70 million years ago. Drive about 35 miles north of Taos, near the village of Cerro, to the Rio Grande Wild River Area. From the lip of the canyon you descend through millions of years of geologic history and land inhabited by Indians since 16,000 B.C. You're liable to encounter raccoons and other wildlife, and once you reach the river you'll see fishermen, rafters, and kayakers. There you can dip your toes in the fabled *Rio.*

- **Maria Benitez Teatro Flamenco.** Flamenco dancing originated in Spain, strongly influenced by the Moors; it is a cultural expression held sacred by Spanish Gypsies. The passionate dance is characterized by intricate toe and heel clicking, sinuous arm and hand gestures, *cante hondo* or "deep song," and expressive guitar solos. A native New Mexican, María Benitez was trained in Spain, a country to which she returns each year to find dancers and prepare her show. Through her fluid movement and aggressive solos, she brings a contemporary voice to this ancient art. This world-class dancer and her troupe perform at the Radisson Hotel in Santa Fe from July through early September.

- **Old Town.** Albuquerque's commercial center until about 1880, Old Town still gives a remarkable sense of what life was once like in a southwestern village. You can meander down crooked streets and narrow alleys and rest on the cottonwood-shaded Plaza. Though many of the shops are now very touristy, you can still happen upon some interesting shopping and dining finds here. Native Americans sell jewelry, pottery, and weavings under a portal on the plaza.

2 Northern New Mexico Today

On rock faces throughout northern New Mexico, you'll find circular symbols carved in sandstone, the wavy mark of Avanu the river serpent, or the ubiquitous Kokopelli playing his magic flute. These petroglyphs are constant reminders of the enigmatic history of the Anasazi, the Indians that inhabited this area from A.D. 800 until as late as 1600. Part of my fascination with this land is the mysterious presence of these ancient people even today. Excavations continue in the area; the other day, right on a highway shoulder between Santa Fe and Española, I saw archaeologists excavating two sites, brushing away dirt to expose kivas and ancient walls.

The Spanish conquistadors, in their inimitable fashion, imposed a new, foreign order upon these people and their land. They brought with them a rich culture bolstered by a fervent belief in spreading Catholicism. As an inevitable component of conquest, they changed most Native American names—today you'll find a number of Native Americans with Hispanic names—and renamed the villages "pueblos." The Spaniards' most far-reaching legacy, however, was the forceful conversion of Indian populations to Catholicism, a religion that many Indians still practice today. In each of the pueblos you'll see a large, often beautiful Catholic church, usually made with sculpted adobe. The churches, set against the ancient adobe dwellings, are symbolic of the melding of two cultures. During the holiday seasons here you'll see Pueblo people perform their ritual dances outside their local Catholic church.

The mix of cultures is today very much apparent in Northern New Mexican cuisine. When the Spaniards came to the New World, they brought cows and sheep. They quickly learned to appreciate the indigenous foods here, most notably corn, beans, squash, and chiles. Look also for such Pueblo dishes as the thin-layered blue piki bread, or chauquehue, a thick corn pudding similar to polenta.

GROWING PAINS Northern New Mexico is experiencing a reconquest of sorts, as the Anglo population soars, and outside money and values again make their way in. The process, called "encroachment" by locals, continues to transform New Mexico's three distinct cultures and their unique ways of life, albeit in a less violent manner than during the Spanish conquest.

Danse Macabre

The Dance of the Matachines, a ritualistic dance performed at Northern New Mexico pueblos and in many Hispanic communities, can be seen as a metaphor for the tribulations and richness of this land. It reveals the cultural miscegenation, identities, and conflicts that characterize Northern New Mexico. It's a dark and vivid ritual in which a little girl, Malinche, is wedded to the church. The dance, depicting the taming of the native spirit, is difficult even for historians to decipher.

Brought to the New World by the Spaniards, the dance has its roots in the painful period during which the Moors were driven out of Spain. However, some symbols seem obvious: At one point, men bearing whips tame "El Toro," a small boy dressed as a bull who has been charging about rebelliously. The whip-men symbolically castrate him and then stroll through the crowd displaying the dismembered body, as if to warn villagers of the consequences of disobedience. At another point, a hunched woman-figure births a small troll-like doll, perhaps representative of the union between Indian and Hispanic cultures.

The Dance of the Matachines ends when two *abuelo* (grandparent) figures dance across the dirt holding up the just-born baby, while the Matachines, adorned with bishoplike headdresses, follow them away in a recessional march. The Matachines' dance, often performed in the early mornings, is so dark and mystical that every time I see it my passion for this area deepens. The image of that baby always stays with me, and in a way represents New Mexico itself: a place born of disparate beliefs that have melded with the sand, sage, and sun and produced incredible richness.

Certainly, the Anglos—many of them from large cities—add a cosmopolitan flavor to life here. The variety of restaurants has greatly improved, as have entertainment options. For their small size, towns such as Taos and Santa Fe offer a broad variety of restaurants and cultural events. Santa Fe has developed a strong dance and drama scene, with treats such as flamenco and opera that you'd expect to find in New York or Los Angeles. And Albuquerque has an exciting nightlife scene downtown; you can walk from club to club and hear a wealth of jazz, rock, country, and alternative music.

Yet many newcomers, attracted by the adobe houses and exotic feel of the place, often bring only a loose appreciation for the area. Some tend to romanticize the lifestyle of the other cultures and trivialize their beliefs. Native American symbology, for example, is employed in ever-popular Southwestern decorative motifs; New Age groups appropriate valued rituals, such as sweats (in which believers sit encamped in a very hot, enclosed space to cleanse their spirits). The effects of cultural and economic change are even apparent throughout the countryside, where land is being developed at an alarming rate, often as lots for new million-dollar homes.

Transformation of the local way of life and landscape is also apparent in the stores continually springing up in the area. For some of us, these are a welcome relief from western clothing stores and provincial dress shops. The down side is that city plazas, which once contained pharmacies and grocery stores frequented by residents, are now crowded with T-shirt shops and galleries appealing to tourists. Many locals in these cities now rarely visit their plazas except during special events such as fiestas.

Environmental threats are another regional reality. Nuclear waste issues form part of an ongoing conflict affecting the entire Southwest, and a section of southern New Mexico has been designated a nuclear waste site. Much of the waste would pass through Santa Fe; the problem necessitated construction of a bypass that will, when completed, direct a great deal of transit traffic around the west side of the city.

Still, new ways of thinking have also brought positive changes to the life here, and many locals have directly benefitted from New Mexico's expansion, influx of wealthy newcomers, and popularity as a tourist destination. Businesses and industries, large and small, have come to the area. In Albuquerque, Intel Corporation now employs over 5,000 workers, and *Outside* magazine recently relocated from Chicago to Santa Fe. Local artists and artisans also benefit from growth. Many craftspeople—furniture makers, tin workers, and weavers—have expanded their businesses. The influx of people has broadened the sensibility of a fairly provincial state. The area has become a refuge for many gays and lesbians, as well as for political exiles, such as Tibetans. With them has developed a level of creativity and tolerance you would generally find in very large cities but not in smaller communities such as these.

CULTURAL QUESTIONS   Faced with new challenges to their ways of life, both Native Americans and Hispanics are marshaling forces to protect their cultural identities. A prime concern is language. Through the years, many Pueblo people have begun to speak more and more English, with their children getting little exposure to their native tongue. In a number of the pueblos, elders are working with school-children in language classes. Some of the pueblos, such as Santa Clara, have even developed written dictionaries, the first time their languages have been presented in this form.

Many pueblos have introduced programs to conserve the environment, preserve ancient seed strains, and protect religious rites. Since their religion is tied closely to nature, a loss of their natural resources would threaten the entire culture. Certain rituals have been closed off to outsiders, the most notable being the Shalako at Zuni, a popular and elaborate series of year-end ceremonies.

Hispanics, through art and observance of cultural traditions, are also embracing their roots. In Northern New Mexico you'll see, adorning many walls, murals that depict important historical events, such as the Treaty of Guadalupe Hidalgo of 1848. The **Spanish Market** in Santa Fe has expanded into a grand celebration of traditional arts—from tin working to santo carving. Public schools in the area have bilingual education programs in place, allowing young people to embrace their Spanish-speaking roots.

Hispanics are also making their voices heard, insisting on more conscientious development of their neighborhoods and rising to positions of power in government. Santa Fe mayor Debbie Jaramillo has made national news as an advocate of the Hispanic people, and Congressman Bill Richardson, Hispanic despite his Anglo surname, was recently appointed U.S. ambassador to the United Nations.

GAMBLING WINS AND LOSSES   Gambling, a fact of life and source of much-needed revenues for Native American populations across the country, is at the center of encroachment-related issues. In 1994, Governor Gary Johnson signed a compact with tribes in New Mexico, ratified by the U.S. Department of the Interior, to allow full-scale gambling. Tesuque Pueblo was one of the first to begin a massive expansion, and many other pueblos followed suit.

In early 1996, however, the State Supreme Court ruled that without legislative action on the matter, the casinos were operating illegally. This was their status through early 1997. Native Americans have remained resolute in their determination to forge ahead with the casinos. Demonstrations of community strength, by the Pojoaque Pueblo and others, ultimately made an impact on legislators. In 1997, law-makers agreed to allow gaming by Indian as well as fraternal and veteran organizations.

Many New Mexicans are concerned about the tone gambling sets in the state. The casinos are for the most part large and unsightly, neon-bedecked buildings that stand out sorely on some of New Mexico's most picturesque land. Though most residents appreciate the boost that gambling can ultimately bring to the Native American economies, many critics wonder where gambling profits actually go—and if the casinos can possibly be a good thing for the pueblos and tribes. Some detractors suspect that profits go directly into the pockets of outside backers.

A number of pueblos and tribes, however, are showing signs of prosperity, and they are using newfound revenues to buy firefighting and medical equipment and to invest in local schools. According to the Indian Gaming Association, casinos directly employ over 4,000 workers and pump $262 million in revenues into the state's economy. Isleta Pueblo recently completed a $3.6 million youth center, which their lieutenant governor says came from gambling revenues. Sandia Pueblo just built a $2 million medical and dental clinic and expanded its police department. Their Governor Alex Lujan calls these projects "totally funded by gaming revenues."

SANTA FE   This is where the splendor of diverse cultures really shines, and it does so in a setting that's unsurpassed. There's a magic in Santa Fe that's difficult to explain, but you'll sense it when you glimpse an old adobe building set against blue mountains and giant billowing thunderheads or when you hear a ranchero song come from a low-rider's radio and you smell chicken and chile grilling at a roadside vending booth. Although it's quickening, the pace of life here is still a few steps slower than the rest of the country. We use the word *mañana* to describe the pace—which doesn't mean "tomorrow" exactly, it just means "not today." There's also a level of creativity here that you'll find in few other places in the world. Artists who have fled big-city jobs are here to follow their passions, as are locals who grew up making crafts and continue to do so. Conversations often center around how to structure one's day so as to take advantage of the incredible outdoors while still making enough money to survive.

Meanwhile, Santa Fe's precipitous growth and enduring popularity with tourists have been the source of conflict and squabbling. Outsiders have bought up land in the hills around the city, building housing developments and sprawling single-family homes. The hills that local populations claimed for centuries as their own are being overrun, while property taxes for all skyrocket. On the positive side, however, local outcry has prompted the city to implement zoning restrictions on where and how development can proceed. Some of the restrictions include banning building on ridgetops and on steep slopes and limiting the size of homes built.

Santa Fe's mayor of the past 4 years, Debbie Jaramillo, draws attention as one of the region's most outspoken opponents of encroachment. A fiery native of Santa Fe, she came into office as a representative of *la gente* (the people), and, contrary to most public officials, set about discouraging tourism and rapid growth. She took a lot of heat for her positions, which to some seemed xenophobic and antibusiness. The mayor initially failed to offer tax breaks or incentive money to some businesses interested in settling in the area, but has since softened her position on development.

TAOS   A funky town in the middle of a beautiful, sage-covered valley, Taos is full of narrow streets dotted with galleries and artisan shops. You might find an artist's studio tucked into a century-old church or a small furniture maker working at the back of his own shop.

More than any other major Northern New Mexico community, Taos has success-fully opposed much of the heavy development slated for the area. In 1987 locals vociferously protested plans to expand the airport so that commercial airlines could fly in; plans have been stalled indefinitely pending environmental impact statements. In 1991 a $40 million golf course and housing development slated for the area was met with such community dissension that its developers eventually desisted. It's hard to say where Taos gets its rebellious strength; the roots may lie in the hippie community that settled here in the '60s, or possibly the Pueblo community around which the city formed. After all, Taos Pueblo was at the center of the 17th-century Pueblo Revolt.

Still, changes are upon Taoseños. The "blinking light" that for years residents used as a reference point has given way to a real traffic light. You'll also see the main route through town becoming more and more like Cerrillos Road in Santa Fe, as fast-food restaurants and service businesses set up shop. Though the town is working to stream-line alternate routes to channel through-traffic around downtown, there's no feasible way of widening the main drag because the street—which started out as a wagon trail—is bordered closely by historic buildings.

ALBUQUERQUE   The largest city in New Mexico, Albuquerque has born the brunt of the state's most massive growth. Currently, the city sprawls over 16 miles, from the lava-crested mesas on the west side of the Rio Grande to the steep alluvial slopes of the Sandia Mountains on the east, and north and south through the Rio Grande Valley. New subdivisions sprout up constantly.

Despite the growth, this town is most prized by New Mexicans for its genuine-ness. You'll find none of the self-conscious artsy atmosphere of Santa Fe here. Instead, there's a traditional New Mexico feel that's evident when you spend some time in the heart of the city. It centers around downtown, a place of shiny skyscrapers built around the original Route 66, which still maintains some of its 1950s charm.

The most emblematic growth problem concerns the **Petroglyph National Monument** on the west side. The area is characterized by five extinct volcanoes. Adjacent lava flows became a hunting and gathering place for prehistoric Native Americans who left a chronicle of their beliefs etched in the dark basalt boulders. Some 15,000 petroglyphs have been found in this archaeological preserve. Now there's a push to carve out a highway corridor right through the center of the monument. Opponents have fought the extension for nearly a decade. Some Native American groups, likening the highway to building a road through a church, oppose the extension on grounds that the petroglyphs are sacred to their culture. U.S. Interior Secretary Bruce Babbitt has refused to give permission for the road to go through. Senator Pete Domenici, however, has introduced a bill into the Senate that would allow the road to be built. The bill must be approved by Congress before construction can begin.

Northern New Mexico's extreme popularity as a tourist destination—it was the "in" place to be through much of the '80s—has dropped off. Though many artists and other businesspeople lament the loss of the crowds, most people are glad that the wave has subsided. It's good news for travelers too, who no longer have to compete so heavily for restaurant seats or space when hiking through ruins. Though parts of Northern New Mexico have lost some of the unique charm that attracted so many to the area, the overall feeling is still one of mystery and a cultural depth unmatched in the world. People here recognize the need to preserve the land and what New Mexicans have traditionally valued as integral to their unique lifestyle.

3   History 101

The Pueblo tribes of the upper Rio Grande Valley are believed to be descendants of the Anasazi, who from the 9th to the 13th centuries lived in the Four Corners region—where the states of New Mexico, Arizona, Colorado, and Utah now meet.

The Anasazi built spectacular structures; you get an idea of their scale and intricacy at the ruins at **Chaco Canyon** and **Mesa Verde.** It isn't known exactly why the Anasazi abandoned their homes (some archaeologists suggest it was due to drought; others claim social unrest), but by the time the Spaniards arrived in the 1500s they were long gone, and the Pueblo culture was well established throughout what would become northern and western New Mexico.

Architectural style was a unifying mark of the otherwise diverse Anasazi and Pueblo cultures. Both built condominium-style communities of stone and mud adobe bricks, three and four stories high. Grouped around central plazas, the villages they constructed incorporated circular spiritual chambers called kivas. As farmers, the Anasazi and Pueblo peoples used the waters of the Rio Grande and its tributaries to irrigate fields of corn, beans, and squash. They were also the creators of elaborate works of pottery.

THE SPANISH OCCUPATION   The Spanish ventured into the upper Rio Grande after conquering Mexico's Aztecs in 1519–21. In 1540 Francisco Vásquez de Coronado led an expedition in search of the fabled Seven Cities of Cíbola, coincidentally introducing horses and sheep to the region. Neither Coronado nor a succession of fortune-seeking conquistadors could locate the legendary cities of gold, so the Spanish concentrated their efforts on exploiting the Native Americans.

Franciscan priests attempted to turn the Pueblo people into model peasants. Their churches became the focal points of every pueblo, with Catholic schools an essential adjunct. By 1625 there were approximately 50 churches in the Rio Grande Valley. (Two of the Pueblo missions, at Isleta and Acoma, are still in use today.) The Pueblo, however, weren't enthused about doing "God's work" for the Spanish—building new adobe missions, tilling fields, and weaving garments for export to Mexico—so soldiers came north to back the padres in extracting labor. In effect, the Pueblo people were forced into slavery.

Santa Fe was founded in 1610 as the seat of Spanish government in the upper Rio Grande. Governor Don Pedro de Peralta named the settlement La Villa Real de la Santa Fe de San Francisco de Asis (The Royal City of the Holy Faith of St. Francis of Assisi). The **Palace of Governors** has been used continuously as a public building ever since—by the Spanish, Mexicans, Americans, Confederate troops (briefly), and Pueblos (1680–92). Today it stands as the flagship of the state museum system.

Decades of resentment against the Spanish colonials culminated in the Pueblo occupation. Uprisings in the 1630s at Taos and Jemez left village priests dead and triggered savage repression. In 1680 a unified Pueblo rebellion, orchestrated from Taos, succeeded in driving the Spaniards from the upper Rio Grande. The leaders of the revolt defiled or destroyed the churches, just as the Spanish had destroyed their religious symbols. They took the Palace of the Governors, where they burned archives and prayer books, and converted the chapel into a kiva. They also burned much of the property in Santa Fe that had been built by the Europeans and laid siege to Spanish settlements up and down the Rio Grande Valley. Forced to retreat to Mexico, the colonists were not able to retake Santa Fe until 12 years later. Bloody battles raged for the next several years, but by the beginning of the 18th century *Nuevo Mexico* was firmly in Spanish hands.

It remained so until Mexico gained its independence from Spain in 1821. The most notable event in the intervening years was the mid-1700s departure of the Franciscans, exasperated by their failure to wipe out all vestiges of traditional Pueblo religion. Throughout the Spanish occupation, eight generations of Pueblos had clung tenaciously to their way of life. However, by the 1750s the number of Pueblo villages had shrunk by half.

ARRIVAL OF THE ANGLOS   The first Anglos to spend time in the upper Rio Grande Valley were mountain men: itinerant hunters, trappers, and traders. Trailblazers of the U.S. westward expansion, they began settling in New Mexico in the first decade of the 19th century. Many married into Pueblo or Hispanic families. Perhaps the best known was Kit Carson, a sometime federal agent, sometime Native American scout, whose legend is inextricably interwoven with that of early Taos. Though he seldom stayed in one place for long, he considered the Taos area his home. He married Josepha Jaramillo, the daughter of a leading Taos citizen. Later, as an Indian agent based in Taos, he became a prime force in the final subjugation of the Plains Indians. The Taos home where he lived off and on for 40 years, until his death in 1868, is now a museum.

Wagon trains and eastern merchants followed Carson and the other early settlers. Santa Fe, Taos, and Albuquerque, already major trading and commercial centers at the end of the Chihuahua Trail (the Camino Real from Veracruz, Mexico, 1,000 miles south), became the western termini of the new **Santa Fe Trail** (from Independence, Missouri, 800 miles east).

Even though independent Mexico granted the Pueblo people full citizenship and abandoned restrictive trade laws instituted by their former Spanish rulers, the subsequent 25 years of direct rule from Mexico City were not peaceful in the upper Rio Grande. Instead, they were marked by ongoing rebellion against severe taxation, especially in Taos. Neither did things quiet down when the United States assumed control of the territory during the Mexican War. Shortly after General Stephen Kearney occupied Santa Fe (in a bloodless takeover) on orders of President James Polk in 1846, a revolt in Taos in 1847 led to the slaying of the new governor of New Mexico, Charles Bent. In 1848 the Treaty of Guadalupe Hidalgo officially transferred title to New Mexico, along with Texas, Arizona, and California, to the United States.

Aside from Kit Carson, perhaps the two most notable personalities of 19th-century New Mexico were priests. Father José Martinez (1793–1867) was one of the first native-born priests to serve his people. Ordained in Durango, Mexico, he jolted the Catholic Church after assuming control of the Taos parish: Martinez abolished the obligatory church tithe because it was a hardship on poor parishioners, published the first newspaper in the territory (in 1835), and fought large land acquisitions by Anglos after the United States annexed the territory.

On all these issues Martinez was at loggerheads with Bishop Jean-Baptiste Lamy (1814–88), a Frenchman appointed in 1851 to supervise the affairs of the first independent New Mexican diocese. Lamy, on whose life Willa Cather based her novel *Death Comes for the Archbishop,* served the diocese for 37 years. Lamy didn't take kindly to Martinez's independent streak and, after repeated conflicts, excommunicated the maverick priest in 1857. But Martinez was steadfast in his preaching. He established an independent church and continued as Northern New Mexico's spiritual leader until his death.

Nevertheless, Lamy made many positive contributions to New Mexico, especially in the fields of education and architecture. Santa Fe's Romanesque **Cathedral of St. Francis** and the nearby Gothic-style **Loretto Chapel,** for instance, were constructed under his aegis. But he was adamant about adhering to strict Catholic religious tenets. Martinez, on the other hand, embraced the folk tradition, including the craft of *santero* (religious icon) carving and a tolerance of the Penitentes, a flagellant sect that flourished after the departure of the Franciscans in the mid–18th century.

With the advent of the **Atchison, Topeka & Santa Fe Railway** in 1879, New Mexico began to boom. Albuquerque in particular blossomed in the wake of a series of major gold strikes in the Madrid valley, close to ancient Native American turquoise mines. By the time the gold lodes began to shrink in the 1890s, cattle and sheep ranching had become well entrenched. The territory's growth culminated in statehood in 1912.

Territorial governor Lew Wallace, who served from 1878 to 1881, was instrumental in promoting interest in the arts, which today flourish in Northern New Mexico. While occupying the Palace of the Governors, Wallace penned the great biblical novel *Ben Hur.* In the 1890s, Ernest Blumenschein, Bert Phillips, and Joseph Sharp launched the Taos art colony; it boomed in the decade following World War I when Mabel Dodge Luhan, D. H. Lawrence, Georgia O'Keeffe, Willa Cather, and many others visited or established residence in the area.

During World War II, the federal government purchased an isolated boys' camp west of Santa Fe and turned it into the **Los Alamos National Laboratory,** where the Manhattan Project and other top-secret atomic experiments were developed and perfected. The science and military legacy continues today; Albuquerque is among the nation's leaders in attracting defense contracts and high technology.

4 Land of Art

It's all in the light—or at least that's what many artists claim drew them to Northern New Mexico. In truth, the light is only part of the attraction: Nature in this part of the country, with its awe-inspiring thunderheads, endless expanse of blue skies, and rugged desert, is itself a canvas. To record the wonders of earth and sky, the early natives of the area, the Anasazi, made petroglyphs and pictographs on the sides of caves and on stones, as well as on the sides of pots they shaped from clay dug in the hills.

Today's Native American tribes carry on that legacy, as do the other cultures that have settled here. Life in Northern New Mexico is shaped by the arts. Everywhere you turn you'll see pottery, paintings, jewelry, and weavings. You're liable to meet an artist whether you're having coffee in a Taos cafe or walking along Canyon Road in Santa Fe.

The area is full of little villages that maintain their own artistic specialties. Each Indian pueblo has a trademark design, such as **Santa Clara's** and **San Ildefonso's** black pottery and **Zuni's** needlepoint silverwork. Bear in mind that the images used often have deep symbolic meaning. When purchasing art or an artifact, you may want to talk to its maker about what the symbols mean.

Hispanic villages are also distinguished by their artistic identities. **Chimayo** has become a center for Hispanic weaving, while the village of **Cordova** is known for its *santo* (icon) carving. Santos, *retablos* (paintings), and *bultos* (sculptures), as well as works in tin, are traditional devotional arts tied to the Roman Catholic faith. Often these works are sold out of artists' homes in these villages, allowing you to glimpse the lives of the artists and the surroundings that inspire them.

Villagers, whether Hispanic or Native American, take their goods to the cities, where for centuries people have bought and traded. Under the portals along the plazas of Santa Fe, Taos, and Albuquerque, you'll find a variety of works in silver, stone, and pottery for sale. In the cities you'll find streets lined with galleries, some very slick, some more modest. At major markets, such as the **Spanish Market** and **Indian Market** in Santa Fe, some of the top artists from the area sell their works. Smaller shows at the pueblos also attract artists and artisans. The **Northern Pueblo Artists and Craftsman show** at Santa Clara Pueblo in July continues to grow each year.

Drawn by the beauty of the local landscape and respect for indigenous art, artists from all over have flocked here, particularly during the 20th century. And they have established locally important art societies; one of the most notable is the **Taos Society of Artists.** An oft-repeated tale explains the roots of this society. The artists Bert Phillips and Ernest L. Blumenschein were traveling through the area from Colorado on a mission to sketch the Southwest when their wagon broke down north of Taos. The scenery so overwhelmed them that they abandoned their journey and stayed. Joseph Sharp joined them, and still later came Oscar Berninghaus, Walter Ufner, Herbert Dunton, and others. You can see a brilliant collection of some of their romantically lit portraits and landscapes at the Van Vechten Lineberry Museum in Taos. The 100th anniversary marking the artists' "broken wheel" will be celebrated in 1998.

A major player in the development of Taos as an artists' community was the arts patron Mabel Dodge Luhan. Herself a writer, she financed the work of many an artist, and in the 1920s she held court for many notables, including Georgia O'Keeffe, Willa Cather, and D. H. Lawrence. This illustrious history goes a long way to explaining how it is that Taos—a town of little more than 8,000—has more than 90 arts and crafts galleries and more than 100 resident painters.

Santa Fe has its own art society, begun in the 1920s by a nucleus of five painters— who became known as **Los Cinco Pintores.** Jozef Bakos, Fremont Ellis, Walter Mruk, Willard Nash, and Will Shuster lived in the area of dusty Canyon Road (now the arts center of Santa Fe, with more than 1,000 artists, countless galleries, and many museums). Despite its small size, Santa Fe is, remarkably, considered the third largest art market in the United States.

Perhaps the most celebrated artist associated with Northern New Mexico was **Georgia O'Keeffe** (1887–1986), a painter who worked and lived most of her later years in the region. O'Keeffe's first sojourn to New Mexico in 1929 inspired her sensuous paintings of the area's desert landscape and bleached animal skulls. The house where she lived in Abiquiu (42 miles northwest of Santa Fe on US 84) is now open for limited public tours (see chapter 10 for details). The **Georgia O'Keeffe Museum** in Santa Fe, the only museum in the United States entirely dedicated to a woman artist, opened in Santa Fe in 1997.

Santa Fe is also home to the **Institute of American Indian Arts** and the **School of Indian Art,** where many of today's leading Native American artists have studied, including the Apache sculptor Allan Houser (whose works you can see near the State Capitol building and in other public areas in Santa Fe). The best-known Native American painter is R. C. Gorman, an Arizona Navajo who has made his home in Taos for more than two decades. Now in his late 50s, Gorman is internationally acclaimed for his bright, somewhat surrealistic depictions of Navajo women. A relative newcomer to national fame is Dan Namingha, a Hopi artist who weaves native symbology into contemporary concerns.

If you look closely, you'll find notable works from a number of local artists. There's Tammy Garcia, a young Taos potter who swept the awards at the 1996 Indian Market with her intricately shaped and carved pots. Cippy Crazyhorse, a Cochiti, has acquired a steady following of patrons of his silver jewelry. All around the area you'll see the frescoes of Frederico Vigil, a noted muralist and Santa Fe native. From the village of Santa Cruz comes a new rising star named Andres Martinez, noted for his Picasso-esque portraits of Hispanic village life.

For the visitor interested in art, however, some caution should be exercised: There's a lot of schlock out there targeting the tourist trade. Yet if you persist, you're likely to find much inspiring work as well. The museums and many of the galleries are excellent repositories of local art. Their offerings range from small-town folk art to works by major artists who show internationally.

5 Architecture: Rich Melting Pot

Northern New Mexico's distinctive architecture reflects the diversity of cultures that have left their imprint on the region. The first people in the area were the Anasazi, who built stone and mud homes at the bottom of canyons and inside caves (which look rather like condominiums to the modern urban eye). When the Spaniards came to the area, they brought adobe bricks. **Pueblo-style adobe architecture** evolved and became the basis for traditional New Mexican homes: sun-dried clay bricks mixed with grass for strength, mud-mortared, and covered with additional protective layers of mud. Roofs are supported by a network of *vigas*—long beams whose ends protrude through the outer facades—and *latillas,* smaller stripped branches layered between the vigas. Other adapted Pueblo architectural elements include plastered adobe-brick kiva fireplaces, *bancos* (adobe benches that protrude from walls), and *nichos* (small indentations within a wall in which religious icons are placed). These adobe homes are characterized by flat roofs and soft, rounded contours.

To Pueblo style the Spaniards wedded other elements, such as portals (porches held up with posts, often running the length of a home) and enclosed patios, as well as the simple, dramatic sculptural shapes of Spanish mission arches and bell towers. They also brought elements from the Moorish architecture found in southern Spain: heavy wooden doors and elaborate *corbels*—carved wooden supports for the vertical posts.

With the opening of the Santa Fe Trail in 1821 and later the 1860s gold boom, both of which brought more Anglo settlers, came the next wave of building. New arrivals contributed architectural elements such as neo-Grecian and Victorian influences popular in the middle part of the United States at the time. Distinguishing features of what came to be known as **Territorial-style** architecture can be seen today; they include brick facades and cornices as well as porches, often placed on the second story. You'll also note millwork on doors and wood trim around windows and doorways, double-hung windows, as well as Victorian bric-a-brac.

Santa Fe Plaza is an excellent example of the convergence of these early architectural styles. On the west side is a Territorial-style balcony, while the Palace of Governors is marked by Pueblo-style vigas and oversized Spanish/Moorish doors.

Nowhere else in the United States are you likely to see such extremes of architectural style as in Northern New Mexico. In Santa Fe you'll see the Romanesque architecture of the **St. Francis Cathedral** and **Loretto chapel,** brought by Archbishop Lamy from France, as well as the railroad station built in the **Spanish Mission style**—popular in the early part of this century.

Since 1957, strict state building codes have required that all new structures within the circumference of the Paseo de Peralta conform to one of two revival styles: Pueblo or Territorial. The regulation also limits the height of the buildings and restricts the types of signs permitted. It also requires buildings to be topped by flat roofs. In 1988 additional citywide standards were established in an effort to impose some degree of architectural taste on new developments.

Albuquerque also has a broad array of styles, most evident in a visit to Old Town. There you'll find the large Italianate brick house known as the **Herman Blueher home** built in 1898; throughout Old Town, you'll find little *placitas,* homes and haciendas built around a courtyard, a strategy developed not only for defense purposes but also as a way to accommodate several generations of the same family in different wings of a single dwelling. The **Church of San Felipe de Neri** at the center of Old Town is centered between two "folk Gothic" towers. This building was begun in a cruciform plan in 1793; subsequent architectural changes resulted in an interesting mixture of styles.

Most notable architecturally in Taos is the **Taos Pueblo,** the site of two structures emulated in homes and business buildings throughout the Southwest. Built to resemble Taos Mountain, which stands behind it, the two structures are pyramidal in form, with the different levels reached by ladders. Much Taos architecture echoes colonial hacienda style. What's nice about Taos is that you can see historic homes inside and out. You can wander through **Kit Carson's old home;** built in 1825, it's an excellent example of a hacienda and is filled with a fine collection of 19th-century Western Americana. Taos Society artist **Ernest Blumenschein's home** is also a museum. Built in 1797 and restored by Blumenschein in 1919, it represents another New Mexico architectural phenomenon: homes that were added onto year after year. Doorways are typically low, and floors rise and fall at the whim of the earth beneath them. The **Martinez Hacienda** is an example of a hacienda stronghold. Built without windows facing outward, it originally had 20 small rooms, many with doors opening out to the courtyard. One of the few refurbished examples of colonial New Mexico architecture and life, the hacienda is on the National Historic Registry.

As you head into villages in the north, you'll see steep pitched roofs on most homes. This is because the common flat-roof style doesn't shed snow; the water builds up and causes roof problems. In just about any town in Northern New Mexico, you may detect the strong smell of tar, a sure sign that another resident is laying out thousands to fix his enchanting but frustratingly flat roof.

Today very few new homes are built of adobe. Instead, most are constructed with wood frames and plasterboard, and then stuccoed over. Several local architects are currently employing innovative architecture to create a Pueblo-style feel. They incorporate straw bails, pumice-crete, rammed earth, old tires, even aluminum cans in the construction of homes. Most of these elements are used in the same way bricks are used, stacked and layered, then covered over with plaster and made to look like adobe. Often it's difficult to distinguish homes built with these materials from those built with wood-frame construction. West of Taos a number of "earthships" have been built. Many of these homes are constructed with alternative materials, some bermed into the sides of hills, utilizing the earth as insulation and the sun as an energy source.

A visitor could spend an entire trip to New Mexico focusing on the architecture. As well as relishing the wealth of architectural styles, you'll find more subtle elements everywhere. You may encounter an oxblood floor, for example. An old Spanish tradition, oxblood is spread in layers and left to dry, hardening into a glossy finish that's known to last centuries. You're also likely to see coyote fences—narrow cedar posts lined up side by side—a system early settlers devised to ensure safety of their animals. Winding around homes and buildings you'll see *acequias,* ancient irrigation canals still maintained by locals for watering crops and trees. Throughout the area you'll notice that old walls are whimsically bowed and windows and floors are often crooked, constant reminders of the effects time has had upon even these stalwart structures.

6 Anthropology 101: Beliefs & Rituals

Religion has always been a central, defining element in the life of the Pueblo people. Within the cosmos, which they view as a single whole, all living creatures are mutually dependent. Thus every relationship a human being may have, whether with a person, animal, or even plant, has spiritual significance. A hunter prays before killing a deer, asking the creature to sacrifice itself to the tribe. A slain deer is treated as a guest of honor, and the hunter performs a ritual in which he sends the animal's soul back to its community, so that it may be reborn. Even the harvesting of plants requires prayer, thanks, and ritual.

The Pueblo people believe that their ancestors originally lived underground— which, as the place from which plants spring, is the source of all life. According to their beliefs, the original Pueblos, encouraged by burrowing animals, entered the world of humans—the so-called "fifth" world—through a hole, a *sipapu.* The ways in which this came about and the deities that the Pueblo people revere vary from tribe to tribe. Most, however, believe this world is bounded by four sacred mountains, where four sacred colors—coral, black, turquoise, and yellow or white—predominate.

There is no single great spirit ruling over this world; instead it is watched over by a number of spiritual elements. Most common are Mother Earth and Father Sun. In this desert land, the sun is an element of both life and death. The tribes watch the skies closely, tracking solstices and planetary movements, to determine the optimal time for crop planting.

Ritualistic dances are occasions of great symbolic importance. Usually held in conjunction with the feast days of Catholic saints (including Christmas Eve), Pueblo ceremonies demonstrate the parallel absorption of Christian elements without the surrendering of traditional beliefs. To this day, communities enact medicine dances, fertility rites, and prayers for rain and for good harvests. The spring and summer corn, or *tablita,* dances are among the most impressive. Ceremonies begin with an early-morning mass and procession to the Plaza; the image of the saint is honored at the forefront. The rest of the day is devoted to song, dance, and feasting, with performers masked and clad as deer, buffalo, eagles, or other creatures.

Visitors are usually welcome to attend Pueblo dances, but they should respect the tribe's requests not to be photographed or recorded. It was exactly this lack of respect that led the Zunis to ban outsiders from attending many of their famous Shalako ceremonies.

Catholicism, imposed by the Spaniards, has infused Northern New Mexico with an elaborate set of beliefs. This is a Catholicism heavy with iconography, expressed in carved santos (statues) and beautiful retablos (paintings) that adorn the altars of many cathedrals. Catholic churches are the focal point of most Northern New Mexico villages. When you take the high road to Taos, be sure to note the church in **Las Trampas,** as well as the one in **Ranchos de Taos;** both have 3- to 4-foot-thick walls sculpted from adobe and inside have an old-world charm, with beautiful retablos decorating the walls and vigas (roof beams) holding up the roofs.

Hispanics in Northern New Mexico, in particular, maintain strong family and Catholic ties, and they continue to honor traditions associated with both. Communities plan elaborate celebrations such as the *quinciniera* for young girls reaching womanhood, weddings with big feasts, and dances in which well-wishers pin money to the bride's elaborately laced gown.

If you happen to be in the area during a holiday, you may even get to see a religious procession or pilgrimage. Most notable is the **pilgrimage to the Santuario de Chimayo,** an hour's drive north of the state capital. Constructed in 1816, the sanctuary has long been a pilgrimage site for Catholics who attribute miraculous healing powers to the earth found in the chapel's anteroom. Several days before Easter, fervent believers begin walking the highway headed north or south to Chimayo, some carrying large crosses, others carrying nothing but a small bottle of water, most praying for a miracle.

In recent years, New Mexico has become known (and in some circles, ridiculed) for **New Age pilgrims and celebrations.** The roots of the local movement are hard to trace. It may have something to do with Northern New Mexico's centuries-old reputation as a place where rebel thinkers come to enjoy the freedom to believe what they want. Pueblo spirituality and deeply felt connection to the land are also factors that have drawn New Agers. At any rate, the liberated atmosphere here has given rise to a thriving New Age network, one that now includes alternative churches, healing centers, and healing schools. You'll find all sorts of alternative medicine and fringe practices here, from aromatherapy to rolfing—a form of massage that realigns the muscles and bones in the body—and chelation therapy, in which an IV drips ethylene diamine tetra-acetic acid into your blood to remove heavy metals. If those sound too invasive, you can always try psychic surgery.

New Age practices and beliefs have given rise to a great deal of local humor targeting their supposed pyschobabble. One pointed joke asks: "How many 12-steppers does it take to change a lightbulb?" Answer: "None. They just form a support group and learn to live in the dark." For many, however, there's much good to be found in the movement. The Dalai Lama visited Santa Fe because the city is seen as a healing center and has become a refuge for Tibetans; notable speakers such as Ram Das and John Bradshaw frequently talk in the area. Many practitioners find the alternatives— healing resources and spiritual paths—they are looking for in the receptive Northern New Mexico desert and mountains.

7 Chiles, Sopaipillas & Other New Mexican Specialties

Northern New Mexicans are serious about eating, and the area's cuisine reflects the amalgam of cultural influences found here. Locals have given their unique blend of Hispanic and Pueblo recipes a rather prosaic, but direct, label: "Northern New Mexico Cuisine."

Food here isn't the same as Mexican cuisine or even those American variations of "Tex-Mex" and "Cal-Mex." New Mexican cooking is a product of Southwestern history: Native Americans taught the Spanish conquerors about corn—how to roast it and how to make corn pudding, stewed corn, cornbread, cornmeal, and posole (hominy)—and they also taught the Spanish how to use chile peppers, a crop indigenous to the New World, having been first harvested in the Andean highlands as early as 4000 B.C. The Spaniards brought the practice of eating beef to the area.

Waves of newcomers have introduced other elements to the food here. From Mexico came the interest in seafood. You'll find fish tacos on many menus as well as shrimp enchiladas and ceviche. New Southwestern cuisine combines elements from various parts of Mexico, such as sauces from the Yucatán Peninsula, and fried bananas served with bean dishes, typical of Costa Rica and other Central American locales. You'll also find Asian elements mixed in, such as pot stickers in a tortilla soup.

The basic ingredients of Northern New Mexico cooking are three indispensable, locally grown vegetables: **chile, beans,** and **corn.** Of these, perhaps the most crucial is the chile, whether brilliant red or green and with various levels of spicy bite. Green chile is hotter if the seeds are left in; red chile is green chile at its ripest stage. Chile forms the base for the red and green sauces that top most Northern New Mexico dishes such as enchiladas and burritos. One is not necessarily hotter than the other; spiciness depends on where and during what kind of season (dry or wet) the chiles were grown. You'll also find salsas, generally made with jalapeños, tomatoes, onions, and garlic, used for chip dipping and as a spice on tacos.

You Say Chili, We Say Chile

You'll never see "chili" on a menu in New Mexico. New Mexicans are adamant that *chile,* the Spanish spelling of the word, is the only way to spell it—no matter what your dictionary might say.

I'm inclined to think chile with an "e" is listed as a secondary spelling only as a courtesy to New Mexicans. We have such a personal attachment to this small agricultural gem that in 1983 we directed our senior U.S. senator, Pete Domenici, to enter New Mexico's official position on the spelling of chile into the *Congressional Record.* That's taking your *chiles* seriously.

Chiles are grown throughout the state, in a perfect climate for cultivating and drying the small but powerful red and green New Mexican varieties. But it is the town of Hatch, New Mexico, that bills itself as the "Chile Capital of the World." Regardless of where you travel in the state, chiles appear on the menu. Virtually anything you order in a restaurant is topped with a chile sauce. If you're not accustomed to spicy foods, certain varieties of red or green chiles will make your eyes water, your sinuses drain, and your palate feel as if it's on fire—all after just one forkful. *Warning:* No amount of water or beer will alleviate the sting. (Drink milk. A sopaipilla drizzled with honey is also helpful.)

But don't let these words of caution scare you away from genuine New Mexico chiles. The pleasure of eating them far outweighs the pain. Start slow, with salsas and chile sauces first, perhaps *relleños* (stuffed peppers) next, followed by *rajas* (roasted peeled chiles cut into strips). Before long, you'll be buying chile ristras and hanging the dried chiles up for decoration. Perhaps you'll be so smitten that you'll purchase bags of chile powder or a chile plant to take home. If you happen to be in New Mexico in the fall, you'll find fresh roasted green chile sold. In the parking lots of most grocery stores and at some roadside stands, you'll smell the scent of roasting chile and see large metal baskets full of peppers rotating over flames. If you have a means of freezing the chile before transporting it home, you can sample the delicacy throughout the year. This will certainly make you an expert on the difference between chile and chili.

Beans—spotted or painted pinto beans with a nutty taste—are simmered with garlic, onion, cumin, and red chile powder and served as a side dish. When mashed and refried in oil, they become *frijoles refritos.* **Corn** supplies the vital dough for tortillas called *masa.* New Mexican corn comes in six colors, of which yellow, white, and blue are the most common.

Even if you are familiar with Mexican cooking, the dishes you know and love are likely to be prepared differently here. The following is a rundown of some regional dishes, a number of which are not widely known outside the Southwest:

biscochito   A cookie made with anise.

carne adovada   Tender pork marinated in red chile sauce, herbs, and spices, and then baked.

chile relleños   Peppers stuffed with cheese, deep-fried, then covered with green chile sauce.

chorizo burrito (also called a "breakfast burrito")   Mexican sausage, scrambled eggs, potatoes, and scallions wrapped in a flour tortilla with red or green chile sauce and melted Jack cheese.

empañada   A fried pie with nuts and currants.

enchiladas   Tortillas filled with peppers or other foods.

fajitas   Strips of beef or chicken sautéed with onions, green peppers, and other vegetables and served on a sizzling platter.

green chile stew   Locally grown chiles cooked in a stew with chunks of meat, beans, and potatoes.

huevos rancheros   Fried eggs on corn tortillas, topped with cheese and red or green chile, served with pinto beans.

pan dulce   A Native American sweetbread.

posole   A corn soup or stew (called hominy in other parts of the South), sometimes prepared with pork and chile.

sopaipillas   A lightly fried puff pastry served with honey as a dessert or stuffed with meat and vegetables as a side dish. Sopaipillas with honey are also often served with your meal—the honey has a cooling effect on your palate after you've eaten a spicy dish.

tacos   More often served as soft rolled tortillas than as crispy shells.

tamales   Made from cornmeal mush, wrapped in husks and steamed.

vegetables and nuts   Despite the prosaic name, unusual local ingredients, such as piñon nuts, jicama, and prickly pear cactus, will often be a part of your meals.

8 Recommended Books

Many well-known writers have made their home in Northern New Mexico in the 20th century. In the 1920s, the most celebrated were **D. H. Lawrence** and **Willa Cather,** both short-term Taos residents. Lawrence, the romantic and controversial English novelist, spent time here between 1922 and 1925; he reflected on his sojourn in *Mornings in Mexico* and *Etruscan Places.* Lawrence's Taos period is described in *Lorenzo in Taos,* which his patron, Mabel Dodge Luhan, wrote. Cather, a Pulitzer prize winner famous for her depictions of the pioneer spirit, penned *Death Comes for the Archbishop;* her fictionalized account of the 19th-century Santa Fe bishop, Jean-Baptiste Lamy, grew out of her stay in the region.

Many contemporary authors also live in and write about New Mexico. John Nichols, of Taos, whose *Milagro Beanfield War* was made into a Robert Redford movie in 1987, writes insightfully about the problems of poor Hispanic farming communities. Albuquerque's Tony Hillerman has for two decades weaved mysteries around Navajo tribal police in books such as *Listening Woman* and *A Thief of Time.* The Hispanic novelist Rudolfo Anaya's *Bless Me, Ultima* and Pueblo writer Leslie Marmon Silko's *Ceremony* capture the lifestyles of their respective peoples. Of the desert environment and politics, no one wrote better than the late Edward Abbey; his *Fire on the Mountain,* set in New Mexico, was one of his most powerful works.

OTHER SUGGESTED READING   Excellent works about Native Americans of New Mexico include *The Pueblo Indians of North America* (Holt, Rinehart & Winston, 1970) by Edward P. Dozier and *Living the Sky: The Cosmos of the American Indian* (University of Oklahoma Press, 1987) by Ray A. Williamson.

For general histories of the state, try Myra Ellen Jenkins and Albert H. Schroeder's *A Brief History of New Mexico* (University of New Mexico Press, 1974) and Marc Simmons's *New Mexico: An Interpretive History* (University of New Mexico Press, 1988). In addition, Claire Morrill's *A Taos Mosaic: Portrait of a New Mexico Village* (University of New Mexico Press, 1973) does an excellent job of portraying the history of that small New Mexican town. I have also enjoyed Tony Hillerman's (ed.) *The Spell of New Mexico* (University of New Mexico Press, 1976) and John Nichols and William Davis's *If Mountains Die: A New Mexico Memoir* (Alfred A. Knopf, 1979).

THE ARTS *Enduring Visions: 1,000 Years of Southwestern Indian Art,* by the Aspen Center for the Visual Arts (Publishing Center for Cultural Resources, 1969), and Roland F. Dickey's *New Mexico Village Arts* (University of New Mexico Press, 1990) are both excellent resources for those interested in Indian art. If you become intrigued with Spanish art during your visit to New Mexico, you'll find E. *Boyd's Popular Arts of Spanish New Mexico* (Museum of New Mexico Press, 1974) to be quite informative.

Planning a Trip to Santa Fe, Taos & Albuquerque

As with any trip, a little preparation is essential before you start. This chapter will provide you with a variety of planning tools, including information on when to go and how to get there.

1 Visitor Information

Numerous agencies can assist you with planning your trip. The Tourism and Travel Division of the **New Mexico Department of Tourism** is located in Room 751, 491 Old Santa Fe Trail, Santa Fe, NM 87503 (☎ **800/545-2040**). Santa Fe, Taos, and Albuquerque each has its own information service for visitors (see the "Orientation" sections in chapters 4, 11, and 15, respectively).

Information about Northern New Mexico is also available on the Internet at the following World Wide Web addresses:

For general New Mexico information, try: **http://www.swcp.com/nm/**.

For Internet addresses of individual cities, see chapters 4, 11, and 15.

2 When to Go

THE CLIMATE   Forget any preconceptions you may have about the New Mexico "desert." The high desert climate of this part of the world is generally dry but not always warm. Santa Fe and Taos, at 7,000 feet above sea level, have midsummer highs in the 80s and lows in the 50s. Spring and fall highs run in the 60s, with lows in the 30s. Typical midwinter daytime temperatures are in the low 40s, and overnight lows are in the teens. Temperatures in Albuquerque, at 5,300 feet, often run about 10°F warmer.

The average annual precipitation ranges from 8 inches at Albuquerque to 12 inches at Taos and 14 at Santa Fe, most of it coming in July and August as afternoon thunderstorms. Snowfall is common from November through March and sometimes as late as May, though the snow seldom lasts long. Santa Fe averages 32 inches total annual snowfall. At the high-mountain ski resorts, as much as 300 inches (25 feet) of snow may fall in a season—and stay.

Average Temperatures (°F) and Annual Rainfall (inches)

	Jan High–Low	Apr High–Low	July High–Low	Oct High–Low	Rainfall (inches)
Albuquerque	47–28	70–41	91–66	72–45	8.9
Santa Fe	40–18	59–35	80–57	62–38	14.0
Taos	40–10	62–30	87–50	64–32	12.1

NORTHERN NEW MEXICO CALENDAR OF EVENTS

January
- **New Year's Day.** Parades, traditional dances, and masses at several pueblos, including Picuris, San Ildefonso, and Taos. January 1. Call ☎ **800/732-8267** for more information.
- **Winter Wine Festival.** A variety of wine offerings and food tastings prepared by local chefs takes place mid-January in the Taos Ski Valley. Call ☎ **505/776-2291** for details.

February
- **Candelaria Day Celebration, San Felipe Pueblo.** Traditional dances. February 2. Call ☎ **505/843-7270** for more information.
- ✪ **Winter Fiesta.** Santa Fe's annual retreat from the midwinter doldrums appeals to skiers and nonskiers alike. Highlights include the Great Santa Fe Chile Cookoff; ski races, both serious and frivolous; snow-sculpture contests; snowshoe races; and hot-air balloon rides.

  Where: Santa Fe Ski Area. When: The last weekend in February. How: Most events are free. Call ☎ **505/982-4429** for information.

March
- **Fiery Food Show.** This annual trade show, which takes place in Albuquerque in early March, features chiles and an array of products that can be made from them. Call ☎ **505/873-9103** for details.
- **Rio Grande Arts and Crafts Festival.** A juried show featuring 200 artists and craftspeople from around the country takes place at the State Fairgrounds in Albuquerque during the second week of March. Call ☎ **505/292-7457** for more information.

April
- **Easter Weekend Celebration,** Nambe, Picuris, and San Ildefonso Pueblos. Celebrations include masses, parades, and corn and other dances. Call ☎ **505/843-7270** for information.
- **Easter Sunday Celebration,** Indian Pueblo Cultural Center, Albuquerque. Traditional dances are performed by Native Americans. Call ☎ **505/843-7270.**
- **American Indian Week,** Indian Pueblo Cultural Center, Albuquerque. A celebration of Native American traditions and culture. Begins late in the second week of April. Call ☎ **505/843-7270.**
- **Gathering of Nations Powwow,** University Arena, Albuquerque. Dance competitions, arts-and-crafts exhibitions, and Miss Indian World contest. Mid- to late April. Call ☎ **505/836-2810.**
- ✪ **Taos Talking Picture Festival.** Filmmakers and film enthusiasts gather to view a variety of films, from serious documentaries to lighthearted comedies. You'll see

locally made films as well as those involving Hollywood big-hitters. In 1997, Philip Kaufman and Louis Gossett Jr. showed their films.

Where: Venues throughout Taos and Taos County. When: Mid-April. How: Taos Talking Picture Festival, 216M N. Pueblo Rd., Taos, NM 87571. Call ☎ **505/751-0637.** Fax 505/751-7385. E-mail: ttpix@taosnet.com; web site: http://www.taosnet.com/ttpix/.

May

✪ **Taos Spring Arts Festival.** Contemporary visual, performing, and literary arts are highlighted during 2 weeks of gallery openings, studio tours, performances by visiting theatrical and dance troupes, live musical events, traditional ethnic entertainment, a film festival, fashion shows, literary readings, and more.

Where: Venues throughout Taos and Taos County. When: The first 2 weeks in May. How: Tickets are available from the Taos County Chamber of Commerce, P.O. Box 1691, Taos, NM 87571 (☎ **800/732-TAOS** or 505/ 758-3873).

• **¡Magnifico! Albuquerque Festival of the Arts.** A 17-day celebration featuring more than 200 special events, attractions, and exhibits. The visual, performing, literary, and culinary arts are honored throughout the city. Early to mid-May. Call ☎ **800/284-2282** or 505/842-9918 for a schedule.

June

• **Rodeo de Taos,** County Fairgrounds, Taos. Fourth weekend in June.

✪ **New Mexico Arts and Crafts Fair.** This is the second-largest event of its type in the United States. More than 200 New Mexico artisans exhibit and sell their crafts, and there is nonstop entertainment for the whole family. Hispanic arts and crafts are also on display.

Where: State Fairgrounds, Albuquerque. When: The last weekend in June (on Friday and Saturday from 10am to 10pm and on Sunday from 10am to 6pm). How: Admission varies. For information, call ☎ **505/884-9043.**

• **Taos Poetry Circus.** Poetry readings, lectures, seminars, and several poetry bouts highlight this spirited event the second week in June. It culminates in the world champion poetry bout.

July

• **Fourth of July celebrations** (including fireworks displays) are held all over New Mexico. Call the chambers of commerce in specific towns and cities for information.

• **Picuris Arts and Crafts Fair,** Picuris Pueblo. Traditional dances and other events. Proceeds go to the restoration of the San Lorenzo Mission. The first weekend in July. Call ☎ **505/843-7270** for details.

• **Santa Fe Wine Festival at Rancho de las Golondrinas.** This event in early July boasts live entertainment and wine tastings presided over by hosts dressed in period clothing. Call ☎ **505/892-4178.**

✪ **Rodeo de Santa Fe.** This 4-day event features a western parade, rodeo dance, and four rodeo performances. It attracts hundreds of cowboys from all over the Southwest who compete for a sizable purse in such events as Brahma bull and bronco riding, calf roping, steer wrestling, barrel racing, trick riding, clown and animal acts, and a local version of bullfighting in which neither the bull nor the matador is hurt.

Where: Rodeo grounds, 4801 Rodeo Rd., off Cerrillos Road, 5$^{1}/_{2}$ miles south of the Plaza. When: The first weekend following the Fourth of July (starting at 7:30pm Wednesday through Saturday; also at 2pm on Saturday). How: For tickets and information, call ☎ **505/471-4300.**

- **Taos Pueblo Powwow.** Intertribal competition in traditional and contemporary dances. The second weekend in July. Call ☎ **505/758-9593** for more information.
- **Eight Northern Pueblos Artist and Craftsman Show.** More than 600 Native American artists exhibit their work at one of the eight northern pueblos. Traditional dances and food booths. The third weekend in July. Call ☎ **505/852-4265** for location and exact dates.
- ✪ **Fiestas de Santiago y Santa Ana.** The celebration begins with a Friday night mass at Our Lady of Guadalupe Church, where the fiesta queen is crowned. During the weekend there are candlelight processions, special masses, music, dancing, parades, crafts, and food booths.

  Where: Taos Plaza. When: The third weekend in July. How: Most events are free. For information, contact the Taos Fiesta Council, P.O. Box 3300, Taos, NM 87571 (☎ **800/732-8267**).
- ✪ **The Spanish Markets.** More than 300 Hispanic artists from New Mexico and southern Colorado exhibit and sell their work in this lively community event. Artists are featured in special demonstrations, while an entertaining mix of traditional Hispanic music, dance, foods, and pageantry create the ambience of a village celebration. Artwork for sale includes painted and carved saints, textiles, tin work, furniture, straw appliqué, and metalwork.

  Where: Santa Fe Plaza, Santa Fe. When: The last full weekend in July. How: Markets are free. For information, contact the Spanish Colonial Arts Society, P.O. Box 1611, Santa Fe, NM 87504 (☎ **505/983-4038**).

August

- ✪ **The Indian Market.** This is the largest all–Native American market in the country. About 800 artisans display their baskets and blankets, jewelry, pottery, wood carvings, rugs, sand paintings, and sculptures at rows of booths. Sales are brisk. Costumed tribal dancing and crafts demonstrations are scheduled in the afternoon.

  Where: Santa Fe Plaza, surrounding streets, and de Vargas Mall. When: The third weekend in August. How: The market is free and hotels are booked months in advance. For information, contact the Southwestern Association on Indian Affairs, P.O. Box 1964, Santa Fe, NM 87501 (☎ **505/983-5220**).
- **Music from Angel Fire,** Angel Fire. World-class musicians perform classical and chamber music. Last week in August to first week in September. Call ☎ **505/ 377-3233** for information and schedules.

September

- **Gourmet Jubilee,** Angel Fire. Farmer's market highlighting New Mexico–made products as well as wines. Cooking classes and dinners are available. Early September. Call ☎ **800/446-8117** for information.
- **New Mexico Wine Festival at Bernalillo,** near Albuquerque. New Mexico wines are showcased at this annual event. Wine tastings, art show, and live entertainment. Labor Day weekend. For schedule of events, call ☎ **505/892-4178.**
- ✪ **La Fiesta de Santa Fe.** An exuberant combination of spirit, history, and general merrymaking, La Fiesta is the oldest community celebration in the United States. The first fiesta was celebrated in 1712, 20 years after the peaceful resettlement of New Mexico by Spanish conquistadors in 1692, following the Pueblo revolt of 1670. La Conquistadora, a carved Madonna credited with the victory, is the focus of the celebration, which includes masses, a parade for children and their pets, a historical/hysterical parade, mariachi concerts, dances, food, and arts, as well as local entertainment on the Plaza. Zozobra, "Old Man Gloom," a 40-foot-tall

effigy of wood, canvas, and paper, is burned at dusk on Friday to revitalize the community.

Where: Santa Fe. When: The first Friday after Labor Day. How: For information, contact the Santa Fe Fiesta Council, P.O. Box 4516, Santa Fe, NM 87502-4516 (☎ **505/988-7575**).

✪ **New Mexico State Fair and Rodeo.** One of America's top 10 state fairs, it features pari-mutuel horse racing, a nationally acclaimed rodeo, entertainment by top country artists, Native American and Spanish villages, the requisite midway livestock shows, and arts and crafts.

Where: State Fairgrounds, Albuquerque. When: 17 days in September. How: Advance tickets can be ordered. Call ☎ **505/265-1791** for information.

✪ **Taos Fall Arts Festival.** Highlights include arts-and-crafts exhibitions and competitions, studio tours, gallery openings, lectures, films, concerts, dances, and stage plays. Simultaneous events include the Old Taos Trade Fair, the Wool Festival, and San Geronimo Day at Taos Pueblo.

Where: Throughout Taos and Taos County. When: 17 days, from mid-September (or the third weekend) through the first week in October. How: Events, schedules, and tickets (where required) can be obtained from the Taos County Chamber of Commerce, P.O. Drawer I, Taos, NM 87571 (☎ **800/732-8267**).

✪ **Old Taos Trade Fair,** Martinez Hacienda, Lower Ranchitos Road, Taos. This 2-day affair reenacts Spanish colonial life of the mid-1820s and features Hispanic and Native American music, weaving and crafts demonstrations, traditional foods, dancing, and "visits" by mountain men. The last full weekend in September. Call ☎ **505/758-0505.**

✪ **San Geronimo Vespers Sundown Dance and Trade Fair,** Taos Pueblo. A mass and procession; traditional corn, buffalo, and Comanche dances; an arts-and-crafts fair; foot races; and pole climbs by clowns. The last weekend in September. Call ☎ **505/758-1028** for details.

October

✪ **Kodak Albuquerque International Balloon Fiesta.** The world's largest balloon rally brings together more than 800 colorful balloons and includes races and contests. There is a sunrise mass ascension. Various special events are staged all week.

Where: Balloon Fiesta Park (at I-25 and Alameda NE) on Albuquerque's northern city limits. When: Second week in October. How: For information, call ☎ **800/733-9918.**

• **Taos Mountain Balloon Rally and Taste of Taos.** The Albuquerque fiesta's "little brother" offers mass dawn ascensions, tethered balloon rides for the public, and a Saturday parade of balloon baskets (in pickup trucks) from Kit Carson Park around the plaza. Taste of Taos includes food and product fairs, chile cookoffs, and the creation of the "world's biggest burrito." The last full weekend in October. Call ☎ **800/732-8267** for more information.

November

• **Weems Artfest,** State Fairgrounds, Albuquerque. Approximately 260 artisans, who work in mixed media, from throughout the world attend this fair. It's one of the top 100 arts-and-crafts fairs in the country. A 3-day weekend in mid-November. For details, call ☎ **505/293-6133.**

December

✪ **Yuletide in Taos.** This pre-Christmas event emphasizes New Mexican traditions, cultures, and arts with carols, festive classical music, Hispanic and Native American songs and dances, historic walking tours, art exhibitions, dance performances, candlelight dinners, and more.

Where: Throughout Taos. When: Throughout December. How: Events are staged by the Taos County Chamber of Commerce, P.O. Drawer I, Taos, NM 87571 (☎ **800/732-TAOS**).

- **Torchlight Procession,** Taos Ski Valley, December 31. Call ☎ **505/776-2291** for information.
- **Winter Spanish Market,** Sweeney Convention Center, Santa Fe. See the Spanish Markets in July (above) for more information. The first full weekend in December. Call ☎ **505/983-4038.**

3 Health & Insurance

HEALTH   One thing that sets New Mexico apart from most other states is its elevation. Santa Fe and Taos are about 7,000 feet above sea level; Albuquerque is more than 5,000 feet above sea level. The reduced oxygen and humidity can precipitate some unique problems, not the least of which is acute mountain sickness. In its early stages you might experience headaches, shortness of breath, loss of appetite and/or nausea, tingling in the fingers or toes, lethargy, and insomnia. It can usually be treated by taking aspirin as well as getting plenty of rest, avoiding large meals, and drinking lots of nonalcoholic fluids (especially water). If it persists or worsens, you must return to a lower altitude. Other dangers of higher elevations include sunburn and hypothermia, and these should be taken seriously. To avoid dehydration, drink water as often as possible.

It is important to monitor your children's health while in New Mexico. They are just as susceptible to mountain sickness, hypothermia, sunburn, and dehydration as you are.

Other things to be wary of are *arroyos,* or creek beds where flash floods can occur without warning in the desert. If water is flowing across a road, *do not* try to drive through it, because chances are the water is deeper and is flowing faster than you think. Just wait it out. Arroyo floods don't last long.

Finally, if you're an outdoorsperson, be on the lookout for snakes—particularly rattlers. Avoid them. Don't even get close enough to take a picture (unless you have a very good zoom lens).

INSURANCE   Before setting out on your trip, check your medical insurance policy to be sure it covers you away from home. If it doesn't, it's wise to purchase a relatively inexpensive traveler's policy, widely available at banks, travel agencies, and automobile clubs. In addition to medical assistance, including hospitalization and surgery, the policy should include the cost of an accident, death, or repatriation; loss or theft of baggage; the cost of trip cancellation; and guaranteed bail in the event of an arrest or other legal difficulties.

4 Tips for Travelers with Special Needs

FOR TRAVELERS WITH DISABILITIES   Throughout the state of New Mexico measures have been taken to provide access for the disabled. Several bed-and-breakfast inns have made one or more of their rooms completely wheelchair accessible, and in Taos there is a completely wheelchair-accessible trail in the state park. If you call the **Developmental Disabilities Planning Council** (☎ **800/552-8195**), they will provide you with free information about traveling with disabilities in New

Mexico. The brochure "Art of Accessibility" lists hotels, restaurants, and attractions in Albuquerque that are accessible to disabled travelers. The **Directory of Recreational Activities for Children with Disabilities** is a list of accessible camps, national forest campgrounds, amusement parks, and individual city services throughout New Mexico. *Access Santa Fe* lists accessible hotels, attractions, and restaurants in the state capital, and the Taos Chamber of Commerce will answer questions regarding accessibility in Taos. No matter what, it is advisable to call hotels, restaurants, and attractions in advance to be sure that they are fully accessible.

FOR SENIORS   Travelers over the age of 65—and in many cases 60, sometimes even 55—may qualify for discounts not available to the younger adult traveler. Some hotels offer rates 10% to 20% lower than the published rate; inquire at the time you make reservations. Many attractions give seniors discounts of up to half the regular adult admission price. Get in the habit of asking about discounts. Seniors who plan to visit national parks and monuments in New Mexico should consider getting a **Golden Age Passport,** which gives anyone over 62 lifetime access to any national park, monument, historic site, recreational area, or wildlife refuge that charges an entrance fee. Golden Age Passports can be obtained at any National Park Office in the country. There is a one-time $10 processing fee. In New Mexico, contact the **National Park Service Office of Communications** (☎ **505/988-6011**) for more information.

If you're retired and are not already a member of the **American Association of Retired Persons (AARP)**, consider joining. The AARP card is valuable throughout North America in your search for travel bargains.

In addition, there are 24 active **Elderhostel** locations throughout the state. For information, call New Mexico Elderhostel at ☎ **505/473-6267.**

A note about health: Senior travelers are often more susceptible to changes in elevation and may experience heart or respiratory problems. Consult your physician before your trip.

FOR FAMILIES   Children are often given discounts that adults, even seniors, would never dream of. For instance, many hotels allow children to stay free with their parents in the same room. The upper age limit may vary from 12 to 18.

Youngsters are almost always entitled to discounts on public transportation and admission to attractions. Though every entrance requirement is different, you'll often find that admission for kids 5 and under is free and for elementary-school-age children it's half price; older students (through college) may also be offered significant discounts.

FOR STUDENTS   Always carry your student identification with you. Tourist attractions, transportation systems, and other services may offer discounts if you have appropriate proof of your student status. Don't be afraid to ask. A high-school or college ID card or International Student Card will suffice.

Student-oriented activities abound on and around college campuses, especially at the University of New Mexico in Albuquerque. In Santa Fe, there are two small 4-year colleges: the College of Santa Fe and the liberal arts school St. John's College.

5 Getting There

BY PLANE   The gateway to Santa Fe, Taos, and other Northern New Mexico communities is the Albuquerque International Sunport (☎ **505/842-4366** for the administrative offices; call the individual airlines for flight information).

Airlines serving Albuquerque include American (☎ **800/433-7300**), America West (☎ **800/235-9292**), Continental (☎ **800/523-3273**), Delta (☎ **800/ 221-1212**), Southwest (☎ **800/435-9792**), TWA (☎ **800/221-2000**), and United (☎ **800/241-6522**).

BY TRAIN   Amtrak (☎ **800/USA-RAIL** or 505/842-9650) passes through Northern New Mexico twice daily. The Southwest Chief, which runs between Chicago and Los Angeles, stops once eastbound and once westbound in Gallup, Grants, Albuquerque, Lamy (for Santa Fe), Las Vegas, and Raton.

You can get a copy of Amtrak's National Timetable from any Amtrak station, from travel agents, or by writing Amtrak, 400 N. Capitol St. NW, Washington, DC 20001.

BY BUS   Because Santa Fe is only about 58 miles northeast of Albuquerque via I-40, most visitors to Santa Fe take the bus directly from the Albuquerque airport. Shuttlejack buses (☎ **505/243-3244** in Albuquerque, **505/982-4311** in Santa Fe) make the 70-minute run between the airport and Santa Fe hotels 7 to 10 times daily each way, from 4:45am to 10:45pm (cost is $20 one-way, payable to the driver). Reservations are required. Three other bus services shuttle between Albuquerque and Taos (via Santa Fe) for $30 to $35 one-way, $55 to $65 round-trip: Pride of Taos Tours/Shuttles (☎ **505/758-8340**), Faust's Transportation (☎ **505/758-3410**), and Twin Heart Express & Transportation (☎ **505/751-1201**).

The public bus depot in Albuquerque is located on 2nd Street at Silver Avenue (300 2nd Street SW). Contact Texas, New Mexico and Oklahoma (T.N.M.& O.; ☎ **505/242-4998**) for information and schedules. Fares run about $11 to Santa Fe and $21 to Taos. However, the bus stations in Santa Fe (858 St. Michael's Dr.; ☎ **505/471-0008**) and Taos (at the Chevron bypass station at the corner of US 64 and NM 68; ☎ **505/758-1144**) are several miles south of each city center. Because additional taxi or shuttle service is needed to reach most accommodations, travelers usually find it more convenient to pay a few extra dollars for an airport-to-hotel shuttle.

BY CAR   The most convenient way to get around the Santa Fe region is by private car. Auto and RV rentals are widely available for those who arrive without their own transportation, either at the Albuquerque airport or at locations around each city.

I have received good rates and service from Avis at the Albuquerque airport (☎ **800/831-2847,** 505/842-4080, or 505/982-4361 in Santa Fe); Thrifty, 2039 Yale Blvd. SE, Albuquerque (☎ **800/367-2277** or 505/842-8733); Hertz, Albuquerque International Airport (☎ **800/654-3131** or 505/842-4235); Dollar, Albuquerque International Airport (☎ **800/369-4226** or 505/842-4304); Budget, Albuquerque International Airport (☎ **505/768-5900**); Alamo, 2601 Yale SE (☎ **800/ 327-9633**); and Rent-A-Wreck of Albuquerque, 501 Yale SE (☎ **800/247-9556** or 505/242-9556).

Drivers who need wheelchair-accessible transportation should call Wheelchair Getaways of New Mexico, 1015 Tramway Lane NE (☎ **800/408-2626** or 505/ 247-2626); it rents vans by the day, week, or month.

If you're arriving by car from elsewhere in North America, Albuquerque is at the crossroads of two major interstate highways. I-40 runs from Wilmington, North Carolina (1,870 miles east), to Barstow, California (580 miles west). I-25 extends from Buffalo, Wyoming (850 miles north), to El Paso, Texas (265 miles south). I-25 skims past Santa Fe's southern city limits. To reach Taos, you'll have to leave I-25 at Santa Fe and travel north 74 miles via US 84/285 and NM 68, or exit I-25

9 miles south of Raton, near the Colorado border, and proceed 100 miles west on US 64.

The following table shows the approximate mileage to Santa Fe from various cities around the United States.

Distances to Santa Fe (in miles)

From	Distance	From	Distance
Atlanta	1,417	Minneapolis	1,199
Boston	2,190	New Orleans	1,181
Chicago	1,293	New York	1,971
Cleveland	1,558	Oklahoma City	533
Dallas	663	Phoenix	595
Denver	391	St. Louis	993
Detroit	1,514	Salt Lake City	634
Houston	900	San Francisco	1,149
Los Angeles	860	Seattle	1,477
Miami	2,011	Washington, D.C.	1,825

PACKAGE TOURS   Unfortunately, you may not find a package tour in New Mexico. The tour companies I spoke to said most visitors to New Mexico have such disparate interests it's difficult to create packages to please them. Still, a few tour companies can help you arrange a variety of day trips during your visit and can also secure lodging. **Southwest Airlines** has begun offering a package tour to Albuquerque and may soon expand to other New Mexico destinations. For information, call **800/ 423-5683.** Otherwise, you may want to contact the following inbound operators:

Destination Southwest, Inc., 20 First Plaza Galeria, Suite 603, Albuquerque, NM 87102 (☎ **800/999-3109** or 505/766-9068; fax 505/766-9065).

Gray Line Tours, 800 Rio Grande NW, Suite 22, Albuquerque, NM 87104 (☎ **800/256-8991** or 505/242-3880; fax 505/243-0692).

Rojotours & Services, P.O. Box 15744, Santa Fe, NM 87506-5744 (☎ **505/ 474-8333;** fax 505/474-2992).

Sun Tours, Ltd., 4300 San Mateo Blvd. NE, Suite B-155, Albuquerque, NM 87110 (☎ **505/889-8888**).

CWT-A to Z Travelink, 6020 Indian School Rd., Albuquerque, NM 87110 (☎ **800/366-0282** or 505/883-5865; fax 505/883-0038).

3 For Foreign Visitors

Although American fads and fashions have spread across Europe and other parts of the world so much that the United States may seem like familiar territory before your arrival, there are still a number of peculiarities and uniquely American situations that any foreign visitor could encounter.

1 Preparing for Your Trip

ENTRY REQUIREMENTS

DOCUMENT REQUIREMENTS   Canadian citizens may enter the United States without passports or visas; they need only proof of residence.

Citizens of the United Kingdom, New Zealand, Japan, and most (but not all) Western European countries traveling on valid passports may not need a visa for fewer than 90 days of holiday or business travel to the United States, provided that they hold a round-trip or return ticket and enter the United States on an airline or cruise line participating in the visa waiver program. (Note that citizens of these visa-exempt countries who first enter the United States may then visit Mexico, Canada, Bermuda, and/or the Caribbean Islands and then reenter the United States by any mode of transportation without needing a visa. Further information is available from any U.S. embassy or consulate.)

Citizens of countries other than those stipulated above, including citizens of Australia, must present two documents: (1) a valid passport with an expiration date at least 6 months beyond the scheduled end of a visit to the United States; and (2) a tourist visa, available without charge from the nearest U.S. consulate.

To obtain a visa, the traveler must submit a completed application form (either in person or by mail) with a 1 1/2-inch-square photo and must demonstrate binding ties to a residence in his or her home country or country of residence. Usually you can obtain a visa immediately or within 24 hours, but it may take longer during the summer rush from June to August. If you cannot go in person, contact the nearest U.S. embassy or consulate for directions on applying by mail. Your travel agent or airline office may also be able to provide you with visa applications and instructions. The U.S. embassy or consulate that issues your visa will determine whether you will be given a multiple- or single-entry visa and any restrictions regarding the length of your stay.

MEDICAL REQUIREMENTS   No inoculations are needed to enter the United States unless you're coming from, or have stopped over in, areas known to be suffering from epidemics, particularly cholera or yellow fever.

If you have a disease that needs treatment with medications containing narcotics or drugs requiring a syringe, carry a valid signed prescription from your physician to allay any suspicions that you are smuggling drugs.

CUSTOMS REQUIREMENTS   Every adult visitor may bring in free of duty: 1 liter of wine or hard liquor; 200 cigarettes or 100 cigars (although no Cuban cigars) or 3 pounds of smoking tobacco; and $100 worth of gifts. These exemptions are offered to travelers who spend at least 72 hours in the United States and who have not claimed them within the preceding 6 months. It's strictly forbidden to bring into the country foodstuffs (particularly cheese, fruit, cooked meats, and canned goods) and plants (vegetables, seeds, tropical plants, and so on). Foreign tourists may bring in or take out up to $10,000 in U.S. or foreign currency with no formalities; larger sums must be declared to Customs on entering or leaving the country.

INSURANCE

Unlike in Canada and Europe, there is no national health-care system in the United States. Because the cost of medical care is extremely high, I strongly advise every traveler to secure health insurance coverage before setting out; check your home policy to verify its coverage, if any, while you are abroad.

You may want to take out a comprehensive travel policy that covers (for a relatively low premium) sickness or injury costs (medical, surgical, and hospital); loss or theft of your baggage; trip-cancellation costs; guarantee of bail in case you are arrested; costs of accidents, repatriation, or death. Such packages (for example, "Europe Assistance Worldwide Services" in Europe) are sold by automobile clubs at attractive rates, as well as by insurance companies and travel agencies.

MONEY

CURRENCY & EXCHANGE   The U.S. monetary system has a decimal base: one American dollar ($1) = 100 cents (100¢).

Dollar bills commonly come in $1 ("a buck"), $5, $10, $20, $50, and $100 denominations (the last two are not welcome when paying for small purchases and are not accepted in taxis or at subway ticket booths). There are also $2 bills (though today seldom encountered).

There are six denominations of coins: 1¢ (one cent or "penny"), 5¢ (five cents or "nickel"), 10¢ (ten cents or "dime"), 25¢ (twenty-five cents or "quarter"), 50¢ (fifty cents or "half dollar"), and the rare $1 piece.

Note: The "foreign-exchange bureaus" so common in Europe are rare even at airports in the United States and nonexistent outside major cities. Try to avoid having to change foreign money (or traveler's checks denominated in a currency other than U.S. dollars) at a small-town bank or even a branch bank in a big city. In fact, you might want to leave any currency other than U.S. dollars at home—it may prove a greater nuisance to you than it's worth.

TRAVELER'S CHECKS   Traveler's checks denominated in U.S. dollars are readily accepted at most hotels, motels, restaurants, and large stores. The best place to change traveler's checks is at a bank. Do not bring traveler's checks denominated in other currencies.

CREDIT & CHARGE CARDS   The method of payment most widely used is credit and charge cards: Visa (BarclayCard in Britain), MasterCard (EuroCard in

Europe, Access in Britain, Chargex in Canada), American Express, Diners Club, Discover, and Carte Blanche. You can save yourself trouble by using "plastic money" rather than cash or traveler's checks in most hotels, motels, restaurants, and retail stores (a growing number of food and liquor stores now accept credit/charge cards). You must have a credit or charge card to rent a car. It can also be used as proof of identity (often carrying more weight than a passport) or as a "cash card," enabling you to draw money from banks and automated-teller machines (ATMs) that accept it.

SAFETY

GENERAL   Tourist areas are generally safe, but, despite recent reports of decreases in violent crime in many cities, it would be wise to check with the tourist offices in Santa Fe, Taos, and Albuquerque if you are in doubt about which neighborhoods are safe. (See the "Orientation" sections in chapters 4, 11, and 15 for the names and addresses of the specific tourist bureaus.)

Remember that hotels are open to the public, and in a large hotel, security may not be able to screen everyone who enters. Always lock your room door; don't assume that once inside your hotel you are automatically safe and no longer need to be aware of your surroundings.

DRIVING   Question your rental agency about personal safety, or ask for a brochure of traveler safety tips when you pick up your car. Obtain written directions, or a map with the route clearly marked, from the agency to show you how to get to your destination. And, if possible, arrive and depart during daylight hours.

In recent years, "car-jacking," a crime that targets both cars and drivers, has been on the rise in all U.S. cities. Incidents involving German and other international tourists in Miami made news around the world. Rental cars are especially targeted. If you exit a highway into a questionable neighborhood, leave the area as quickly as possible. If you have an accident, even on the highway, stay in your car with the doors locked until you are able to assess the situation or until the police arrive. If you are bumped from behind by another car on the street or are involved in a minor accident with no injuries and the situation appears to be suspicious, motion to the other driver to follow you to the nearest police precinct, a well-lit service station, or an all-night store. Never get out of your car in such situations.

If you see someone on the road who indicates a need for help, do not stop. Take note of the location, drive to a well-lighted area, and telephone the police by dialing ☎ 911.

Also, make sure that you have enough gasoline in your tank to reach your intended destination, so that you're not forced to look for a service station in an unfamiliar and possibly unsafe neighborhood—especially at night. These warnings cannot be overemphasized; failure to do any of these things could be exceedingly dangerous or even fatal.

2 Getting to the U.S.

Travelers from overseas can take advantage of the **APEX (Advance-Purchase Excursion) fares** offered by all the major international carriers. Aside from these, attractive values are offered by Icelandair on flights from Luxembourg to New York and by Virgin Atlantic Airways from London to New York/Newark and to Los Angeles. To reach Northern New Mexico from Europe, you'll probably have to stop at one of these airports anyway to make a connecting flight.

British travelers should check out British Airways (☎ **0345/222-111** in the U.K. or **800/247-9297** in the U.S.), which offers direct flights from London to New York and to Los Angeles, as does Virgin Atlantic Airways (☎ **0293/747-747** in the U.K.

or **800/862-8621** in the U.S.). Canadian readers might book flights on Air Canada (☎ **800/268-7240** in Canada or **800/776-3000** in the U.S.), which offers service from Toronto, Montreal, and Calgary to New York and to Los Angeles. In addition, many other international carriers serve the New York and Los Angeles airports, including: Air France (☎ **800/237-2747**), Alitalia (☎ **800/223-5730**), Japan Airlines (☎ **800/525-3663**), Lufthansa (☎ **800/645-3880**), Quantas (☎ **008/177-767** in Australia), Swissair (☎ **800/221-4750**), and SAS (☎ **800/221-2350**).

Visitors arriving by air, no matter the port of entry, should reserve patience and resignation before setting foot on U.S. soil. Getting through Immigration control may take as long as 2 hours on some days, especially summer weekends. Add the time it takes to clear Customs, and you'll see that you should allow extra time for delays when planning connections between international and domestic flights—an average of 2 to 3 hours at least.

In contrast, travelers arriving by car or by rail from Canada will find that the border-crossing formalities have been streamlined to the point that they are practically nonexistent. Air travelers from Canada, Bermuda, and some places in the Caribbean can sometimes go through Customs and Immigration at the point of departure, which is much quicker and less tedious.

For further information about transportation to Santa Fe, Taos, and Albuquerque, see "Orientation" in chapters 4, 11, and 15, respectively.

3 Getting Around the U.S.

BY PLANE   On transatlantic or transpacific flights, some prominent American airlines (for example, American Airlines, Delta, Northwest, TWA, and United) offer travelers special discount tickets under the name **Visit USA,** allowing travel between a number of U.S. destinations at minimum rates. These tickets are not for sale in the United States and must, therefore, be purchased before you leave your foreign point of departure. This system is the best, easiest, and fastest way to see the United States at low cost. You should obtain information well in advance from your travel agent or the office of the airline concerned, since the conditions attached to these discount tickets can be changed without advance notice.

BY TRAIN   Long-distance trains in the United States are operated by Amtrak, the national rail passenger corporation. International visitors can buy a **USA Railpass,** good for 15 or 30 days of unlimited travel on Amtrak (☎ **800/872-7245**). The pass is available through many foreign travel agents. In 1997 prices for a 15-day pass were $260 off-peak, $375 peak; a 30-day pass costs $350 off-peak, $480 peak. (With a foreign passport, you can also buy passes at some Amtrak offices in the United States, including locations in Boston, Chicago, Los Angeles, Miami, New York, San Francisco, and Washington, D.C.) Reservations are generally required and should be made for each part of your trip as early as possible. Even cheaper than the above are regional USA Railpasses, allowing unlimited travel through a specific section of the United States. Reservations are generally required and should be made for each part of your trip as early as possible.

Visitors should be aware of the limitations of long-distance rail travel in the United States. With a few notable exceptions, service is rarely up to European standards: Delays are common, routes are limited and often infrequently served, and fares are rarely much lower than discount airfares. Thus, cross-country train travel should be approached with caution.

BY BUS   The cheapest way to travel around the United States is by bus. Greyhound/Trailways (☎ **800/231-2222**), the sole nationwide bus line, offers an

Ameripass for unlimited travel for 7 days (for $179), 15 days (for $289), 30 days (for $399), and 60 days (for $599). Bus travel in the United States can be both slow and uncomfortable, so this option is not for everyone. Furthermore, bus stations are often situated in undesirable neighborhoods.

BY CAR   Travel by car gives visitors the freedom to make—and alter—their itineraries to suit their own needs and interests. And it offers the opportunity to visit some of the off-the-beaten-path locations, places that cannot be reached easily by public transportation. For many foreign travelers, traveling the wide-open roads of the western U.S. by car is the stuff of legend. For information on renting cars in the United States, see "Automobile Organizations" and "Automobile Rentals" in "Fast Facts: For the Foreign Traveler," below; "By Car" in "Getting There," in chapter 2; and "By Car" in "Getting Around," in chapters 4, 11, and 15.

FAST FACTS: For the Foreign Traveler

Automobile Organizations   Auto clubs supply maps, suggested routes, guidebooks, accident and bail-bond insurance, and emergency road service. The major auto club in the United States, with 955 offices nationwide, is the **American Automobile Association (AAA).** Members of some foreign auto clubs have reciprocal arrangements with AAA and enjoy its services at no charge. If you belong to an auto club in your home country, inquire about AAA reciprocity before you leave. You may be able to join AAA even if you're not a member of a reciprocal club; to inquire, call AAA (☎ **800/881-7585**). AAA can provide you with an **International Driving Permit,** validating your foreign license.

In addition, some automobile-rental agencies now provide many of these same services. Inquire about their availability when you rent your car.

Automobile Rentals   To rent a car you will need a major credit or charge card and a valid driver's license. In addition, you usually need to be at least 25 years old (some companies do rent to younger people but add a daily surcharge). Be sure to return your car with the same amount of gas you started out with; rental companies charge excessive prices for gasoline. See "By Car" in "Getting Around," in chapters 4, 11, and 15, for the phone numbers of car-rental companies in Santa Fe, Taos, and Albuquerque, respectively.

Business Hours   See "Fast Facts," in chapters 4, 11, and 15.

Climate   See "When to Go," in chapter 2.

Currency Exchange   You'll find currency-exchange services at major airports with international service. Elsewhere, they may be quite difficult to come by. In the United States, a very reliable choice is **Thomas Cook Currency Services, Inc.** They sell commission-free foreign and U.S. traveler's checks, drafts, and wire transfers; they also do check collections (including Eurochecks). Their rates are competitive, and the service is excellent. Thomas Cook maintains several offices in New York City, including one at 511 Madison Ave. (☎ **800/287-7362**), and at the JFK Airport International Arrivals Terminal (☎ **718/656-8444**).

For Santa Fe, Taos, and Albuquerque banks that handle foreign-currency exchange, see the "Fast Facts" sections in chapters 4, 11, and 15, respectively.

Drinking Laws   You must be 21 to purchase alcoholic beverages in New Mexico.

Electric Current   The United States uses 110–120 volts A, 60 cycles, compared with 220–240 volts A, 50 cycles, as in most of Europe. In addition to a 100-volt transformer, small appliances of non-American manufacture, such as hair dryers or shavers, will require a plug adapter, with two flat, parallel pins.

Embassies/Consulates   All embassies are located in the national capital, Washington, D.C.; some consulates are located in major U.S. cities, and most countries maintain a mission to the United Nations in New York City. The embassies and consulates of the major English-speaking countries—Australia, Canada, the Republic of Ireland, New Zealand, and the United Kingdom—are listed below. If you are from another country, you can get the telephone number of your embassy by calling "Information" in Washington, D.C. (☎ 202/555-1212).

The embassy of **Australia** is at 1601 Massachusetts Ave. NW, Washington, DC 20036 (☎ 202/797-3000). There is an Australian consulate at Century Plaza Towers, 19th Floor, 2049 Century Park East, Los Angeles, CA 90067 (☎ 310/229-4800). The consulate in New York is located at the International Building, 630 Fifth Ave., Suite 420, New York, NY 10111 (☎ 212/408-8400).

The embassy of **Canada** is at 501 Pennsylvania Ave. NW, Washington, DC 20001 (☎ 202/682-1740). There's a Canadian consulate in Los Angeles at 550 S. Hope St., 9th Floor, Los Angeles, CA 90071 (☎ 213/346-2700). The one in New York is located at 1251 Ave. of the Americas, New York, NY 10020 (☎ 212/596-1600).

The embassy of the **Republic of Ireland** is at 2234 Massachusetts Ave. NW, Washington, DC 20008 (☎ 202/462-3939). The consulate in New York is located at 345 Park Ave., 17th Floor, New York, NY 10022 (☎ 212/319-2555). The consulate in San Francisco is at 655 Montgomery St., Suite 930, San Francisco, CA 94111 (☎ 415/392-4214).

The embassy of **New Zealand** is at 37 Observatory Circle NW, Washington, DC 20008 (☎ 202/328-4800). The consulate in New York is located at 780 Third Ave., Suite 1904, New York, NY 10017-2024 (☎ 212/832-4038). The consulate in Los Angeles is located at 12400 Wilshire Blvd., Suite 1150, Los Angeles, CA 90025 (☎ 310/207-1605).

The embassy of the **United Kingdom** is at 3100 Massachusetts Ave. NW, Washington, DC 20008 (☎ 202/462-1340). The consulate in New York is located at 845 Third Ave., New York, NY 10022 (☎ 212/745-0200). The consulate in Los Angeles is located at 11766 Wilshire Blvd., Suite 400, Los Angeles, CA 90025 (☎ 310/477-3322).

Emergencies   Call ☎ **911** to report a fire, call the police, or get an ambulance. This is a toll-free call (no coins are required at a public telephone).

If you encounter traveler's problems, check the local telephone directory to find an office of the **Traveler's Aid Society,** a nationwide, nonprofit, social-service organization geared toward helping travelers in difficult straits. Their services might include reuniting families separated while traveling, providing food and/or shelter to people stranded without cash, or even offering emotional counseling. If you're in trouble, seek them out.

Gasoline (Petrol)   One U.S. gallon equals 3.8 liters or 0.83 Imperial gallons. There are usually several grades (and price levels) of gasoline available at most gas stations, and their names change from company to company. Unleaded gas with the highest octane ratings is the most expensive; however, most rental cars take the least expensive—"regular" unleaded gas. Furthermore, the price is often lower if you pay in cash rather than by credit or charge card.

Most gas stations are essentially self-service, although a number of them now offer higher-priced full service as well. Late- or all-night stations are usually self-service only.

Holidays   On the following legal national holidays, banks, government offices, post offices, and many stores, restaurants, and museums are closed: January 1 (New

Year's Day), the third Monday in January (Martin Luther King Jr.'s Birthday [observed]), the third Monday in February (Presidents' Day, Washington's Birthday), the last Monday in May (Memorial Day), July 4 (Independence Day), the first Monday in September (Labor Day), the second Monday in October (Columbus Day), November 11 (Veterans Day/Armistice Day), the fourth Thursday in November (Thanksgiving Day), and December 25 (Christmas). Also, the Tuesday following the first Monday in November is Election Day and is a legal holiday in presidential-election years (next in 2000).

Legal Aid   The well-meaning foreign visitor will probably never become involved with the American legal system. However, there are a few things you should know just in case. If you are stopped for a minor infraction of the highway code (for example, speeding), never attempt to pay the fine directly to a police officer; you may wind up arrested on the much more serious charge of attempted bribery. Pay fines by mail or directly to the clerk of the court. If you're accused of a more serious offense, it's wise to say and do nothing before consulting a lawyer. Under U.S. law, an arrested person is allowed one telephone call to a party of his or her choice. Call your embassy or consulate.

Mail   If you want your mail to follow you on your vacation and you aren't sure of your address, your mail can be sent to you, in your name, **c/o General Delivery** (Poste Restante) at the main post office of the city or region where you expect to be. (For the addresses and telephone numbers in Santa Fe, Taos, and Albuquerque, see the "Fast Facts" sections in chapters 4, 11, and 15, respectively.) The addressee must pick up the mail in person and must produce proof of identity (driver's license, credit or charge card, passport, etc.).

Domestic **postage rates** are 20¢ for a postcard and 32¢ for a letter. Check with any local post office for current international postage rates to your home country.

Generally found at intersections, **mailboxes** are blue with a red-and-white stripe and carry the designation **U.S. Mail.** If your mail is addressed to a U.S. destination, don't forget to add the five-digit **postal code,** or ZIP (Zone Improvement Plan) code, after the two-letter abbreviation of the state to which the mail is addressed (CA for California, NM for New Mexico, NY for New York, and so on).

Newspapers & Magazines   National newspapers include *The New York Times, USA Today,* and the *Wall Street Journal.* National news weeklies include *Newsweek, Time,* and *U.S. News and World Report.* In large cities most newsstands offer a small selection of the most popular foreign periodicals and newspapers, such as the *Economist, Le Monde,* and *Der Spiegel.* For information on local publications, see the "Fast Facts" sections in chapters 4, 11, and 15.

Radio/Television   Audiovisual media, with four coast-to-coast networks—ABC, CBS, NBC, and Fox—joined in recent years by the Public Broadcasting System (PBS) and the cable network CNN, play a major part in American life. In big cities, viewers have a choice of several dozen channels (including basic cable), most of them transmitting 24 hours a day, not counting the pay-TV channels that show recent movies or sports events. All options are usually indicated on your hotel TV set. You'll also find a wide choice of local radio stations, both AM and FM, each broadcasting particular kinds of talk shows and/or music—classical, country, jazz, pop, gospel—punctuated by news broadcasts and frequent commercials.

Rest Rooms   Visitors can usually find a rest room in a bar, restaurant, hotel, museum, department store, service station, or train station.

Safety   See "Safety" in "Preparing for Your Trip," earlier in this chapter.

Taxes In the United States there is no VAT (value-added tax) or other indirect tax at a national level. Every state, and each county and city in it, has the right to levy its own local tax on purchases, including hotel and restaurant checks, airline tickets, and so on. Taxes are already included in the price of certain services, such as public transportation, cab fares, telephone calls, and gasoline. The amount of sales tax varies from about 4% to 12%, depending on the state and city, so when you're making major purchases, such as photographic equipment, clothing, or stereo components, it can be a significant part of the cost.

Telephone/Telegraph/Telex The telephone system in the United States is run by private corporations, so rates, especially for long-distance service and operator-assisted calls, can vary widely—even on calls made from public telephones. Local calls in the United States usually cost 25¢ (they're 25¢ throughout New Mexico).

Generally, hotel surcharges on long-distance and local calls are astronomical. It's usually cheaper to call collect, use a telephone charge card, or use a public pay telephone, which you'll find clearly marked in most public buildings and private establishments as well as on the street. Outside metropolitan areas, public telephones are more difficult to find. Stores and gas stations are your best bet.

Most long-distance and international calls can be dialed directly from any phone (stock up on quarters if you're calling from a pay phone or use a telephone charge card). For calls to Canada and other parts of the United States, dial 1 followed by the area code and the seven-digit number. For international calls, dial 011 followed by the country code (Australia, 61; Republic of Ireland, 353; New Zealand, 64; United Kingdom, 44), then the city code (for example, 171 or 181 for London, 121 for Birmingham) and the telephone number of the person you wish to call.

For reversed-charge (collect calls) and person-to-person calls, dial 0 (zero, not the letter "O") followed by the area code and number you want; an operator will then come on the line, and you should specify that you are calling collect, or person-to-person, or both. If your operator-assisted call is international, ask for the overseas operator.

For local directory assistance ("information"), dial **1-411;** for long-distance information, dial 1, then the appropriate area code and 555-1212.

Like the telephone system, telegraph and telex services are provided by private corporations such as ITT, MCI, and above all, Western Union, the most important. You can bring your telegram in to the nearest Western Union office (there are hundreds across the country) or dictate it over the phone (☎ **800/325-6000**). You can also telegraph money (using a major credit or charge card) or have it telegraphed to you, very quickly over the Western Union system. (Note, however, that this service can be very expensive—the charge can run as high as 15% to 25% of the amount sent.)

Most hotels have fax machines available for guest use (be sure to ask about the charge to use it), and many hotel rooms are even wired for guests' fax machines. You'll probably also see signs for public faxes in the windows of local shops.

Telephone Directory There are two kinds of telephone directories available to you. The general directory is the so-called White Pages, in which private and business subscribers are listed in alphabetical order. The inside front cover lists the emergency numbers for police, fire, and ambulance, and other vital numbers (such as the poison-control center, crime-victims hotline, and so on). The first few pages are devoted to community-service numbers, including a guide to long-distance and international calling, complete with country codes and area codes.

The second directory, printed on yellow paper (hence its name, Yellow Pages), lists all local services, businesses, and industries by type of activity, with an index at the back. The listings cover not only such obvious items as automobile repairs by make of car or drugstores (pharmacies), often by geographical location, but also restaurants by type of cuisine and geographical location, bookstores by special subject and/or language, places of worship by religious denomination, and other information that the tourist might otherwise not readily find. The Yellow Pages also often include city plans or detailed area maps, often showing ZIP codes and public transportation routes.

Time   The United States is divided into six time zones: From east to west, Eastern standard time (EST), Central standard time (CST), Mountain standard time (MST), Pacific standard time (PST), Alaska standard time (AST), and Hawaii standard time (HST). Always keep the changing time zones in mind if you are traveling (or even telephoning) long distances in the United States. For example, noon in New York City (EST) is 11am in Chicago (CST), 10am in Santa Fe (MST), 9am in Los Angeles (PST), 8am in Anchorage (AST), and 7am in Honolulu (HST).

New Mexico is on Mountain standard time (MST), 7 hours behind Greenwich Mean Time. Daylight saving time is in effect from the first Sunday in April through the last Saturday in October (actually, the change is made at 2am on Sunday), except in Arizona, Hawaii, part of Indiana, and Puerto Rico. Daylight saving time moves the clock 1 hour ahead of standard time. (Americans use the adage "Spring ahead, fall back" to remember which way to change their clocks and watches.)

Tipping   This is part of the American way of life, based on the principle that one should pay for any special service received. (Often service personnel receive little direct salary and depend almost entirely on tips for their income.) Here are some rules of thumb:

In hotels, tip bellhops $1 per piece of luggage carried, and tip the chamber staff $1 per day. Tip the doorman or concierge only if he or she has provided some additional service (for example, calling a cab for you or obtaining difficult-to-get theater tickets).

In restaurants, bars, and nightclubs, tip the service staff 15% to 20% of the check, tip bartenders 10% to 15%, tip checkroom attendants $1 per garment, and tip valet-parking attendants $1 per vehicle. Tip the doorman only if he or she has provided some specific service (such as calling a cab for you). Tipping is not expected in cafeterias and fast-food restaurants.

Tip cab drivers 15% of the fare.

As for other service personnel, tip redcaps at airports or railroad stations $1 per piece of luggage, and tip hairdressers and barbers 15% to 20%.

Tipping ushers in cinemas, movies, and theaters and gas-station attendants is not expected.

THE AMERICAN SYSTEM OF MEASUREMENTS

Length

1 inch (in.)			=	2.54cm			
1 foot (ft.)	=	12 in.	=	30.48cm	=	.305m	
1 yard (yd.)	=	3 ft.			=	.915m	
1 mile	=	5,280 ft.					= 1.609km

To convert miles to kilometers, multiply the number of miles by 1.61 (for example, 50 mi. × 1.61 = 80.5km). Note that this conversion can be used to convert speeds from miles per hour (m.p.h.) to kilometers per hour (kmph).

To convert kilometers to miles, multiply the number of kilometers by .62 (example, 25km × .62 = 15.5 mi.). Note that this same conversion can be used to convert speeds from kilometers per hour to miles per hour.

Capacity

1 fluid ounce (fl. oz.)			=	.03 liter	
1 pint (pt.)	=	16 fl. oz.	=	.47 liter	
1 quart (qt.)	=	2 pints	=	.94 liter	
1 gallon (gal.)	=	4 quarts	=	3.79 liters	= .83 Imperial gal.

To convert U.S. gallons to liters, multiply the number of gallons by 3.79 (example, 12 gal. × 3.79 = 45.48 liters).

To convert liters to U.S. gallons, multiply the number of liters by .26 (example, 50 liters × .26 = 13 U.S. gal.).

To convert U.S. gallons to Imperial gallons, multiply the number of U.S. gallons by .83 (example, 12 U.S. gal. × .83 = 9.96 Imperial gal.).

To convert Imperial gallons to U.S. gallons, multiply the number of Imperial gallons by 1.2 (example, 8 Imperial gal. × 1.2 = 9.6 U.S. gal.).

Weight

1 ounce (oz.)			=	28.35g			
1 pound (lb.)	=	16 oz.	=	453.6g	=	.45kg	
1 ton	=	2,000 lb.	=		907kg	=	.91 metric ton

To convert pounds to kilograms, multiply the number of pounds by .45 (example, 90 lb. × .45 = 40.5kg).

To convert kilograms to pounds, multiply the number of kilos by 2.2 (example, 75kg × 2.2 = 165 lb.).

Area

1 acre			=	.41 ha		
1 square mile	=	640 acres	=	2.59 ha	=	2.6 sq. km

To convert acres to hectares, multiply the number of acres by .41 (example, 40 acres × .41 = 16.4ha).

To convert hectares to acres, multiply the number of hectares by 2.47 (example, 20ha × 2.47 = 49.4 acres).

To convert square miles to square kilometers, multiply the number of square miles by 2.6 (example, 80 sq. mi × 2.6 = 208 sq. km).

To convert square kilometers to square miles, multiply the number of square kilometers by .39 (example, 150 sq. km × .39 = 58.5 sq. mi.).

Temperature

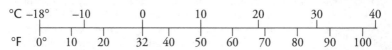

To convert degrees Fahrenheit to degrees Celsius, subtract 32 from °F, multiply by 5, then divide by 9 (example, 85°F − 32 × $^5/_9$ = 29.4°C).

To convert degrees Celsius to degrees Fahrenheit, multiply °C by 9, divide by 5, and add 32 (example, 20°C × $^9/_5$ + 32 = 68°F).

4 Getting to Know Santa Fe

After visiting Santa Fe, Will Rogers reportedly once said, "Whoever designed this town did so while riding on a jackass backwards and drunk." You, too, may find yourself perplexed maneuvering through the city. The meandering lanes and one-way streets can frustrate your best intentions. That's why people call it a walking town. Truly, that is the best way to get a feel for the idiosyncrasies of the place.

Like most cities of Hispanic origin, Santa Fe contains a plaza in the center of the city. Here you'll find tall shade trees and lots of grass, a nice place to sit and make travel plans. The area is full of restaurants, shops, art galleries, and museums, many within centuries-old buildings, and is dominated by the beautiful St. Francis Cathedral, a French Romanesque structure to the east.

In the plaza you'll notice the variety of people that inhabit and visit this city. Here you'll see Native Americans selling jewelry under the portal of the Palace of the Governors, teenagers in souped-up low riders cruising along, and people young and old hanging out in the ice-cream parlor. Such diversity, coupled with the variety of architecture, prompted the tourism department here to begin calling Santa Fe "The City Different."

Not far away is the Canyon Road district, a narrow, mostly one-way street packed with galleries and shops. Once the home of many artists, today you'll still find some who work within gallery studios. There are a number of fine restaurants in this district as well.

Farther to the east slopes the rugged Sangre de Cristo range. Locals spend a lot of time in these mountains picnicking, hiking, and skiing; for many, these mountains are why they choose to live in the region. When you look up at the mountains you'll see the peak of Santa Fe Baldy (with an elevation of more than 12,600 feet) as well as other peaks more than 12,000 feet high. Back in town, to the south of the plaza, is the Santa Fe River; it's a tiny tributary of the Rio Grande that is little more than a trickle for much of the year.

North is the Española Valley, and beyond that, the village of Taos, about 66 miles away. South of the city are ancient Native American turquoise mines in the Cerrillos Hills, and to the southwest is metropolitan Albuquerque, some 58 miles away. To the west, across the Caja del Rio Plateau, is the Rio Grande, and beyond that, the 11,000-foot Jemez Mountains and Valle Grande—an ancient and massive volcanic caldera. Pueblos dot the entire Rio Grande Valley an hour's drive in any direction.

1 Orientation

ARRIVING

BY PLANE The **Santa Fe Municipal Airport** (☎ **505/473-7243**), just outside the southwestern city limits on Airport Road off Cerrillos Road, has three paved runways, used primarily by private planes. In conjunction with United Airlines, commuter flights are offered by United Express, which is operated by Mountain West Airlines (☎ 800/241-6522). There are four daily departures from Denver during the week and three on weekends. When departing from Santa Fe, passengers can connect to other United Airlines flights in Denver. Call for schedules and fares.

Getting to and from the airport: Virtually all air travelers to Santa Fe arrive in Albuquerque, where they either rent a car or take one of the bus services. See "Getting There," in chapter 2, for details.

BY TRAIN & BUS For detailed information about train and bus service to Santa Fe, see "Getting There," in chapter 2.

BY CAR I-25 skims past Santa Fe's southern city limits, connecting it along one continuous highway from Billings, Montana, to El Paso, Texas. I-40, the state's major east–west thoroughfare, which bisects Albuquerque, affords coast-to-coast access to "The City Different." (From the west, motorists leave I-40 in Albuquerque and take I-25 north; from the east, travelers exit I-40 at Clines Corners and continue 52 miles to Santa Fe on US 285.) For those coming from the northwest, the most direct route is via Durango, Colorado, on US 160, entering Santa Fe on US 84.

For information on car rentals in Albuquerque, see "Getting There," in chapter 2; and for agencies in Santa Fe, see "Getting Around," later in this chapter.

VISITOR INFORMATION

The **Santa Fe Convention and Visitors Bureau** is located at 201 W. Marcy St., in Sweeney Center at the corner of Grant Street downtown (P.O. Box 909), Santa Fe, NM 87504-0909 (☎ **800/777-CITY** or 505/984-6760). If you would like information before you leave home but don't want to wait for it to arrive by mail, try this web site address: **http://www.santafe.org**. It will take you directly to the Santa Fe Convention and Visitors Bureau's home page.

CITY LAYOUT

MAIN ARTERIES & STREETS The limits of downtown Santa Fe are demarcated on three sides by the horseshoe-shaped Paseo de Peralta and on the west by St. Francis Drive, otherwise known as US 84/285. Alameda Street follows the north shore of the Santa Fe River through downtown, with the State Capitol and other federal buildings on the south side of the river, and most buildings of historic and tourist interest on the north, east of Guadalupe Street.

The plaza is Santa Fe's universally accepted point of orientation. Its four diagonal walkways meet at a central fountain, around which a strange and wonderful assortment of people of all ages, nationalities, and lifestyles can be found at nearly any hour of the day or night.

If you stand in the center of the plaza looking north, you'll be gazing directly at the Palace of the Governors. In front of you is Palace Avenue; behind you, San Francisco Street. To your left is Lincoln Avenue and to your right is Washington Avenue, which divides the downtown avenues into "east" and "west." St. Francis Cathedral is the massive Romanesque structure a block east, down San Francisco Street. Alameda Street is 2 full blocks behind you.

Santa Fe

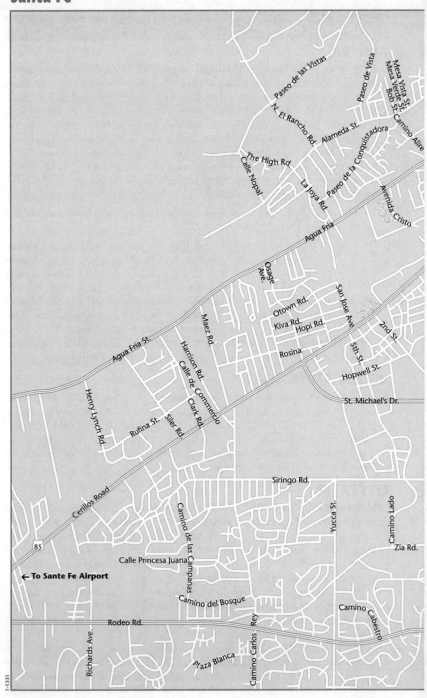

← To Sante Fe Airport

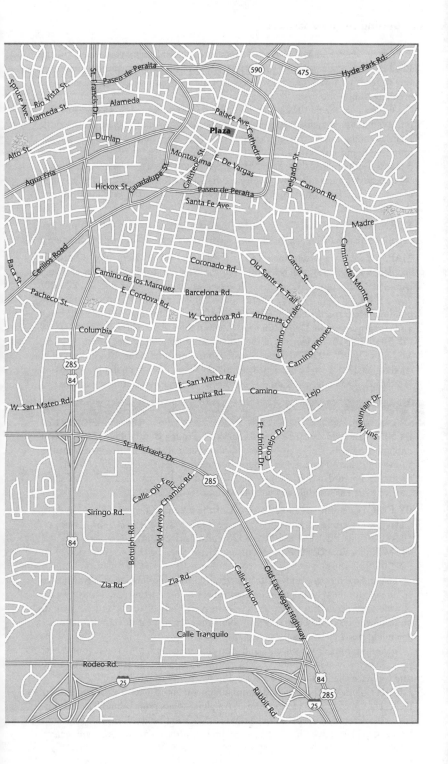

Near the intersection of Alameda Street and Paseo del Peralta, you'll find Canyon Road running east toward the mountains. Much of this street is one-way. The best way to see it is to walk up or down, taking time to explore shops and galleries and even have lunch or dinner.

Diagonally to the southwest from the downtown area, beginning opposite the state office buildings on Galisteo Avenue, is Cerrillos Road. Once the main north–south highway connecting New Mexico's state capital with its largest city, it is now a 6-mile-long motel and fast-food strip. St. Francis Drive, which crosses Cerrillos Road 3 blocks south of Guadalupe Street, is a far less tawdry byway, linking Santa Fe with I-25, located 4 miles southeast of downtown. The Old Pecos Trail, on the east side of the city, also joins downtown and the freeway. St. Michael's Drive interconnects the three arteries.

FINDING AN ADDRESS   Because of the city's layout, it's often difficult to know exactly where to look for a particular street address. It's best to call ahead for directions.

MAPS   Free city and state maps can be obtained at tourist information offices. An excellent state highway map is published by the **New Mexico Department of Tourism,** 491 Old Santa Fe Trail (P.O. Box 20002), Santa Fe, NM 87504 (☎ **800/ 733-6396** or 505/827-7336). There's also a Santa Fe Visitors Center in the same building. More specific county and city maps are available from the **State Highway and Transportation Department,** 1120 Cerrillos Rd., Santa Fe, NM 87504 (☎ **505/827-5100**). Members of the **American Automobile Association,** 1644 St. Michael's Dr. (☎ **505/471-6620**), can obtain free maps from the AAA office. Other good regional maps can be purchased at area bookstores. Gousha publishes a laminated "FastMap" of Santa Fe and Taos that has proved indispensable during my travels.

2 Getting Around

BY BUS   In 1993, Santa Fe opened **Santa Fe Trails** (☎ **505/984-6730**), its first public bus system. There are six routes, and visitors can pick up a map from the Convention and Visitors Bureau. Buses operate Monday through Friday from 6am to 9pm and Saturday from 8am to 8pm. There is no service on Sunday or holidays. Call for a current schedule and fare information.

BY CAR   Cars can be rented from any of the following firms in Santa Fe: Avis, Garrett's Desert Inn, 311 Old Santa Fe Trail (☎ **505/982-4361**); Budget, 1946 Cerrillos Rd. (☎ **505/984-8028**); Enterprise, 2641A Cerrillos Rd. and 4450 Cerrillos Rd. (☎ **505/473-3600**); and Hertz, Santa Fe Hilton, 100 Sandoval St. (☎ **505/982-1844**).

If Santa Fe is merely your base for an extended driving exploration of New Mexico, be sure to give the vehicle a thorough road check before starting out. There are a lot of wide-open desert and wilderness spaces in New Mexico, and if your car were to break down you could be stranded for hours in extreme heat or cold before someone might pass by.

Make sure your driver's license and auto club membership (if you're a member) are valid before you leave home. Check with your auto insurance company to make sure you're covered when out of state and/or when driving a rental car.

Street parking is difficult to find during summer months. There's a parking lot near the federal courthouse, 2 blocks north of the plaza; another one behind Santa Fe Village, a block south of the plaza; and a third at Water and Sandoval streets. If you stop by the Santa Fe Convention and Visitors Bureau, at the corner of Grant and Marcy streets, you can pick up a wallet-size guide to Santa Fe parking areas. The map shows both street and lot parking.

Unless otherwise posted, the speed limit on freeways is 75 mph; on most other two-lane open roads it's 65 mph. The minimum age for drivers is 16. Seat belts are required for drivers and all passengers age 5 and over; children under 5 must use approved child seats.

Since Native American reservations enjoy a measure of self-rule, they can legally enforce certain designated laws. For instance, on the Navajo reservation (New Mexico's largest), it is forbidden to transport alcoholic beverages, leave established roadways, or go without a seat belt. Motorcyclists must wear helmets. If you are caught breaking reservation laws, you are subject to *reservation* punishment.

The **State Highway and Transportation Department** has a toll-free hotline (☎ **800/432-4269**) providing up-to-the-hour information on road closures and conditions.

A word of warning: New Mexico has the highest per capita rate of traffic deaths of any American state. Drive carefully!

BY TAXI   It's best to telephone for a cab, because they are difficult to flag from the street. Taxis have no meters, but fares are set for given distances. Expect to pay a standard fee of $1.85 for the service and an average of about $1.50 per mile. **Capital City Cab** (☎ **505/438-0000**) is the main company in Santa Fe.

BY BICYCLE/ON FOOT   A bicycle is an excellent way to get around town. Check with **Palace Bike Rentals,** 409 E. Palace Ave. (☎ **505/984-2151**), for rentals.

The best way to see downtown Santa Fe is on foot. Free walking-tour maps are available at the tourist information center in Sweeney Center, 201 W. Marcy St. (☎ **800/777-CITY** or 505/984-6760), and several walking tours are included in chapter 7.

FAST FACTS: Santa Fe

Airport   See "Orientation," above.

American Express   There is no office in Santa Fe; the nearest one is in Albuquerque (see "Fast Facts: Albuquerque," in chapter 15).

Area Code   All of New Mexico is in area code **505.**

Baby-sitters   Most hotels can arrange for sitters on request. Alternatively, call the **Santa Fe Kid Connection** at ☎ **505/471-3100.**

Business Hours   **Offices** and **stores** are generally open Monday through Friday from 9am to 5pm, with many stores also open Friday night, Saturday, and Sunday in the summer season. Most **banks** are open Monday through Thursday from 10am to 3pm and Friday from 10am to 6pm; drive-up windows may be open later. Some may also be open Saturday morning. Most branches have cash machines available 24 hours. See also "Liquor Laws," below.

Car Rentals   See "Getting Around," above.

Climate   Santa Fe is consistently 10°F cooler than the nearby desert but has the same sunny skies, averaging more than 300 days of sunshine out of 365. Midsummer (July and August) days are dry and sunny (around 80°F), often with brief afternoon thunderstorms; evenings are typically in the upper 50s. Winters are mild and fair, with occasional and short-lived snow (average annual snowfall is 32 inches, although the ski basin gets an average of 225 inches). The average annual rainfall is 14 inches, most of it in summer; the relative humidity is 45%. (See also "When to Go," in chapter 2.)

Currency Exchange You can exchange foreign currency at two banks in Santa Fe: Sun West Bank, 1234 St. Michael's Dr. (☎ **505/471-1234**), and First Security Bank, 121 Sandoval St. (☎ **505/983-4312**).

Dentists Located in the geographic center of the city is Dr. Leslie E. La Kind, at 400 Botulph Lane (☎ **505/988-3500**). Dr. La Kind offers emergency service.

Doctors The Lovelace Alameda Clinic, 901 W. Alameda St. (☎ **505/995-2900**), is in the Solano Center near St. Francis Drive. It's open daily from 8am to 8pm. For Physicians and Surgeons Referral and Information Services, call the American Board of Medical Specialties at ☎ **800/776-2378.**

Embassies/Consulates See "Fast Facts: For the Foreign Traveler," in chapter 3.

Emergencies For police, fire, or ambulance emergency, dial ☎ **911.**

Eyeglass Repair The **Quintana Optical Dispensary,** 109 E. Marcy St. (☎ **505/988-4234**), provides 1-hour prescription service Monday through Friday from 9am to 5pm and Saturday from 9am to noon. They will also repair your current glasses.

Hospitals St. Vincent Hospital, 455 St. Michael's Dr. (☎ **505/983-3361**), is a 268-bed regional health center. Patient services include urgent and emergency-room care and ambulatory surgery. Other health services include the Women's Health Services Family Care and Counseling Center (☎ **505/988-8869**). Lovelace Health Systems has a walk-in office at 901 W. Alameda St. (☎ **505/995-2900**).

Hotlines The following hotlines are available in Santa Fe: battered families (☎ **505/473-5200**), poison control (☎ **800/432-6866**), psychiatric emergencies (☎ **505/983-3361**), and sexual assault (☎ **505/473-7818**).

Information See "Visitor Information," above.

Libraries The Santa Fe Public Library is half a block from the plaza at 145 Washington Ave. (☎ **505/984-6780**). There are branch libraries at Villa Linda Mall and at 1713 Llano St., just off St. Michael's Drive. The New Mexico State Library is at 325 Don Gaspar Ave. (☎ **505/827-3800**). Specialty libraries include the Archives of New Mexico, 404 Montezuma St., and the New Mexico History Library, 110 Washington Ave.

Liquor Laws The legal drinking age is 21 throughout New Mexico. Bars may remain open until 2am Monday through Saturday and until midnight on Sunday. Wine, beer, and spirits are sold at licensed supermarkets and liquor stores, but there are no package sales on election days until after 7pm. It is illegal to transport liquor through most Native American reservations.

Lost Property Contact the city police at ☎ **505/473-5000.**

Newspapers & Magazines *The New Mexican*—Santa Fe's daily paper—is the oldest newspaper in the West. Its offices are at 202 E. Marcy St. (☎ **505/983-3303**). The weekly *Santa Fe Reporter,* published on Wednesday, is often more willing to be controversial; its entertainment listings are excellent. Regional magazines published locally are *New Mexico Magazine* (monthly, statewide interest) and the *Santa Fean Magazine* (monthly, Southwestern lifestyles).

Pharmacies The **R&R Professional Pharmacy,** at 1691 Galisteo St. (☎ **505/988-9797**), is open Monday through Friday from 9am to 6pm and Saturday from 9am to noon. Emergency and delivery service is available.

Photographic Needs   Everything from film purchases to camera repairs to 1-hour processing can be handled by the **Camera Shop,** 109 E. San Francisco St. (☎ **505/ 983-6591**). Twenty-four–hour processing is available at **Camera & Darkroom,** 216 Galisteo St. (☎ **505/983-2948**).

Police   In case of emergency, dial ☎ **911.**

Post Offices   The **Main Post Office** is at 120 S. Federal Place (☎ **505/988-6351**), 2 blocks north and 1 block west of the plaza. The **Coronado Station branch** is at 2071 S. Pacheco St. (☎ **505/438-8452**). Both are open Monday through Friday from 8am to 4pm and Saturday from 9am to 1pm. Most of the major hotels have stamp machines and mailboxes with twice-daily pickup. The ZIP code for central Santa Fe is 87501.

Radio   Santa Fe's radio stations include KSFR-FM 90.7 (classical and jazz), KNYN-FM 95.5 (country), KBAC-FM 98.1 (adult contemporary), KLSK-FM 104.1 (adult alternative), KBOM-FM 106.7 (oldies), and KVSF, 1260 AM (news and talk). Albuquerque stations are easily received in Santa Fe.

Safety   Though the tourist district appears very safe, Santa Fe is not on the whole a safe city; theft is common (see "Northern New Mexico Today," in chapter 1), and the number of reported rapes has risen. The good news is that Santa Fe's over-all crime statistics do appear to be falling. Still, when walking the city streets, guard your purse carefully, because there are many bag-grab thefts, particularly during the summer tourist months. Also, be as aware of your surroundings as you would in any other major city.

Taxes   A tax of 10.5625% is added to all lodging bills.

Taxis   See "Getting Around," above.

Television   The three Albuquerque network affiliates—KOB-TV (Channel 4, NBC), KOAT-TV (Channel 7, ABC), and KQRE-TV (Channel 13, CBS)—all have offices at the State Capitol.

Time Zone   New Mexico is on Mountain standard time, 1 hour ahead of the West Coast and 2 hours behind the East Coast. When it's 10am in Santa Fe, it's noon in New York, 11am in Chicago, and 9am in San Francisco. Daylight saving time is in effect from early April to late October.

Useful Telephone Numbers   Information on **road conditions** in the Santa Fe area can be obtained from the state police (☎ **505/827-5594**). For **time and temperature,** call ☎ **505/473-2211.**

Weather   For weather forecasts, call ☎ **505/988-5151.**

5 Where to Stay in Santa Fe

There may not be a bad place to stay in Santa Fe. From downtown hotels to Cerrillos Road motels, ranch-style resorts to quaint bed-and-breakfasts, the standard of accommodation is universally high.

You should be aware of the seasonal nature of the tourist industry in Santa Fe. Accommodations are often booked solid through the summer months, and most places raise their prices accordingly. Rates increase even more during Indian Market, the third weekend of August. During these periods, it's essential to make reservations well in advance.

Still, there seems to be little agreement on what constitutes the tourist season; one hotel may raise its rates July 1 and lower them again in mid-September, while another may raise its rates from May to November. Some hotels raise their rates again over the Christmas holidays or recognize a shoulder season. It pays to shop around during the "in-between" seasons of May through June and September through October.

No matter the season, discounts are often available to seniors, affiliated groups, corporate employees, and others. If you have any questions about your eligibility for these lower rates, be sure to ask.

A combined city-state tax of 10.75% is added to every hotel bill in Santa Fe. And, unless otherwise indicated, all recommended accommodations come with private bath. All hotels listed offer rooms for nonsmokers and for travelers with disabilities. For the B&Bs, I've indicated in the text whether they do this or not.

ACCOMMODATIONS CATEGORIES   In this chapter, hotels/motels are listed first by geographical area (downtown, Northside, or Southside) and then by price range, based on midsummer rates for doubles: **Very Expensive** refers to rooms that average $150 or more per night; **Expensive** rooms are those that go for $110 to $150; **Moderate** encompasses rooms that range from $75 to $110; and **Inexpensive,** those that cost up to $75.

Following hotel/motel and bed-and-breakfast recommendations, you'll find suggestions for campgrounds and RV parks.

RESERVATIONS SERVICES   Although Santa Fe has more than 50 hotels, motels, bed-and-breakfast establishments, and other accommodations, it can still be difficult to find available rooms at the peak of the tourist season. Year-round assistance is available from **Santa Fe Central Reservations,** 320 Artist Rd., Suite 10 (☎ **800/776-7669** or 505/983-8200; fax 505/984-8682). This service will

also book tickets for the Santa Fe Opera, Chamber Music Festi~~~
Teatro Flamenco, and the Desert Chorale, as well as jeep trips int~
white-water rafting, horseback riding, mountain bike tours, and gou~~
gency Lodging Assistance—especially helpful around fiesta time in Septem~~
is available free after 4pm daily (☎ **505/986-0043**).

1 Best Bets

- **Best Historic Hotel: La Fonda,** 100 E. San Francisco St. (☎ **505/982-5511**), is the oldest (and one of the nicest) hotels in Santa Fe. It has hosted a long list of notables, including Ulysses S. Grant and Kit Carson. Billy the Kid is rumored to have been a dishwasher there.
- **Best for a Romantic Getaway:** Bed-and-breakfast inns are always my first choice for a romantic getaway, and in Santa Fe I would recommend one of the new suites at the **Water Street Inn,** 427 Water St. Eclectic Southwestern decor and a private patio with a fountain make for an elegantly romantic stay (☎ **505/984-1193**).
- **Best for Families: Bishop's Lodge,** Bishop's Lodge Road (☎ **505/983-6377**), offers a wide variety of activities for children (pony ring, trail rides, pool, and day program) and is a good bet for families who can afford to spend a little extra money. Otherwise, **Homewood Suites,** 400 Griffin St. (☎ **505/988-3000**), which offers units with full kitchens, a 24-hour convenience store, and video rental shop, and features a swimming pool, is a true bargain for families.
- **Best Location:** There are a number of centrally located hotels. Among them, **La Fonda** (see address and telephone above) is the only hotel right on the Plaza; the **Hotel Plaza Real,** 125 Washington Ave. (☎ **505/988-4900**), and the **Inn of the Anasazi,** 113 Washington Ave. (☎ **505/988-3030**), are both just a half block from the plaza.
- **Best Fitness Facilities:** The **Eldorado Hotel,** 309 W. San Francisco St. (☎ **505/ 988-4455**), has a wonderful heated rooftop pool and Jacuzzi, exercise room with a view, professional masseuse, and his-and-hers saunas.
- **Best Southwestern Bed-and-Breakfast:** If you want to stay in a Southwestern-style bed-and-breakfast that reflects the unique style of its owner, try **Adobe Abode,** 202 Chapelle St. (☎ **505/983-3133**), which is within walking distance of the Plaza. Each of the rooms is individually and creatively decorated. In addition to traditional Southwestern features, you'll find everything from Balinese puppets to Oaxacan hand-loomed fabrics, as well as cowboy paraphernalia.

2 Downtown

Everything within the horseshoe-shaped Paseo de Peralta and east a few blocks on either side of the Santa Fe River, is considered downtown Santa Fe. All of these accommodations are within walking distance of the plaza.

VERY EXPENSIVE

Eldorado Hotel

309 W. San Francisco St., Santa Fe, NM 87501. ☎ **800/955-4455** or 505/988-4455. Fax 505/995-4544. 219 rms, 18 suites, 8 condo suites. A/C MINIBAR TV TEL. Jan 1–Jan 31, $129–$239 single or double; Feb 1–May 1, $159–$269 single or double; May 2–June 25 and Aug 24–Oct 25, $209–$319 single or double; June 26–Aug 23, $239–$349 single or double; Oct 26–Dec 18, $169–$279 single or double; Dec 19–Dec 31, $189–$309 single or double. Year-round, $249–$975 suite. Ski and other package rates are available. AE, DC, DISC, MC, V. 24-hour valet parking, $8 per night.

...nce its opening in 1986, the Eldorado has stood like a monolith at the center of town. Locals wonder how the five-story structure bypassed the two-story zoning restrictions. Still, the architects did manage to meld Pueblo revival style with an interesting cathedral feel, inside and out. The lobby is grand, with a high ceiling that continues into the court area and the cafe.

Take your time while wandering through, since well over a million dollars' worth of art, most of it from Northern New Mexico, adorns the place. Most notable in the entry to the court is an *olla*, or pot, made in the early 1920s by a Zia potter, and decorated with a parrot, a bird sacred to the Anasazi as well as the Pueblo people.

The rooms follow along with the artistic Southwestern motif, which the Eldorado manages to pull off better than most. There's a warmth here, created in particular by the kiva fireplaces in many of the rooms as well as the tapestries supplied by Seret and Sons, a local antique dealer. This, along with the Inn of the Anasazi, are the hotels in town for those who expect consistency and fine service. The suites here come with a "butler" who will do "anything legal" for you, including walking your dog.

You'll find small families and businesspeople staying here, as well as conference-goers. Most of the rooms have views of downtown Santa Fe, many from balconies. The hotel rightfully prides itself on the quietness of the rooms.

If you're really indulging, join the ranks of Mick Jagger, Geena Davis, and King Carlos of Spain and try the penthouse five-room Presidential suite. Just down the street from the main hotel is Zona Rosa, which houses two-, three-, and four-bedroom condo suites with full kitchens.

Dining/Entertainment: The innovative and elegant Old House restaurant was built on the preserved foundation of an early 1800s Santa Fe house. The viga-latilla ceiling, polished wood floor, and pottery and *kachinas* (Pueblo Indian carved dolls) in nichos give it a distinct regional touch, found also in its creative Southwestern cuisine. More casual meals are served in the spacious Eldorado Court. The lobby lounge offers low-key entertainment.

Services: Concierge, room service, butlers, dry cleaning and laundry service, nightly turndown, twice-daily maid service, safe-deposit boxes.

Facilities: Heated rooftop swimming pool and Jacuzzi, medium-sized health club, his-and-hers saunas, professional masseuse, business center, beauty salon, boutiques.

Hilton of Santa Fe

100 Sandoval St. (P.O. Box 25104), Santa Fe, NM 87504-2387. ☎ **800/336-3676**, 800/HILTONS, or 505/988-2811. Fax 505/986-6439. 158 rms, 3 suites, 3 casitas. A/C MINIBAR TV TEL. Jan 1–May 1 and Oct 29–Dec 21, $89–$219 double; May 2–June 29 and Sept 4–Oct 28, $119–$239 double; June 30–Sept 3, $139–$260 double; Dec 22–Dec 31, $160–$260 double. Year-round, $380–$520 suite/casita. Extra person $20. AE, CB, DC, DISC, MC, V. Free parking.

With its city-landmark bell tower, the Hilton encompasses a full city block (a few-minutes' walk from the plaza) and incorporates most of the historic landholdings of the 350-year-old Ortiz family estate. It's built around a central pool and patio area and is a fine blend of ancient and modern styles.

Rooms are fairly standard, not nearly as refined as the Eldorado, but many visitors like this hotel because it offers all the amenities of a fine hotel at a fairly reasonable price. It also has an intimacy that some of the other large downtown hotels lack. The lobby is cozy, with huge vigas and a big fireplace; it's decorated in a refined Southwestern style.

Remodeled in 1997 in a warm Aztec-Southwestern style, the guest rooms are large, most with a small patio or balcony. All rooms now have hair dryers, honor bars, safes, and coffeemakers.

Downtown Santa Fe Accommodations

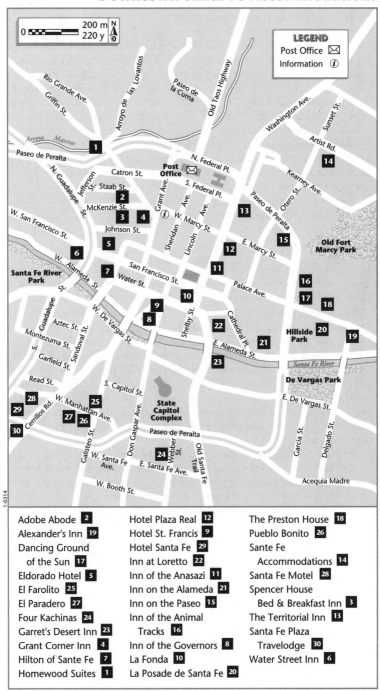

Adobe Abode **2**
Alexander's Inn **19**
Dancing Ground
   of the Sun **17**
Eldorado Hotel **5**
El Farolito **25**
El Paradero **27**
Four Kachinas **24**
Garret's Desert Inn **23**
Grant Corner Inn **4**
Hilton of Sante Fe **7**
Homewood Suites **1**

Hotel Plaza Real **12**
Hotel St. Francis **9**
Hotel Santa Fe **29**
Inn at Loretto **22**
Inn of the Anasazi **11**
Inn on the Alameda **21**
Inn on the Paseo **15**
Inn of the Animal
   Tracks **16**
Inn of the Governors **8**
La Fonda **10**
La Posade de Santa Fe **20**

The Preston House **18**
Pueblo Bonito **26**
Sante Fe
   Accommodations **14**
Santa Fe Motel **28**
Spencer House
   Bed & Breakfast Inn **3**
The Territorial Inn **13**
Santa Fe Plaza
   Travelodge **30**
Water Street Inn **6**

In June 1994 the Hilton opened Casa Ortiz de Santa Fe, a small building adjacent to the main hotel, which houses three *casitas*. The building was once the coach house (ca. 1625) of Nicholas Ortiz III. Today the thick adobe walls encompass elegantly Southwestern-style suites. Each has a living room with kiva fireplace, fully stocked kitchenette (with microwave, stove, and minirefrigerator), and bathroom with whirlpool tub. One of the one-bedroom units offers a fireplace in the bedroom.

Dining/Entertainment: Two restaurants occupy the premises of the early 18th-century Casa Ortiz. The Piñon Grill serves a variety of wood-fire grilled items in a casual atmosphere. The Chamisa Courtyard, which serves breakfast, features casual garden-style tables amid lush greenery under a large skylight; it's built on the home's enclosed patio. El Cañon wine and coffee bar specializes in fine wines by the glass and gourmet coffees. El Cañon also serves breakfast, lunch, and dinner. Specialties include freshly baked breads and pastries as well as sandwiches.

Services: Concierge, room service, courtesy van, dry cleaning, laundry service.

Facilities: VCRs, Spectravision, outdoor swimming pool, Jacuzzi, car-rental agency, travel agency, gift shop.

Homewood Suites

400 Griffin St., Santa Fe, NM 87501. ☎ **800/225-5466** or 505/988-3000. Fax 505/988-4700. 105 suites. A/C TV TEL. $99–$150 1-bedroom suite; $109–$160 2-bedroom suite; $119–$170 1-bedroom suite with gas fireplace; $179–$299 2-bedroom, 2-bathroom suite with gas fireplace. Rates include continental breakfast. AE, DC, DISC, MC, V. Free parking.

This is upscale practicality. Within walking distance from the Plaza, tucked within a residential neighborhood, Homewood is as it sounds—the place you go when you want to be able to cook and eat as you do at home. Built in 1994, its guest rooms on 3 stories are decorated in Southwestern style, with a cold efficiency that marks the place. The rooms are consistently comfortable, with full kitchens that include microwaves, stoves, refrigerators, and dishwashers, as well as amenities such as ironing boards and irons, recliners, and VCRs. Some have balconies and patios, and some have gas fireplaces. There is a homey feel in the main room, where an extended continental breakfast and evening hors d'oeuvres are served around a "kitchen" environment.

Services: Complimentary grocery shopping and local shuttle service.

Facilities: Fully stocked kitchens; VCRs and video rentals; year-round heated outdoor pool and two outdoor Jacuzzis; well-equipped health club; activity court; 24-hour business center with a computer, copier, and business supplies; Laundromat; Suite Shop (for sundries, food, and video rentals); picnic area with gas grill.

Inn at Loretto

211 Old Santa Fe Trail (P.O. Box 1417), Santa Fe, NM 87501. ☎ **800/727-5531** or 505/988-5531. Fax 505/984-7988. 137 rms, 3 suites. A/C TV TEL. Jan 3–June 25, $145–$200 double; June 26–Nov 4, $170–$235 double; Nov 5–Dec 22, $145–$200 double; Dec 23–Jan 2, $170–$220 double. Year-round, $500–$1,500 suite. Extra person $15. Children 12 and under stay free in parents' room. AE, CB, DC, DISC, MC, V. Free parking.

This much-photographed hotel, just 2 blocks from the plaza, was built in 1975 to resemble the Taos Pueblo. Light and shadow dance upon the five-level structure as the sun crosses the sky. Two years ago it came under new ownership and is now undergoing a $3.7 million makeover. The outdated decor is being replaced by a Southwest/Montana ranch style throughout that reeks of Ted Turner and Jane Fonda—as it should since they use the same decorator. The new owners, Noble House Hotels and Resorts, are determined to compete with other large hotels such as the Eldorado. With faux painted walls and an interesting and cozy new lobby lounge, they may manage to catch up. Still, the hotel's rooms are pretty standard, a bit small, and the bathrooms are basic; but the prices are pretty reasonable. With the renovation, the hotel hopes to attract more groups. Overall, it is fairly quiet and has nice views—especially on the northeast side, where you'll see both the historic St. Francis Cathedral and the Loretto Chapel (with its "miraculous" spiral staircase; see "More Attractions" in chapter 7).

Dining and Entertainment: Named for a carved wooden serpent above the bar, Nellie's (see chapter 6) serves fine nouveau Southwestern cuisine in a hip-folk-artsy environment.

Services: Concierge, room service, complimentary valet laundry, airport shuttle service.

Facilities: Outdoor heated swimming pool (mid-May to mid-October), tennis and golf privileges nearby can be arranged, shopping arcade (with a fine-art gallery, four boutiques, gold designer, bookstore, three gift shops, sundries shop, and hair salon), audiovisual conferencing equipment.

✪ Inn of the Anasazi

113 Washington Ave., Santa Fe, NM 87501. ☎ **800/688-8100** or 505/988-3030. Fax 505/988-3277. 59 rms. A/C MINIBAR TV TEL. Nov–Mar, $199–$345 double; Apr–Oct, $235–$395 double. Holiday and festival rates may be higher. AE, CB, DC, DISC, MC, V. Valet parking $10 per day.

In an incredible feat, the designers of this fine luxury hotel have managed to create a feeling of grandness in a very limited space. Flagstone floors, vigas, and latillas are enhanced by oversized cacti that evoke the feeling of an Anasazi cliff dwelling and lend a warm and welcoming feel. Accents are appropriately Navajo, in a nod to the fact that the Navajo live in the area the Anasazi once inhabited. A half block off the plaza, this hotel was built in 1991 to cater to travelers who know their hotels. Amenities include stereos and VCRs in all rooms, as well as private safes, coffeemakers with custom blended coffee beans, bathroom telephones, hair dryers, 100 percent cotton linens, and organic bath oils and shampoos, as well as organic food in the restaurant. On the ground floor are the "living room" and "library," with oversized furniture and replica Anasazi pottery and Navajo rugs.

The rooms, even the smallest, are spacious, with diamond-finished walls and decor in cream tones accented by novelties such as iron candle sconces, original art, four-poster beds, gaslit kiva fireplaces, and humidifiers. All the rooms are quiet and comfortable, though none have dramatic views.

Dining/Entertainment: See the Anasazi Restaurant in chapter 6 for a full description.

Services: Concierge, room service (6am to 11pm), laundry service, newspaper delivery, in-room massage, twice-daily maid service, tours of galleries and museums, stationary bicycles available for use in guest rooms, free coffee or refreshments in lobby.

Facilities: VCRs, video rentals, access to nearby health club can be arranged, library/boardroom, audiovisual and communication equipment available.

✪ Inn on the Alameda

303 E. Alameda St., Santa Fe, NM 87501. ☎ **800/289-2122** or 505/984-2121. Fax 505/986-8325. 67 rms, 9 suites. A/C TV TEL. Nov–Feb, $140–$190 single or double; $200–$300 suite. Mar–June, $155–$210 single or double; $225–$315 suite. July–Oct, $170–$225 single or double; $240–$315 suite. Holiday and special-event rates may be higher. Rates include breakfast. AE, CB, DC, DISC, MC, V. Free parking.

Just across the street from the bosque-shaded Santa Fe River and 3 blocks from the Plaza is the Inn on the Alameda, a cozy stop for those who like the services of a hotel with the intimacy of an inn. Begun 10 years ago as a bed-and-breakfast, it now sprawls into four buildings. There are casita suites to the west, two 3-story buildings at the center, and another 1-story that contains suites. All are pueblo-style adobe, ranging in age, but most were built in the late 1980s.

The owner, Joe Schepps, appreciates traditional Southwestern style; he's used red brick in the dining area and Mexican equipal furniture in the lobby. He went all out here in the construction, using thick vigas and shiny latillas, all set around a grand fireplace.

The rooms follow a similar good taste, though the decor is more standard Santa Fe style (boxy with pastel upholstery) than the lobby might lead you to expect. The newer deluxe rooms and suites in the easternmost building are in the best shape. The traditional rooms are quaint, some with interesting angled bed configurations, but the bathrooms are in need of, and I hear are getting, a renovation. Beware of the casitas on the western corner of the property. After some recent tree trimming, they've tended to pick up traffic noise. Still, take note of the trees surrounding the inn—cottonwoods and aspens, which, when you step out on some balconies, make you feel as though you're in a tree house. If you're an art shopper, this is an ideal spot because it's a quick walk to Canyon Road. Amenities include robes, refrigerators, and kiva fireplaces in some rooms. Pets are welcome (the hotel offers a pet program that features pet amenities and a pet-walking map).

Dining/Entertainment: An elaborate continental "Breakfast of Enchantment" is served each morning in the Agoyo Room, outdoor courtyard, or your own room. A full-service bar is open nightly.

Services: Concierge, limited room service, dry cleaning and laundry service, newspaper delivery, baby-sitting can be arranged.

Facilities: Medium-sized fitness facility, massage, two open-air Jacuzzis, Laundromat.

✪ La Fonda

100 E. San Francisco St. (P.O. Box 1209), Santa Fe, NM 87501. ☎ **800/523-5002** or 505/ 982-5511. Fax 505/988-2952. 153 rms, 21 suites. A/C TV TEL. $179 standard double, $189 deluxe double; $200–$500 suite. Extra person $15. Children under 12 stay free in parents' room. AE, CB, DC, DISC, MC, V. Parking $4 per day in a 3-story garage.

Whether you stay in this hotel on the southeast corner of the plaza or elsewhere, it's worth strolling through, just to get a sense of how Santa Fe once was, and still in some ways is. This was the inn at the end of the Santa Fe Trail; it saw trappers, traders, and merchants, as well as notables such as President Rutherford B. Hayes and General Ulysses S. Grant. The original inn was dying of old age in 1920 when it was razed and replaced by the current La Fonda. Its architecture is Pueblo Revival: imitation adobe with wooden balconies and beam ends protruding over the tops of windows. Inside, the lobby is rich, slightly dark, with people bustling about, drinking in the cafe and buying jewelry from Native Americans.

As you head farther into this 4-story building you may come across Ernesto Martinez, who wanders around finding things to paint. You'll see his colorful, playful designs throughout the hallways and in the rooms decorating tin mirror frames and carved wooden headboards.

The hotel has seen some renovation through the years, as well as a whole new wing recently completed to the east—where you'll find deluxe suites and new meeting spaces. Overall, however, this hotel isn't the model of refinement. For that, you'd best go to the Hotel Santa Fe or other newer places. No room is the same here, and while each has its own funky touch, some are more kitsch than quaint. Some rooms have minirefrigerators, fireplaces, and private balconies. Still, if you want a feel of real Santa Fe, this is the place to stay.

Dining and Entertainment: The French Pastry Shop is the place to get cappuccino and crepes; La Fiesta Lounge draws many locals to their economical New Mexican food lunch buffet; and La Plazuela offers what some believe to be the best chile rellenos in town in a skylit garden patio. The Bell Tower Bar, at the southwest corner of the hotel, is the highest point in downtown Santa Fe, a great place for a cocktail and a view of the city.

Services: Concierge, room service, dry cleaning and laundry service, tour desk.

Facilities: Outdoor swimming pool, two indoor Jacuzzis, cold plunge, massage room, ballroom, and shopping arcade.

EXPENSIVE

✪ Hotel St. Francis

210 Don Gaspar Ave., Santa Fe, NM 87501. ☎ **800/529-5700** or 505/983-5700. Fax 505/989-7690. 82 rms, 2 suites. A/C TV TEL. May 1–Oct 31 plus the weeks of Dec 25 and Jan 1, $118–$188 double; $228–$353 suite. Nov–Feb, $88–$138 double; $178–$278 suite. Mar–Apr, $98–$153 double; $178–$278 suite. Children under 12 stay free in parents' room. AE, CB, DC, DISC, MC, V. Free parking.

One block from the plaza, this hotel is on its third incarnation. Built in the 1880s as the Palace Hotel, it burned in 1922 and was rebuilt as the de Vargas Hotel in 1923. Legend has that it rivaled the State Capitol Building as a site for conducting official government business in the '30s and '40s. Later it fell into a lamentable state of disrepair. In 1986 it was renovated—but forget Southwestern style. Here you have a touch of European luxury, including, in the lobby, a Victorian fireplace where cherubs hover, a theme repeated throughout the hotel.

The rooms follow the European decor, each with its own unique bent. You'll find a "fishing room," "golf room," "garden room," and "music room," the themes evoked in the furnishings: a vintage set of golf clubs here, a sheet of music in a dry flower arrangement there. The hotel attracts individual travelers, as well as families and many Europeans, who are well taken care of by their concierge who speaks six languages. If you don't plan to spend much time in your room, you can even get a lovely but small standard room here for $100 in high season, unheard of anywhere else downtown. Request a room facing east, and you'll wake each day to a view of the mountains, seen through lovely lace. All rooms have refrigerators and closet safes.

Dining and Entertainment: A recent renovation has given the restaurant and bar a European gentleman's club ambience. Breakfast, lunch, and dinner specials are served daily with prices worth checking out. The lobby and veranda are favorites for locals to take their afternoon tea. You'll eat scones, pastries, and tea sandwiches—all baked in-house—and drink tea, sherry, port, or champagne.

Services: Concierge, room service, dry cleaning and laundry service, free coffee or refreshments in lobby.

Facilities: Spectravision movie channels, guest membership at nearby health club.

Hotel Plaza Real

125 Washington Ave., Santa Fe, NM 87501. ☎ **800/279-7325** or 505/988-4900. Fax 505/983-9322. 56 rms, 44 suites. A/C MINIBAR TV TEL. Nov–May, $119–$189 double; June–Oct, $149–$219 double. Year-round, $295–$495 suite. Extra person $15. Children under 12 stay free in parents' room. Rates include breakfast. AE, CB, DC, DISC, MC, V. Parking $10 per day.

New Orleans meets Santa Fe in this Territorial-style hotel built in 1990. The lobby is rustically elegant, built around a fireplace with balconies perched above. Given its location just steps from the plaza, this is a real bargain. The rooms are clean and nicely decorated with Southwestern-style furniture and accent amenities such as *bancos* (adobe benches) and French doors opening onto balconies or terraces surrounding a quiet courtyard decorated with chile *ristras* where an occasional pigeon will flap by. Most of the rooms are suites and cover guests needs well: coffeemakers, hair dryers, irons and ironing boards in each. At press time, a restaurant serving three meals was planned. Try the bar for an afternoon drink or the veranda for coffee. Services include concierge (who gives excellent historical walking tours), 24-hour desk, dry cleaning and laundry service, twice daily maid service, and access to a nearby health club.

Hotel Santa Fe

1501 Paseo de Peralta, Santa Fe, NM 87501. ☎ **800/825-9876** or 505/982-1200. Fax 505/984-2211. 131 rms, 91 suites. A/C MINIBAR TV TEL. Jan 2–Feb 10, $79 double; $139 suite. Feb 11–April 30, $99 double; $115 suite. May 1–June 19, $119 double; $179 suite. June 20–Aug 24, $139 double; $199 suite. Aug 25–Nov 1, $119 double; $179 suite. Nov 2–Dec 23, $89 double; $159 suite. Dec 24–Jan 1, $139 double; $199 suite. Extra person $10. Children 17 and under stay free in parents' room. AE, CB, DC, DISC, MC, V. Free parking.

About a 10-minute walk south of the Plaza you'll find this newer 3-story establishment, the only Native American–owned hotel in Santa Fe. The Picuris Pueblo is the majority stockholder here, and part of the pleasure of staying is the culture they bring to your visit. This is not to say that you'll get any sense of the rusticity of a pueblo in your accommodations—this sophisticated 5-year-old hotel is decorated in Southwestern style with a few novel aspects such as an Allan Houser bronze buffalo dancer watching over the front desk and an *horno*-shaped fireplace surrounded by comfortable furniture in the lobby.

The rooms have clean lines and an appreciable newness, accented with pine Taos-style furniture. Where you will get a sense of the Native American presence is in the patio during summer, when Picuris dancers come to perform, and bread bakers uncover the horno and prepare loaves for sale.

Rooms on the north side get less street noise from Cerrillos Road and have better views of the mountains, but you won't have the sun shining onto your balcony.

Dining and Entertainment: The restaurant is another place where the hotel's origins are recognizable. The famed Corn Dance Cafe serves a standard breakfast, but for lunch and dinner you can dine on Native American food from all over the Americas. Expect buffalo and turkey instead of beef and chicken. Many of the dishes are accompanied by what chef Loretta Oden calls the three sisters: corn, beans, and squash.

Services: Concierge, limited room service, dry cleaning and laundry service, in-room massage, twice-daily maid service, baby-sitting, secretarial services, courtesy shuttle to the plaza and Canyon Road.

Facilities: Outdoor heated pool, access to nearby health club, Jacuzzi, conference rooms, Laundromat, car-rental desk, Picuris pueblo gift shop.

Inn of the Governors

234 Don Gaspar Ave., Santa Fe, NM 87501. ☎ **800/234-4534** or 505/982-4333. Fax 505/989-9149. 100 rms. A/C TV TEL. Jan 1–Mar 6, $117–$197 double; $207–$237 suite. Mar 7–May 1, $137–$217 double; $227–$247 suite. May 2–June 26, $157–$257 double; $287 suite. June 27–Sept 13, $167–$277 double; $297 suite. Sept 14–Oct 25, $157–$277 double; $287–$297 suite. Oct 26–Dec 18, $137–$217 double; $227–$247 suite. Dec 19–Dec 31, $167–$277 double; $287–$307 suite. Extra person $10. Children under 18 stay free in parents' room. AE, CB, DC, MC, V. Free parking.

This inn, tucked 2 blocks off the plaza, has an intimate feel despite its 100 rooms. The Southwestern decor lobby is accented with gray weathered wood; outside is a heated pool that steams through the winter. The building is Territorial style, the novelty of which is lost on people outside the compound, but once inside you can see the brick-trimmed roofline and distinctive portals. Built in 1965, it has been continuously renovated, leaving the rooms consistent.

The rooms are accented with Mexican furniture and headboards, wrought-iron lamps, and hand-painted tin mirror frames, which provide a softer feel than is found in some of the newer hotels. As its name would imply, the Governor's mansion used to rest on this spot. Winter sees a fair amount of legislative traffic, but the majority of their clientele are travelers. Rooms on the north side look toward downtown and the mountains, many with balconies. All rooms have minirefrigerators; many of the superior and deluxe rooms have fireplaces (wood provided daily), and some have balconies and stereos.

Dining and Entertainment: The Mañana Bar and Restaurant serves three meals daily in a casual atmosphere. On warm days meals are served in an outdoor courtyard. The menu features light, healthy American cuisine. The adjacent bar is rustic, with cowhide chairs at a copper-topped bar beneath a viga ceiling; a pianist performs 6 nights a week.

Services: Concierge, room service, dry cleaning and laundry service, complimentary newspaper in lobby, in-room massage, twice-daily maid service, baby-sitting, express checkout, courtesy car, free coffee or refreshments in lobby.

Facilities: Spectravision movie channels, outdoor heated pool, access to nearby health club, conference rooms.

La Posada de Santa Fe

330 E. Palace Ave., Santa Fe, NM 87501. ☎ **800/727-5276** or 505/986-0000. Fax 505/982-6850. 119 rms, 40 suites. A/C TV TEL. May–Oct plus the Thanksgiving and Christmas seasons, $110–$295 single or double; $189–$397 suite. Nov–Apr (except holidays), $77–$215 single or double; $125–$285 suite. Various packages available. AE, CB, DC, DISC, MC, V. Free parking.

If you're in the mood to stay in a little New Mexico adobe village, you'll enjoy this hotel, just 3 blocks from the plaza. The main building is an odd mix of architecture. The original part was a Victorian mansion built in 1882 by Abraham Staab, a German immigrant, for his bride, Julia. Later it was "adobeized"—an adobe structure was literally built around it—so that now the Victorian presence is only within the charming bar and a half-dozen rooms, which still maintain the original brick, mahogany, and marble, as well as Italian paintings and French furniture and tapestries. Julia Staab, who died in 1896, continues to haunt the place. Mischievous but good-natured, she is Santa Fe's best-known and most frequently witnessed ghost.

The rest of the hotel follows in the pueblo-style construction and is quaint, especially in the summer when surrounded by acres of green grass. Here you get to experience squeaky maple floors, vigas and latillas, and, in many rooms, kiva fireplaces. However, in terms of design, this hotel has not kept up with the times. Some rooms have dingy gray carpet and poor polyester bedspreads. Some feel a bit cold, without enough floor coverings to make for coziness. Though renovation is ongoing, you may want to see a room before you take one. The hotel attracts travelers and a fair number of families. Most rooms don't have views but have outdoor patios, and most are tucked back into the quiet compound.

Dining and Entertainment: The Staab House Restaurant, open for three meals daily, has been fully restored with ceiling vigas, pueblo weavings, and tin work on the walls. From spring to early fall, meals are also served outside on a big patio. New Mexican cuisine is a house specialty. The lounge has seasonal happy-hour entertainment (usually local musicians).

Services: Concierge, room service, dry cleaning and laundry service.

Facilities: Outdoor swimming pool, guest use of local health club, beauty salon, boutique.

MODERATE

Garrett's Desert Inn

311 Old Santa Fe Trail, Santa Fe, NM 87501. ☎ **800/888-2145** or 505/982-1851. Fax 505/989-1647. 82 rms. A/C TV TEL. $84–$109 depending on season and type of room. AE, DISC, MC, V. Free parking.

Completion of this hotel in 1957 prompted the Historic Design Review Board to implement zoning restrictions throughout downtown. Apparently, residents were appalled by the huge air conditioners adorning the roof. Though they're still unsightly, the hotel makes up for them in other ways. First, with all the focus today on retro fashions, this hotel 3 blocks from the Plaza is totally "in." It's a clean, two-story, concrete block building around a broad parking lot. The hotel underwent a complete remodeling in 1994; it managed to maintain some '50s touches, such as art deco colored tile in the bathrooms and plenty of space in the rooms, while enlarging the windows and putting in sturdy doors and wood accents. Rooms are equipped with tile vanities and hair dryers. Above all, it's centrally located, within walking distance from the plaza and Canyon Road, but also far enough from busy streets to provide needed quiet. Locals frequent the hotel's Le Cafe on the Trail for crepes and pancakes. There's a year-round heated pool.

INEXPENSIVE

Santa Fe Motel

510 Cerrillos Rd., Santa Fe, NM 87501. ☎ **800/999-1039** or 505/982-1039. Fax 505/986-1275. 21 rms, 1 house. A/C TV TEL. May–Oct, $85–$95 double; $180 Thomas House. Nov–Apr, $70–$80 double; $150 Thomas House. Rates from May–Sept include continental breakfast. AE, MC, V. Free parking.

This adobe-style motel south of the Santa Fe River provides a clean, inexpensive stay within walking distance from downtown. Five rooms have kitchenettes, complete with two-burner stoves, minirefrigerators, and microwave ovens. Southwestern motifs predominate in the rooms, spread across four turn-of-the-century buildings. A recent renovation brought a new level of charm to all of the hotel's rooms. One now has a fireplace, others feature skylights, viga ceilings were added in some cases, and several have their own private patio entrances. Fresh-brewed coffee is served each morning in the office, where a bulletin board lists Santa Fe activities. The nearby Thomas House, on West Manhattan Avenue, is a fully equipped rental home with living and dining rooms and off-street parking.

Santa Fe Plaza Travelodge

646 Cerrillos Rd., Santa Fe, NM 87501. ☎ **800/578-7878** or 505/982-3551. Fax 505/983-8624. 48 rms. A/C TV TEL. Nov–Apr, $40–$60 single or double; May–Oct, $60–$75 single or double. AE, CB, DC, DISC, MC, V. Free parking.

You can count on the motel next door to Hotel Santa Fe on busy Cerrillos Road for comfort, convenience, and a no-frills stay. The rooms are very clean, nicely lit, and despite the busy location, relatively quiet. New mattresses and a pretty Southwestern ceiling border added to the decor make the rooms comfortable. Each room has a minirefrigerator, table and chairs, as well as a coffeemaker. The curbside pool, though basic, will definitely provide relief on hot summer days. *Beware, though:* The managers here are not always courteous.

3 Northside

Within easy reach of the plaza, Northside encompasses the area that lies north of the loop of the Paseo de Peralta.

VERY EXPENSIVE

✪ Bishop's Lodge

Bishop's Lodge Rd. (P.O. Box 2367), Santa Fe, NM 87504. ☎ **505/983-6377.** Fax 505/989-8739. 68 rms and 20 suites. A/C TV TEL. European Plan (meals not included), Mar 25–May 26 and Sept 6–Dec 31, $140 standard double, $215 deluxe double, $245 superdeluxe double; $170 standard suite, $285 deluxe suite. May 27–June 30, $160 standard double, $245 deluxe double, $275 superdeluxe double; $210 standard suite, $315 deluxe suite. July 1–Sept 5, $195 standard double, $295 deluxe double, $325 superdeluxe double; $245 standard suite, $355 deluxe suite. Modified American Plan (available May 27–Labor Day), $236–$271 standard double, $321–$371 deluxe double, $351–$401 superdeluxe double; $286–$321 standard suite, $391–$431 deluxe suite. Jan 1–Mar 24, $95 standard double, $175 deluxe double, $210 superdeluxe double; $125 standard suite, $249 deluxe suite. AE, DISC, MC, V. Free parking.

More than a century ago, when Bishop Jean-Baptiste Lamy was the spiritual leader of Northern New Mexico's Roman Catholic population, he often escaped clerical politics by hiking 3¹/₂ miles north over a ridge into the Little Tesuque Valley. There he built a retreat and a humble chapel (now on the National Register of Historic Places) with high-vaulted ceilings and a hand-built altar. Today Lamy's 1,000-acre getaway has become Bishop's Lodge. Purchased in 1918 from the Pulitzer family (of publishing fame) by Denver mining executive James R. Thorpe, it has remained in his family's hands to this day.

The guest rooms, spread through 10 buildings, all feature handcrafted furniture and regional artwork. Guests receive a complimentary fruit basket upon arrival. Standard rooms are spacious and many have balconies, while deluxe rooms feature traditional kiva fireplaces and private decks or patios; some older units have flagstone floors and viga ceilings. Superdeluxe rooms offer a combination bedroom/sitting room. The deluxe suites are extremely spacious, with living rooms, separate bedrooms, private patios and decks, and artwork of near-museum quality. All deluxe units come with fireplaces, refrigerators, and in-room safes. The Lodge is an active resort three seasons of the year; in the winter, it takes on the character of a romantic country retreat. Bishop's Lodge is exceptionally well cared for—in fact, each year approximately 10 rooms are renovated.

Dining and Entertainment: Three large adjoining rooms with wrought-iron chandeliers and wall-size Native American–theme oil paintings comprise Bishop's Lodge dining room. Santa Feans flock here for breakfast, lunch, Sunday brunch, and dinner (featuring creative regional cuisine with continental flair). Attire is casual at breakfast and lunch but more formal at dinner (men generally wear a sport coat, although they aren't required). There's a full vintage wine list, and El Rincon Bar serves before- and after-dinner drinks.

Services: Concierge, room service, laundry service, newspaper delivery, in-room massage, twice-daily maid service, baby-sitting, express checkout, courtesy shuttle three times daily, free coffee or refreshments in the lobby in the mornings, seasonal cookouts.

Facilities: Outdoor pool with lifeguard, small health club, aerobics classes, four tennis courts, pro shop and instruction, Jacuzzi, sauna, hiking and self-guided nature walk (the Lodge is a member of the Audubon Cooperative Sanctuary System), daily guided horseback rides, introductory riding lessons, children's pony ring, supervised skeet and trap shooting, stocked trout pond for children, Ping-Pong, summer daytime program with counselors for children.

Rancho Encantado

Route 4, Box 57C, Santa Fe, NM 87501. ☎ **800/722-9339** or 505/982-3537. Fax 505/983-8269. 90 rms, 36 suites. A/C TV TEL. Jan 1–June 26, $115–$235 lodge room, suite, or villa; June 27–Oct 12, Nov 26–29, and Dec 19–31, $175–$375 lodge room, suite, or villa; Oct 13–Nov 25 and Nov 30–Dec 18, $115–$235 lodge room, suite, or villa. AE, DC, MC, V. Free parking.

Located 8 miles north of Santa Fe in the foothills of the Sangre de Cristo Mountains, Rancho Encantado, with its sweeping panoramic views, claims to be "where the magic of New Mexico comes to life." Betty Egan, a former World War II captain in the Women's Army Corps, purchased the property in the mid-1960s. Mrs. Egan, then recently widowed, was determined to begin a new and prosperous life with her family, and so in 1968 the 168-acre ranch became Rancho Encantado. The property came under new ownership in 1995, and with it came long-needed upgrades to sleeping rooms, the main lodge, and grounds.

The handsome main lodge is comfortable and unassuming, decorated in traditional Southwestern style with hand-painted tiles, ceiling vigas, tile floors, antique furnishings, Pueblo rugs, and Hispanic art objects hanging on stuccoed walls. The large fireplace in the living room/lounge is a focal point, especially on cold winter afternoons. In the main lodge and adjoining area the rooms are quite cozy with a Victorian bed-and-breakfast feel. All rooms provide coffeemakers plus coffee for those who couldn't make it out the front door otherwise.

Surrounding the lodge are clusters of cottages and casitas. These are comfortable units with a homey feel, though they could use a decorator's touch. Across the street from the main building are two-bedroom/two-bath villas. These newer split-level adobe units are equipped with fireplaces in the living room and master bedroom plus a full kitchen. Whatever you choose, you're sure to find the accommodations here more than adequate; satisfied guests have included Princess Anne, Robert Redford, Jimmy Stewart, Whoopi Goldberg, and John Wayne.

The new owners are working to expand special programs at the resort. Watch for summer sunset margarita horseback rides, barbecues, barn dances, and western movie screenings. Winter activities are in the planning stages.

Dining and Entertainment: Rancho Encantado's restaurant is a favorite dinner spot for opera-goers. The food is good and the atmosphere relaxing. The west wall of the dining room has picture windows that overlook the Jemez Mountains, offering diners a first-rate view of the spectacular New Mexico sunset. Specialties include tenderloin of beef, served with a Jack Daniels sauce, and a smoked salmon burrito. You can also get a good egg salad sandwich. The Cantina, with its big-screen TV, is a popular gathering spot; there is also a snack bar on the premises.

Services: Laundry service, baby-sitting, courtesy limo.

Facilities: Outdoor pool, small health club, Jacuzzi, tennis courts (tennis pro in summer), hiking trails, horseback riding, sand volleyball, basketball, horseshoes, pool table, library.

EXPENSIVE

Radisson Deluxe Hotel Santa Fe

750 N. St. Francis Dr., Santa Fe, NM 87501. ☎ **800/333-3333** or 505/982-5591. Fax 505/988-2821. 116 rms, 12 suites, 32 condo units. A/C TV TEL. $89–$149 double, $218–$436 suite, $119–$209 condo, depending on time of year. AE, CB, DC, DISC, MC, V. Free parking.

🏨 Family-Friendly Hotels

Bishop's Lodge *(see page 59)* A children's pony ring, riding lessons, tennis courts with instruction, a pool with lifeguard, stocked trout pond just for kids, a summer daytime program, horseback trail trips, and more make this a veritable day camp for all ages.

El Rey Inn *(see page 64)* Kids will enjoy the play area, table games, and pool; parents will appreciate the kitchenettes and laundry facilities.

Rancho Encantado *(see page 60)* Horseback riding (on trails or pony ring), pool, tennis courts, and many indoor and outdoor games will keep kids happily busy here.

Set on a hill as you head north toward the Santa Fe Opera, this 3-story hotel, like the Inn at Loretto, provides a decent stay. The lobby is unremarkable, with tile floors and aged wood trim. Previously remodeled in 1994, the hotel recently came under new ownership; more renovation is ongoing. In some ways it needs it; the rooms' door frames need repairs, and many hallways need repainting. I'd reserve a stay here for summer months when the country club–feeling pool is open. The rooms are decorated in blond furniture with a Southwestern motif, some with views of the mountains, others overlooking the pool. All are equipped with hair dryers, irons, and ironing boards. There's free drop-off and pick-up to the opera, plaza, and elsewhere in town. Premium rooms are more spacious, some with large living rooms and private balconies. Each parlor suite has a Murphy bed and kiva fireplace in the living room, a big dining area, a wet bar and refrigerator, and a jetted bathtub. Cielo Grande condo units nearby come with fully equipped kitchens, fireplaces, and private decks.

Dining/Entertainment: The Santa Fe Salsa Company Restaurant and Bar serves three meals a day. Dinner main courses are international with a Southwestern flair. A jazz combo plays nightly in the bar, and the nightclub features the quick Spanish steps of New Mexico's best-known flamenco dancer, María Benitez, and her Estampa Flamenca troupe.

Services: Room service; dry cleaning and laundry service; complimentary *USA Today;* child care can be arranged; complimentary shuttle is available.

Facilities: Outdoor swimming pool, Jacuzzi, access to health club next door, Laundromat.

Santa Fe Accommodations

320 Artist Rd., Santa Fe, NM 87501. ☎ **800/745-9910** or 505/98-CONDO. Fax 505/ 984-8682. 140 units. A/C. Jan 6–Feb 28, Apr 1–May 14, and Nov 1–Dec 14, from $90 1-bedroom; $178 2-bedroom; $240 3-bedroom. Mar 1–Mar 31, May 15–June 30, and Sept 15–Oct 31, from $135 1-bedroom; $198 2-bedroom; $272 3-bedroom. July 1–Sept 14 and Dec 15–Jan 5, from $155 1-bedroom; $218 2-bedroom; $293 3-bedroom. Extra person $20. Children 18 and under stay free in parents' room. Rates include extended continental breakfast. AE, DC, MC, V. Free on- and off-street parking.

Santa Fe Accommodations is not a hotel, but a company that manages a number of properties around Santa Fe. If you're interested in a condominium-type stay, they have a few options. The first is **Fort Marcy Hotel Suites.** Located about an 8-minute walk from the plaza, these condominiums climb up a hill north of town. They are privately owned, so decor varies, although Santa Fe Accommodations gives incentives to encourage remodeling among owners. Logically, some are better decorated than others. All have full kitchens, with microwave ovens, stoves, ovens, refrigerators, and most have dishwashers. They also have irons and ironing boards. The units h~

plenty of room, and the grounds are well kept, though some of the units are showing their age (built in 1975). The grounds have a rural feel with chamisa and pine trees. They've recently remodeled the club house and meeting space. There's an indoor pool and Jacuzzi, nature trails, business center, conference rooms, and a Laundromat.

Two blocks from the plaza, you'll find **Seret's 1001 Nights.** If it's a trip to a foreign land and the lush comforts of Middle Eastern decor that you seek, choose these one- and two-bedroom units. These buildings, built in the historic Barrio de Analco, have wonderful touches such as arched windows and doorways. They're filled with Middle Eastern antiques and tapestries. Each has a fully stocked kitchen with stove, refrigerator, and microwave. *Beware, however:* Parking is scarce in this part of town, though you can arrange to park at the supplier's store, Seret and Sons, one block away.

4 Southside

Santa Fe's major strip, Cerrillos Road, is US 85, the main route to and from Albuquerque and the I-25 freeway. It's about 5¼ miles from the plaza to the Villa Linda Mall, which marks the southern boundary of the city. Most motels are on this strip, although several of them are east, closer to St. Francis Drive (US 84) or the Las Vegas Highway.

EXPENSIVE

Residence Inn
1698 Galisteo St., Santa Fe, NM 87505. ☎ **800/331-3131** or 505/988-7300. Fax 505/988-3243. 120 suites. A/C TV TEL. $89–$169 studio suite; $99–$190 penthouse suite. Rates vary according to season and include continental breakfast. AE, CB, DC, DISC, JCB, MC, V. Free parking.

Designed to look like a neighborhood, this inn provides the efficient stay that you'd expect from a Marriott. It's located about 10 minutes from the Plaza, a quiet drive through a few neighborhoods. The lobby and breakfast area are warmly decorated in red tile, with a fireplace and Southwest accents such as bancos and drums. There are two sizes of suites, both roomy, with fully equipped kitchens that include microwave, dishwasher, stove top, oven, refrigerator and coffeemaker. All rooms have fireplaces and balconies and are decorated in a Southwestern rose color. Outside there is a sport court where guests can play basketball, tennis, and volleyball, as well as a grill, a pool open in summer and three Jacuzzis. Access to a nearby health club can be arranged. A newspaper is delivered to your room and free coffee and refreshments are available in the lobby. Most who stay here are tourists, but you'll also encounter some government workers as well as business travelers and film crews.

Dining/Entertainment: A social hour with complimentary hors d'oeuvres takes place on Monday through Thursday from 5 to 6:30pm.

Services: Dry cleaning and laundry service, complimentary local newspaper, grocery shopping service.

Facilities: Outdoor swimming pool, three Jacuzzis, guest membership at nearby health club, sports court, jogging trail, Laundromat, barbecue grills on patio.

MODERATE

Days Inn
3650 Cerrillos Rd., Santa Fe, NM 87501. ☎ **800/325-2525** or 505/438-3822. Fax 505/438-3795. 96 rms, 20 suites. A/C TV TEL. Apr–Sept, $95–$105 double; $135–$155 suite. Oct–Mar, $45–$75 double; $85–$95 suite. Children 12 and under stay free in parents' room. Rates include continental breakfast. AE, CB, DC, DISC, MC, V. Free parking.

Accommodations & Dining on Cerrillos Road

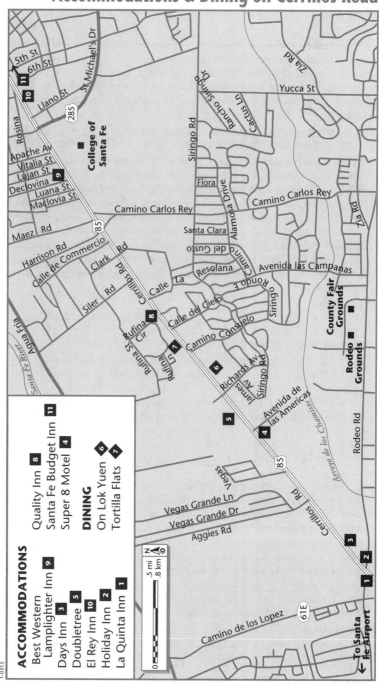

ACCOMMODATIONS

Best Western Lamplighter Inn **9**
Days Inn **3**
Doubletree **5**
El Rey Inn **10**
Holiday Inn **2**
La Quinta Inn **1**
Quality Inn **8**
Santa Fe Budget Inn **11**
Super 8 Motel **4**

DINING

On Lok Yuen **6**
Tortilla Flats **7**

63

A 3-story palace in pink, the Days Inn opened in 1990 as an impressive new motel on Santa Fe's south side. It's a security-conscious hostelry; rooms can be entered only from interior corridors and there are private safes in each room. Typical rooms, decorated in a soft Southwestern style, feature a queen-size bed or two double beds plus a sofa or easy chair. Deluxe suites and 14 upgraded "king" rooms offer wet bars and refrigerators. Some suites also boast Jacuzzi tubs and private balconies. Amenities include a 24-hour desk, Laundromat, and a hot tub and swimming pool.

Doubletree

3347 Cerrillos Rd., Santa Fe, NM 87505. ☎ **800/777-3347** or 505/473-2800. Fax 505/473-4905. 213 rms. AC TV TEL. $59–$249 double. AE, DC, DISC, MC, V. Free parking.

This is a good choice if you don't mind mixing business with pleasure. Since the Doubletree caters to a lot of conference traffic, there's a definite business feel to this hotel, built in 1986 and remodeled in 1996. The decor is tasteful Southwestern with rooms opening onto cavelike balconies or walkways bordered by grass. All rooms have hair dryers, vanity mirrors, irons and ironing boards, minirefrigerators, and coffeemakers. Though its situated on busy Cerrillos Road, the rooms are placed so that they are quiet. A 24-hour business center provides fax, copy machine, and computer facilities. The Cafe Santa Fe serves New Mexican, Italian, and American food. There are two Jacuzzis and a beautiful indoor pool, large enough for laps, as well as a small health club.

✪ El Rey Inn

1862 Cerrillos Rd. (P.O. Box 4759), Santa Fe, NM 87502. ☎ **800/521-1349** or 505/982-1931. Fax 505/989-9249. 86 rms, 8 suites. A/C TV TEL. $56–$99 single or double; $99–$207 suite. Rates include continental breakfast. AE, DC, DISC, MC, V. Free parking.

Staying at "The King" makes you feel like you're traveling the old Route 66 through the Southwest. The white stucco buildings of this court motel are decorated with bright trim around the doors and hand-painted Mexican tiles on the walls. Opened in the 1930s, additions were made in the 1950s; remodeling is ongoing. The lobby has vigas and tile floors decorated with Oriental rugs and dark Spanish furniture. No two rooms are alike. The oldest section, nearest the lobby, feels a bit cramped, though the rooms have style, with art-deco tile in the bathrooms and vigas on the ceilings. Some have little patios. Be sure to request to be as far back as possible from Cerrillos Road.

The two stories of suites around the Spanish colonial courtyard are the sweetest deal I've seen in all of Santa Fe. These feel like a Spanish inn, with carved furniture and cozy couches. Some rooms have kitchenettes. The owners recently purchased the motel next door and have now added 10 deluxe units around the courtyard. The new rooms offer more upscale amenities and gas log fireplaces, as well as distinctive furnishings and artwork. Complimentary continental breakfast is served in a sunny room or on a terrace in the warmer months. There's also a sitting room with a library and games tables, outdoor swimming pool, Jacuzzi, sauna, picnic area, children's play area, and Laundromat. For its cheaper rooms, El Rey is Santa Fe's best moderately (and even inexpensively) priced accommodation.

Holiday Inn

4048 Cerrillos Rd., Santa Fe, NM 87505. ☎ **800/465-4329** or 505/473-4646. Fax 505/473-2186. 112 rooms, 18 minisuites. AC TV TEL. Jan 1–May 21, $79–$109 double; May 22–Sept 30, $99–$119 double; Oct 1–Dec 31, $79–$109 double. Minisuites are $10 more than regular rooms. AE, DC, DISC MC, V. Free parking.

Though the decor, done in pastel blues, is a bit cheesy for my taste, this is a good place to get a standard, reliable room. It's located about 5 miles from the historic district, about a 15-minute drive. Built in the early 1980s, the hotel was remodeled in

1996. The lobby has clean lines, though it's dominated by a cumbersome elevator shaft at the center. Judging by the number of pickup trucks in the parking lot, this seems to be the place where a lot of local ranchers and traders stay, as well as plenty of businesspeople. The sunny rooms are brightly wallpapered and have plenty of amenities: hair dryers, irons, and ironing boards; minisuites have refrigerators and wet bars. There's a small pool heated year-round as well as a sauna, Jacuzzi, and health club. Bobby Rubino's offers ribs and continental cuisine. Room service is available, as are shuttles to and from the Santa Fe airport.

INEXPENSIVE

Best Western Lamplighter Inn
2405 Cerrillos Rd., Santa Fe, NM 87505. ☎ **800/767-5267** or 505/471-8000. Fax 505/471-1397. 80 rms. A/C TV TEL. $49–$79 double; $62–$89 double with kitchenette, depending on the season and type of room. AE, CB, DC, DISC, MC, V. Free parking.

This motel, built in the early 1970s, is doing its best to keep up with the times. Renovated in 1991–92, it comprises three buildings. The oldest faces busy Cerrillos Road and offers the smallest rooms and probably the most street noise. I recommend the back building where noise is diminished and rooms are larger with vaulted ceilings. In a third building are the doubles with kitchenettes, each with a VCR and well-stocked kitchen with microwave, stove, and oven. All rooms are decorated in Southwestern-style decor and have coffeemakers and refrigerators. An indoor pool and Jacuzzi area has panels that open to let in sunlight in summer.

La Quinta Inn
4298 Cerrillos Rd., Santa Fe, NM 87505. ☎ **800/531-5900** or 505/471-1142. Fax 505/438-7219. 130 rms. A/C TV TEL. June to mid-Oct, $88 double; late Oct–May, $66 double. Large discount for AAA members. Rates include continental breakfast. AE, CB, DC, DISC, MC, V. Free parking.

Though it's a good 15-minute drive from the plaza, this is my choice of economical Cerrillos Road chain hotels. Built in 1986, it was just fully remodeled in a very comfortable and tasteful way. The rooms within the 3-story white brick buildings have an unexpectedly elegant feel, with lots of deep colors and art-deco tile in the bathrooms. There's plenty of space in these rooms, and they're lit for mood as well as for reading. Each has a coffeemaker. A continental breakfast is served in the intimate lobby. The kidney-shaped pool has a nice lounging area. If you're a shopper or moviegoer, this hotel is just across a parking lot from the Villa Linda Mall. The Kettle, a 24-hour coffee shop, is adjacent. La Quinta also has a 24-hour desk, Laundromat (as well as valet laundry), and complimentary coffee at all times in the lobby. Continental breakfast includes fresh fruit, danish, cereal, bagels, coffee, tea, and juice. Pets are accepted.

Quality Inn
3011 Cerrillos Rd., Santa Fe, NM 87505. ☎ **800/228-5151** or 505/471-1211. Fax 505/438-9535. 99 rms. A/C TV TEL. $55–$95 single or double. AE, CB, DC, DISC, ER, JCB, MC, V. Free parking.

This hotel on the south end of town, about a 10-minute drive from the plaza, provides good standard rooms at reasonable prices. Built in 1970, it's a 2-story white stucco building with a red-tile roof. Its most recent remodeling took place in 1996, mostly replacing carpet, drapes, and some furnishings. Expect nothing fancy here, even with the remodeling. Rooms, though, are fairly quiet, with big windows to let in lots of light. Request a room looking in toward the courtyard and pool, and you'll get a bit of a resort feel. All rooms have coffeemakers; deluxe rooms have wet bars and minirefrigerators. There is limited room service and valet laundry. Baby-sitting is available, and pets are accepted. The coffee shop and dining room serves Mexican fare.

Santa Fe Budget Inn

725 Cerrillos Rd., Santa Fe, NM 87501. ☎ **800/288-7600** or 505/982-5952. Fax 505/984-8879. 160 rms. A/C TV TEL. July 4–Oct 25 and Dec 25–Jan 2, $75–$86 double ($10–$20 higher during Indian Market). Rest of the year, $50–$58 double. A Sun–Thurs "supersaver" rate may apply in the off-season. AAA and AARP members receive $3 discounts. AE, CB, DC, MC, V. Free parking.

About a 10-minute walk to the plaza, this is a decent economical motel. Santa Fe Opera and Fiesta posters add a splash of color to the predominantly brown rooms. Each of the modest units is furnished with one queen-size or two double beds, striped bedspreads, reading lamps, and a desk with built-in dresser. Each room has satellite TV. Guests can enjoy an outdoor swimming pool, lots of parking space, and two restaurants adjacent to the motel.

Super 8 Motel

3358 Cerrillos Rd., Santa Fe, NM 87501. ☎ **800/800-8000** or 505/471-8811. Fax 505/471-3239. 96 rms. A/C TV TEL. May–Sept, $49.88 double; Oct–Apr, $42.88 double. AE, CB, DC, DISC, MC, V. Free parking.

It's nothing flashy, but this pink stucco, boxy motel attracts regulars who know precisely what to expect—a clean, comfortable room with standard furnishings: double beds, working desk, in-room safes, and free local phone calls. Rooms for nonsmokers are available. The motel has a 24-hour desk and continental breakfast is included.

5 Bed-&-Breakfasts

If you prefer a homey, intimate setting to the sometimes impersonal ambience of a large hotel, one of Santa Fe's bed-and-breakfast inns may be right for you. All those listed here are located in or close to the downtown area and offer comfortable accommodations at moderate to inexpensive prices.

✪ Adobe Abode

202 Chapelle St., Santa Fe, NM 87501. ☎ **505/983-3133.** Fax 505/986-0972. 6 rms. TV TEL. $110–$150 double. Rates include breakfast. DISC, MC, V. Free parking.

A short walk from the plaza, in a quiet residential neighborhood, Adobe Abode is one of Santa Fe's most imaginative B&Bs. The living room is filled with everything from Mexican folk art and pottery to Buddhas and ethnic masks. The open kitchen features a country pine table as well as Balinese puppets. The creativity of the owner/innkeeper, Pat Harbour, shines in each of the guest rooms as well. The Out of Africa Room, in the main house, has elegant fabrics and tribal art. The Texas Hill Country Room features antiques and designer denim and plaid bedding. Both are in the main house, which was built in 1907 and renovated in 1989. In back are casitas occupying newer buildings, designed with flair. The Bronco Room is filled with cowboy paraphernalia: hats, Pendleton blankets, pioneer chests, and my favorite—an entire shelf lined with children's cowboy boots. Finally, Pat has added a new 2-room suite—the Provence Suite—which she decorated in sunny yellow and bright blue. Two rooms have fireplaces, while several have private patios. All rooms have coffeemakers and terry-cloth robes. Complimentary sherry, fruit, and Santa Fe cookies are served daily in the living room. Every morning a healthy breakfast of fresh-squeezed orange juice, fresh fruit, homemade muffins, scones or pastries, and a hot dish is served in the kitchen.

Alexander's Inn

529 East Palace Ave., Santa Fe, NM 87501. ☎ **505/986-1431.** Fax 505/982-8572. 6 rms (4 with bath), 4 cottages. A/C TV TEL. $75–$160 double. Rates include continental breakfast. MC, V. Free parking.

Located not far from central downtown—in a quiet residential area—Alexander's Inn is quite different from other accommodations here; most of its decor is not Southwestern style. Instead, the 10-year-old inn (whose building dates back to 1903) has a Victorian/New England style. The rooms in the main house have stenciling on the walls, hook and Oriental rugs, muted colors such as apricot and lilac, and white iron or four-poster queen-size beds (there are some king-size beds as well). There are also cottages complete with kitchens (equipped with stove, oven, and refrigerator, some with microwave) and living rooms with kiva fireplaces. Some of these don't have quite the charm of rooms in the main house; others have plenty of Southwestern charm, so discuss your desires when making reservations. Fresh flowers adorn all the rooms throughout the year, and all have robes, hair dryers, and makeup mirrors. Guests can enjoy privileges at El Gancho Tennis Club as well as the hot tub in the back garden. Mountain bikes are available for guest use. An extended continental breakfast of homemade baked goods is served on the veranda every morning, and afternoon tea and cookies are available. Pets are accepted.

Dancing Ground of the Sun

711 Paseo de Peralta, Santa Fe, NM 87501. ☎ **800/645-5673** or 505/986-9797. Fax 505/986-8082. 8 rooms. TV TEL. Nov–Apr, $75–$210 double; May–Oct and all holidays, $95–$255 double. Rates include breakfast. MC, V. Free parking.

A great deal of thought and energy went into decorating these units—and it shows. Each of the eight rooms, five of which are casitas, has been outfitted with handcrafted Santa Fe–style furnishings made by local artisans, and the decor of each room focuses on a mythological Native American figure, whose likeness has been hand-painted on the walls of that unit. There's Corn Dancer, who represents the anticipation of an abundant harvest; Kokopelli, a flute player believed to bring good fortune and abundance to the Native American people; and other themes. Many have ceiling vigas and all have nice touches, such as *nichos* (little niches) that come with an older adobe building such as this, constructed in the 1930s. Each of the five casitas has a fireplace and a fully equipped kitchen with microwave, refrigerator, stove, and coffeemaker; two are equipped with washers and dryers. Spirit Dancer and Deer Dancer, completed in 1996, are the inn's newest rooms; they have kitchenettes. Each evening, the next day's breakfast of healthful, fresh-baked food is delivered to your door for you to enjoy at your leisure. The two casitas closest to the street may have some street noise, but it should die down by bedtime. Smoking and pets are not permitted.

✪ Dos Casas Viejas

610 Agua Fría St., Santa Fe, NM 87501. ☎ **505/983-1636.** Fax 505/983-1749. 8 rms. TV TEL. $165–$245 single or double. Free off-street parking.

These two old houses *(dos casas viejas),* located not far from the plaza in what some call the *barrio* on Agua Fría Street, are hidden behind an old wooden security gate. You enter into a meandering brick lane along which are the elegant guest rooms. The innkeepers, Susan and Michael Strijek, manage to provide guests with most of the luxuries one would expect in a fine hotel. The grounds are nicely manicured, and the rooms, each with a patio and private entrance, are finely renovated and richly decorated. All rooms have Mexican-tile floors and kiva fireplaces; most have diamond-finished stucco walls and embedded vigas. They're furnished with Southwestern antiques and original art. Some have canopy beds and one has a beautiful sleigh bed;

all are covered with fine linens and down comforters. Each room is supplied with robes, a hair dryer, and complimentary gourmet treats and refreshments. Valet laundry, newspaper delivery, in-room massage, and free coffee or refreshments in the lobby are available. Guests can use the library and dining area (where a European breakfast is served each morning) in the main building. Breakfast can also be enjoyed on the main patio alongside the beautiful lap pool or on your private patio (after you collect it in a basket).

El Farolito

514 Galisteo St., Santa Fe, NM 87501. ☎ **888/634-8782** or 505/988-1631. Fax 505/ 988-4589. 7 casitas. TV TEL. $85–$135 casita for two. Rates include breakfast. AE, DISC, MC, V. Free parking.

Before press time, the new owners of this inn, within walking distance to the Plaza, hope to create a fantasy experience for guests in each room. Already they've filled the common area with works by notable New Mexico artists. The planned room themes include a Native American Room, decorated with rugs and pottery; a South-of-the-Border room, decorated with Mexican folk art; and others. The walls of most of the rooms are stylishly rubbed with beeswax during plastering to give them a smooth, golden finish, and all have kiva fireplaces and private patios. Part of the inn was built before 1912 and the rest is new, but the old-world elegance carries through. A full breakfast is included with the price of the room.

El Paradero

220 W. Manhattan Ave., Santa Fe, NM 87501. ☎ **505/988-1177.** E-mail: elpara@trail.com. 14 rms (10 with bath), 2 suites. A/C TEL. May 15–Oct 31, $75–$110 double; $135 suite. Nov 1–May 14, $65–$100 double; $120 suite. Rates include breakfast. MC, V. Free parking.

Located a few blocks south of the Santa Fe River, El Paradero (The Stopping Place) began in about 1810 as a Spanish adobe farmhouse. It doubled in size in 1878, when Territorial-style details were added; in 1912 Victorian touches were incorporated in the styling of its doors and windows. Innkeepers Ouida MacGregor and Thom Allen opened this as one of Santa Fe's first bed-and-breakfasts 16 years ago. They're deeply involved in the city and can direct visitors to unexpected sights and activities. Nine ground-level rooms surround a central courtyard; they offer a clean, white decor, hardwood or brick floors, folk art, and handwoven textiles on the walls. Three more luxurious upstairs rooms feature tile floors and baths as well as private balconies. Five rooms have fireplaces. Two suites occupying a brick 1912 coachman's house, a railroad-era Victorian building, are elegantly decorated with period antiques and provide living rooms with fireplaces, kitchen nooks, TVs, and phones. The ground floor of the main building has a parlor, a living room with a piano and fireplace, and a Mexican-style breakfast room, where a full gourmet breakfast and afternoon tea are served daily. Also, if you'd like to take a trip up the mountain and have a picnic, they'll prepare a lunch for you.

Four Kachinas Inn

512 Webber St., Santa Fe, NM 87501. ☎ **800/397-2564** or 505/982-2550. 4 rms. TV TEL. $68–$125 double. Rates include breakfast. DISC, MC, V. Free parking.

Located on a quiet, residential street, but still well within walking distance of downtown, the Four Kachinas Inn is a wonderful little bed-and-breakfast. The rooms are sparsely decorated with Southwestern artwork, including antique Navajo rugs, kachinas, and handmade furniture (in fact, some of the furniture was made from the wood salvaged from the old barn that once stood on the property here). As you might have guessed, each of the rooms is named for a Hopi kachina: The Koyemsi Room is named for the "fun-loving, mudhead clown kachina"; the Poko Room, for the dog kachina that "represents the spirits of domestic animals"; the Hon Room, for the "powerful healing bear kachina"; and the Tawa Room, for the sun god kachina. Three of the rooms are on the ground floor. The upstairs room offers a beautiful view of the Sangre de Cristo Mountains. In a separate building, constructed of adobe bricks that were made on the property, there is a lounge where guests can gather at any time of day to enjoy complimentary beverages and snacks. The snacks are a treat—one of the owners won "Best Baker of Santa Fe" at the county fair a couple of years ago. In the guest lounge there is a library of art and travel books, and works of art are for sale. An extended continental breakfast of juice, coffee or tea, pastries, yogurt, and fresh fruit is brought to guests' rooms each morning. One of the rooms here is completely accessible for disabled travelers.

✪ Grant Corner Inn

122 Grant Ave., Santa Fe, NM 87501. ☎ **800/964-9003** or 505/983-6678 for reservations, 505/984-9001 for guest rooms. Fax 505/983-1526. 12 rms (10 with bath), 1 hacienda. A/C TV TEL. $80–$120 single or double without bath; $100–$155 single or double with bath. Hacienda, $105–$130 guest rooms rented separately; $215–$255 for entire house. Rates include full gourmet breakfast. MC, V. Free parking.

This early 20th-century manor at the corner of Grant Avenue and Johnson Street is just 2 blocks west of the plaza. Each room is furnished with antiques, from brass or four-poster beds to armoires and quilts, and monogrammed terry-cloth robes are available for those staying in rooms with shared baths. All rooms have ceiling fans, and some are equipped with small refrigerators. Each room has its own character. For example, no. 3 has a hand-painted German wardrobe closet dating from 1772 and a washbasin with brass fittings in the shape of a fish; no. 8 has an exclusive outdoor deck that catches the morning sun; and no. 11 has an antique collection of dolls and stuffed animals. Two rooms have kitchenettes, and two also have laundry facilities. The inn's office doubles as a library and gift shop. Children age 6 and under are accepted in specific rooms. In addition to the rooms mentioned above, Grant Corner Inn now offers accommodations in their Hacienda, located at 604 Griffin Street. It's a Southwestern-style condominium with two bedrooms, living and dining rooms, and a kitchen. It can be rented in its entirety or the rooms can be rented separately, depending on your needs.

Breakfast, for both the inn and the Hacienda, is served each morning in front of the living-room fireplace or on the front veranda in summer. The meals are so good that an enthusiastic public arrives for brunch here every Saturday and Sunday (the inn is also open to the public for weekday breakfasts).

Inn of the Animal Tracks

707 Paseo de Peralta, Santa Fe, NM 87501. ☎ **505/988-1546.** 5 rms. A/C TV TEL. $90–$130. Rates include continental breakfast. AE, MC, V. Free parking.

If you tend to miss your animals while you're away, this is the place to stay. The main sitting room is set around a kiva fireplace, and you may share the couch with a resident dog. Each room is decorated with pictures and memorabilia of animals such as an eagle, deer, otter, wolf, and rabbit. The rooms have a cozy feel and are well lit. I recommend the Otter Room with its old-fashioned tub, even though the room is near the street, with traffic noise throughout the day and quieter at night. Inhabiting a 90-year-old adobe home, this inn maintains that old character, with important renovations added in 1989. Guests tend to like the frequently served breakfast burrito, as well as other warm breakfasts and afternoon snacks such as apple pie and carrot cake. Situated $2^{1}/_{2}$ blocks from both the plaza and Canyon Road, you'll hardly need a car.

Inn on the Paseo

630 Paseo de Peralta, Santa Fe, NM 87501. ☎ **800/457-9045** or 505/984-8200. Fax 505/989-3979. 18 rms, 2 suites. A/C TV TEL. $85–$165 double. Rates include extended continental breakfast. AE, DC, MC, V. Free parking.

Located just a few blocks from the plaza, this is a good choice for travelers who want to be able to walk to the shops, galleries, and restaurants but would rather not stay at a larger hotel. As you enter the inn you'll be welcomed by the warmth of the large fireplace in the foyer. Southwestern furnishings dot the spacious public areas and the work of local artists adorns the walls. The guest rooms are medium-sized, meticulously clean, and very comfortable. They have a hotel feel, the arrangements are fairly consistent from room to room, but the bathrooms are a bit stark. Still, one room boasts a fireplace and many feature four-poster beds and private entrances. The focal point of each room is an original handmade patchwork quilt. The owner is a third-generation quilter, and she made all the quilts you'll see hanging throughout the inn (more than 25 of them). A breakfast buffet is served on the sundeck in warmer weather and indoors by the fire on cooler days. It consists of muffins, breads, granola, fresh fruit, and coffee or tea. Complimentary refreshments are served every afternoon.

Preston House

106 Faithway St., Santa Fe, NM 87501. ☎ **505/982-3465.** Fax 505/982-3465. 15 rms (13 with bath), 2 cottages, 1 adobe house. A/C TV TEL. High season, $75–$85 double without bath; $106–$160 double with bath; $150 cottage or adobe house. Low season, $48–$58 double without bath; $78–$98 double with bath; $115 cottage or adobe house. Extra person $20. Rates include continental breakfast and afternoon tea. AE, MC, V. Free parking.

This is a different type of building for "The City Different"—a century-old Queen Anne home. This style of architecture is rarely seen in New Mexico; it has been painted sky blue with white trim. The house's owner, noted silk-screen artist and muralist Signe Bergman, adores its original stained glass. Located 5 blocks east of the plaza, off Palace Avenue near La Posada hotel, the Preston House has several types of rooms. Six in the main house are furnished with period antiques and decorated with floral wallpaper and lace drapes. Many rooms feature brass beds covered with quilts; some have decks, several offer fireplaces, and only two require sharing a bath. All rooms are equipped with phones. Seven more rooms with bath are located in an adobe building catercorner from the house. Two private cottages behind Preston House and an adobe home across the street provide more deluxe facilities. All rooms are stocked with terry-cloth robes; sherry is served in a common area. An expanded continental buffet breakfast is served daily, as is afternoon tea with homemade cookies, cakes, and pies. Children 10 and over are welcome.

Pueblo Bonito

138 W. Manhattan Ave., Santa Fe, NM 87501. ☎ **800/461-4599** or 505/984-8001. Fax 505/984-3155. 20 rms, 6 suites. TV TEL. Jan–Feb, $70–$110 double; March–April and Nov–Dec, $80–$120 double. May–Oct, holidays, and special events, $95–$140 double. Rates include continental breakfast. MC, V. Free parking.

Private courtyards and narrow flagstone paths give a look of elegance for this former circuit judge's 19th-century adobe hacienda and stables, located a few blocks south of the Santa Fe River. Adobe archways lead to hybrid rose gardens shaded by prolific apricot and pear trees; guests are invited to help themselves to the fruit. Each guest room is named for a Pueblo tribe of the surrounding countryside. Every room—decorated with Native American rugs on wood or brick floors—has a queen-size bed and fireplace. A 1997 renovation added some new furniture and mirrors. Bathrooms are small but attractively tiled. Six rooms are suites, each with a fully stocked kitchen, with microwave, refrigerator, stove, dishwasher, and coffeemaker as well as a living/dining room with fireside seating and bedroom. A couple of other rooms offer refrigerators and wet bars. The remainder are standard units with locally made willow headboards and couches, dining alcoves, and old Spanish-style lace curtains. Continental breakfast is served daily from 8 to 10am in the dining room or on the sundeck; if you prefer, there's room service. Afternoon tea is served daily from 4 to 6pm. There's also a Laundromat and an indoor Jacuzzi.

✪ Spencer House Bed & Breakfast Inn

222 McKenzie St., Santa Fe, NM 87501. ☎ **800/647-0530** (7am to 7pm) or 505/988-3024. 4 rms, 1 cottage. A/C. $90–$150 double. Rates include breakfast. AE, MC, V. Free parking.

The Spencer House is unique among Santa Fe bed-and-breakfasts. Instead of Southwestern-style furnishings, you'll find beautiful antiques from England, Ireland, and colonial America. One guest room features an antique brass bed, another a pencil-post bed, yet another an English panel bed, and all rooms utilize Ralph Lauren fabrics and linens. Each bed is also outfitted with a fluffy down comforter. All bathrooms are completely new, modern, and very spacious. Owner Jan McConnell takes great pride in the Spencer House and keeps it spotlessly clean. From the old Bissell carpet sweeper and drop-front desk in the reading nook to antique trunks in the bedrooms, no detail has been overlooked. In summer, a full breakfast—coffee, tea, yogurt, cereal, fresh fruit, and main course—is served on the outdoor patio. In winter, guests dine indoors by the wood-burning stove. Afternoon tea is served in the breakfast room. In 1995 two new rooms were added. One has a fireplace, private patio, TV, and private telephone. The second is an 800-square-foot cottage with a living room, dining area, full kitchen and bath, private patio, and screened-in porch. Take note of the careful renovation of this 1920s adobe, as it received an award from the Santa Fe Historical Board.

Territorial Inn

215 Washington Ave., Santa Fe, NM 87501. ☎ **505/989-7737.** Fax 505/986-9212, attn: Mary. 10 rms (8 with bath). AC, TV TEL. $80–$160 single or double. Higher rates during special events. Extra person $20. Rates include continental breakfast. AE, DC, MC, V. Free parking.

This 2-story, Territorial-style building, which dates from the 1890s and is situated 1¹⁄₂ blocks from the plaza, has a delightful Victorian feel with plenty of amenities. Constructed of stone and adobe with a pitched roof, it has 2 stories connected by a curving tiled stairway. Eight of its rooms, typically furnished with Early American antiques, offer private baths; the remaining two share a bath. All rooms are equipped with ceiling fans and sitting areas, and two have fireplaces. Each room also has robes. Free coffee or refreshments are available in the lobby and an in-room massage as well as access to a nearby health club can be arranged. An extended continental breakfast is served in a sophisticated common area or in warm months in the back garden, which is shaded by large cottonwoods. There is also a rose garden and a gazebo-enclosed hot tub.

✪ Water Street Inn

427 Water St., Santa Fe, NM 87501. ☎ **800/646-6752** or 505/984-1193. Fax 505/984-6235. 11 rms. A/C TV TEL. $125–$195 single or double. Rates include continental breakfast. AE, DISC, MC, V. Free parking.

An award-winning adobe restoration to the west of the Hilton hotel and 4 blocks from the plaza, this friendly inn features beautiful Mexican-tile baths, several kiva fireplaces or wood stoves, and antique furnishings. Each room is packed with Southwestern art and books. A happy hour, with quesadillas and margaritas, is offered in the living room or on the upstairs portal in the afternoon, where an extended continental breakfast is also served. All rooms are decorated in a Moroccan/Southwestern style. Room 3 features a queen-size hideaway sofa to accommodate families. (Yes, children are welcome, as are pets, with prior approval.) Room 4 provides special regional touches in its decor and boasts a chaise longue, fur rug, built-in seating, and corner fireplace. Four new suites have elegant contemporary Southwestern furnishings and outdoor private patios with fountains. Now there's also an outdoor Jacuzzi. All rooms have VCRs and offer newspaper delivery and twice-daily maid service.

6 RV Parks & Campgrounds

RV PARKS

At least four private camping areas, mainly for recreational vehicles, are located within a few minutes' drive of downtown Santa Fe. Typical rates are $20 for full RV hookups, $15 for tents. Be sure to book ahead at busy times.

Babbitt's Los Campos RV Park

3574 Cerrillos Rd., Santa Fe, NM 87501. ☎ **505/473-1949.** Fax 505/471-9220.

The resort has 95 spaces with full hookups, picnic tables, showers, rest rooms, laundry, and a soda and candy concession. It's just 5 miles south of the plaza, so it's plenty convenient, but keep in mind that it is surrounded by the city.

Camel Rock RV Campground

Route 5, Box 360H (US 84/285), Santa Fe, NM 87501. ☎ **800/TRY-RV-PARK** or 505/455-2661.

Another campground with full hookups, pull-through sites, tent sites, showers, rest rooms, laundry, seasonal swimming pool and hot tub, and a store that sells Native American jewelry and fishing gear. It's approximately 10 miles north of Santa Fe.

Rancheros de Santa Fe Campground

736 Old Las Vegas Hwy. (exit 290 off I-25), Santa Fe, NM 87505. ☎ **505/466-3482.** DISC, MC, V.

Tents, motor homes, and trailers requiring full hookups are welcome here. The park's 130 sites are situated on 22 acres of piñon and juniper forest. Facilities include tables, grills and fireplaces, hot showers, rest rooms, Laundromat, grocery store, nature trails, outdoor swimming pool, playground, games room, free nightly movies, public telephones, and propane. Cabins are also available. It's located about 6 miles southeast of Santa Fe and is open from March 15 to November 1. It's a campground with a web site! **Http://www.roadrunner.com/~ranchero/.**

Santa Fe KOA

934 Old Las Vegas Hwy. (exit 290 or 294 off I-25), Santa Fe, NM 87501. ☎ **505/466-1419** or 505/KOA-1514 for reservations. DISC, MC, V.

This campground offers full hookups, pull-through sites, tent sites, picnic tables, showers, rest rooms, laundry, store, "Santa Fe–style" gift shop, playground, recreation room, propane, and dumping station. It's located about 11 miles northeast of Santa Fe in the foothills of the Sangre de Cristo Mountains.

CAMPGROUNDS

There are three forested sites along NM 475 going toward the Santa Fe Ski Basin. All are open from May to October. Overnight rates start at about $6, depending on the particular site.

Hyde Memorial State Park

P.O. Box 1147 (NM 475), Santa Fe, NM 87503. ☎ **505/983-7175.**

This park is about 8 miles from the city. Its campground includes shelters, water, tables, fireplaces, and pit toilets. Maps of Santa Fe showing where firewood can be found are supplied. Seven RV pads with electrical pedestals and an RV dump station are available. There is also a small pond and nature trails.

Santa Fe National Forest

P.O. Box 1689 (NM 475), Santa Fe, NM 87504. ☎ **505/988-6940.**

Black Canyon campground, with 44 sites, is located just before you reach Hyde State Park. It has potable water and sites for trailers up to 32 feet long. Big Tesuque campground, with 10 newly rehabilitated sites, is about 12 miles from town. Both Black Canyon and Big Tesuque campgrounds, located along the Santa Fe Scenic Byway, NM 475, are equipped with vault toilets.

6 Where to Dine in Santa Fe

Santa Fe may not be a major city, but it abounds in dining options, with hundreds of restaurants of all categories to choose from. Competition among them is steep, and spots are continually opening and closing. Locals watch closely to see which ones will survive. Some chefs create dishes that incorporate traditional Southwestern foods with ingredients not indigenous to the region; their restaurants are referred to in the listings as "creative Southwestern." There is also standard regional New Mexican cuisine, and beyond that diners can opt for excellent steak and seafood, as well as continental, European, Asian, and, of course, Mexican menus.

Especially during peak tourist seasons, dinner reservations may be essential. Reservations are always recommended at better restaurants.

In the listings below, **Very Expensive** refers to restaurants where most dinner main courses are priced above $25; **Expensive** includes those where the main courses generally cost between $18 and $25; **Moderate** means those in the $12 to $18 range; and **Inexpensive** refers to those charging $12 and under.

1 Best Bets

- **Best Value:** If you're looking for good food and large portions for little money, you'll find it at **Tortilla Flats**, 3139 Cerrillos Rd. (☎ 505/471-8685). Portions are gigantic, and the atmosphere is quite friendly.
- **Best for Kids:** Without doubt, the **Cowgirl Hall of Fame,** 319 S. Guadalupe St. (☎ 505/982-2565), is great for children. The food on the kids' menu is simple enough to suit their tastes, and they'll be endlessly amused in the children's play area.
- **Best Continental:** For 31 years, **The Compound,** 653 Canyon Rd. (☎ 505/982-4353), has been serving fine cuisine. In an old adobe home nestled down a hidden driveway off Canyon Road, it will delight your culinary sensibilities.
- **Best Creative American:** Decor, service, and cuisine are top-notch at **Santacafé,** 231 Washington Ave. (☎ 505/984-1788). The chef has a magical way of combining Asian and Southwestern spices.
- **Best Creative Southwestern: Anasazi,** 113 Washington Ave. (☎ 505/988-3236), at the Inn of the Anasazi, provides time-worn Southwestern atmosphere and eclectic food that melds such staples as tortilla soup and wontons into exquisite creations.

Downtown Santa Fe Dining

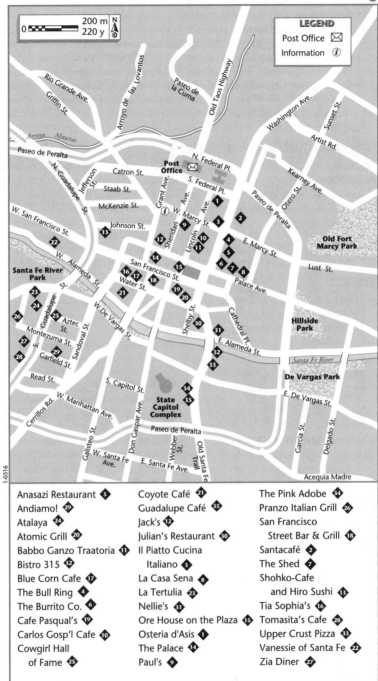

Anasazi Restaurant **5**
Andiamo! **29**
Atalaya **24**
Atomic Grill **20**
Babbo Ganzo Traatoria **11**
Bistro 315 **32**
Blue Corn Cafe **17**
The Bull Ring **4**
The Burrito Co. **6**
Cafe Pasqual's **19**
Carlos Gosp'l Cafe **10**
Cowgirl Hall
  of Fame **25**

Coyote Café **21**
Guadalupe Café **35**
Jack's **12**
Julian's Restaurant **30**
Il Piatto Cucina
  Italiano **3**
La Casa Sena **8**
La Tertulia **23**
Nellie's **31**
Ore House on the Plaza **15**
Osteria d'Asis **1**
The Palace **14**
Paul's **9**

The Pink Adobe **34**
Pranzo Italian Grill **26**
San Francisco
  Street Bar & Grill **18**
Santacafé **2**
The Shed **7**
Shohko-Cafe
  and Hiro Sushi **13**
Tia Sophia's **16**
Tomasita's Cafe **28**
Upper Crust Pizza **33**
Vanessie of Santa Fe **22**
Zia Diner **27**

- **Best Italian:** There are other good Italian restaurants in town, but my personal favorite is **Babbo Ganzo Trattoria,** 130 Lincoln Ave. (☎ **505/986-3835**). Their thin-crust pizzas are the best, and their pastas and specialties are consistently good.
- **Best Spanish: El Farol,** 808 Canyon Rd. (☎ **505/988-9912**), offers great ambience and local color, not to mention the longest and best tapas menu in Santa Fe.
- **Best Northern New Mexico Enchilada: Guadalupe Cafe,** 422 Old Santa Fe Trail (☎ **505/982-9762**), serves its enchiladas flat, the way most New Mexicans like them. Corn tortillas are layered with chicken, beef, or cheese, and smothered with a fiery red or green chile sauce.

2 Restaurants by Cuisine

AMERICAN

Jack's (Downtown, *E*)
San Francisco Street Bar and Grill
(Downtown, *I*)

BARBECUE

Cowgirl Hall of Fame (Downtown, *I*)

CHINESE

Hunan Restaurant (Southside, *M*)
On Lok Yuen (Southside, *I*)
Szechwan Chinese Cuisine
(Southside, *M*)

CONTINENTAL

Bistro 315 (Downtown, *M*)
The Compound (Downtown, *E*)
Geronimo (Downtown, *E*)
Pink Adobe (Downtown, *E*)

CREATIVE SOUTHWESTERN

Atomic Grill (Downtown, *M*)
Coyote Cafe (Downtown, *E*)
La Casa Sena (Downtown, *E*)

DELI/CAFE

Carlos' Gosp'l Cafe (Downtown, *I*)

ECLECTIC

Atalaya (Downtown, *I*)
Nellie's (Downtown, *E*)

INTERNATIONAL

Paul's (Downtown, *M*)
Zia Diner (Downtown, *I*)

ITALIAN

Andiamo! (Downtown, *M*)
Babbo Ganzo Trattoria
(Downtown, *M*)

Il Piatto Cucina Italiano
(Downtown, *M*)
Julian's (Downtown, *M*)
The Palace (Downtown, *M*)
Pranzo Italian Grill (Downtown, *M*)
Upper Crust Pizza (Downtown, *I*)

JAPANESE

Shohko-Cafe and Hiro Sushi
(Downtown, *M*)

MEXICAN

Old Mexico Grill (Southside, *M*)
Toushie's (Southside, *M*)

NEW AMERICAN

Celebrations (Downtown, *M*)

NEW MEXICAN

Anasazi Restaurant (Downtown, *E*)
Blue Corn Cafe (Downtown, *I*)
The Burrito Co. (Downtown, *I*)
Cafe Pasqual's (Downtown, *M*)
Green Onion (Southside, *I*)
Guadalupe Cafe (Downtown, *I*)
La Choza (Downtown, *I*)
La Tertulia (Downtown, *M*)
Maria's New Mexican Kitchen
(Southside, *I*)
Santacafé (Downtown, *E*)
The Shed (Downtown, *I*)
Tecolote Cafe (Southside, *I*)
Tía Sophia's (Downtown, *I*)
Tomasita's Cafe (Downtown, *I*)
Tortilla Flats (Southside, *I*)

SPANISH

El Farol (Downtown, *M*)

Key to abbreviations: *E* = Expensive, *I* = Inexpensive, *M* = Moderate

STEAKS/SEAFOOD

Bobcat Bite (Southside, *I*)
Bull Ring (Downtown, *M*)
El Nido (Northside, *E*)
Legal Tender (Out of Town, *M*)
Ore House on the Plaza
  (Downtown, *M*)

Steaksmith at El Gancho
  (Southside, *M*)
Tiny's Restaurant & Lounge
  (Southside, *I*)
Vanessie of Santa Fe (Downtown, *M*)

VIETNAMESE

Saigon Cafe (Southside, *I*)

3 Downtown

This area includes the circle defined by the Paseo de Peralta and St. Francis Drive, as well as Canyon Road.

EXPENSIVE

✪ Anasazi Restaurant

113 Washington Ave. ☎ **505/988-3236.** Reservations recommended. Breakfast $5.25–$9.50; lunch $8–$11.75; dinner $17.50–$29. AE, DC, DISC, MC, V. Daily 7–10:30am, 11:30am–2:30pm, and 5:30–10pm. NORTHERN NEW MEXICAN/NATIVE AMERICAN.

This ranks right up there with Santa Café as one of Santa Fe's richest dining experiences. And though it's part of the Inn of the Anasazi (see chapter 5), it's a fine restaurant in its own right. You'll dine surrounded by diamond-finished walls decorated with petroglyph symbols. Stacked flagstone furthers the Anasazi feel of this restaurant named for the ancient people who once inhabited the area. There's no pretension here; the waitstaff is friendly but not overbearing, and tables are spaced nicely, making it a good place for a romantic dinner. All the food is inventive, and organic meats and vegetables are used whenever available.

For breakfast try the breakfast burrito with homemade chorizo, green chile potatoes, and refried Anasazi beans. A must with lunch or dinner is the grilled corn tortilla soup with ginger-pork pot stickers. It's thick, served with tortilla strips and thinly sliced scallions, and a chile-spiced bread stick like a snake in the grass. For an entree, I enjoyed grilled swordfish with a roasted corn puree, light enough to enhance the fish flavor rather than diminish it. For dinner I recommend the cinnamon chile–rubbed beef tenderloin with white-cheddar chipotle, chile mashed potatoes, and mango salsa. Desserts are thrilling; try the sour-cream chocolate cake, rich and moist. There are daily specials, as well as a nice list of wines by the glass and special wines of the day.

✪ The Compound

653 Canyon Rd. ☎ **505/982-4353.** Reservations recommended. Main courses $17.50–$25. AE, MC, V. Tues–Sat 6pm until the last diners leave. CONTINENTAL.

The Compound, designed by noted architect Alexander Girard, is set on beautifully landscaped grounds amid tall firs and pines on the south bank of the Santa Fe River. It's reached by a long driveway off Canyon Road, at the rear of an exclusive housing compound. The interior has a minimalist decor, which sets of splashes of color from carefully selected folk artwork. Service is attentive and elegant. Neat, causal attire is fine.

The menu isn't long, but it is carefully selected. Choose from fish or meat dishes, all delicately prepared with sauces that truly enhance flavors. For a starter I had a special salmon mousse; the salad with endive and blue cheese was also excellent. My favorite entree is the boneless medaillons of lamb, very tender, served with a light mint sauce. The filet of beef is also delicious, as are the scallops. A different special is offered nightly. Main courses are served with saffron rice or braised potatoes, and three vegetables. Seafood and vegetables are delivered to the restaurant daily; pastries and French bread are baked on the premises. For dessert try the crème caramel or the Belgian chocolate mousse. There is, of course, an extensive wine list.

Coyote Cafe

132 Water St. ☎ **505/983-1615.** Reservations recommended. Main courses $6.50–$15.95 (Rooftop Cantina); fixed-price dinner $39.50 (Coyote Cafe). AE, DC, DISC, MC, V. Cafe: Oct–Apr, daily 11:30am–2pm and 6–9pm; Apr–Oct, Sat–Sun 11:30am–2pm, daily 6–9pm. Rooftop Cantina: daily 11am–9pm. CREATIVE SOUTHWESTERN.

This is still the trendiest place to dine in Santa Fe. Owner Mark Miller is a master of the palate and enjoys juxtaposing flavors in inventive ways. Tourists throng here: In fact, during summer, reservations are recommended days in advance. The cafe overlooks Water Street from tall windows on the second floor of a downtown building. Beneath the skylight, set in a cathedral ceiling, is a veritable zoo of animal sculptures in modern folk-art forms. The space can be a little noisy. Smoking is not allowed here.

The cuisine, prepared on a pecan-wood grill in an open kitchen, is Southwestern with a modern twist. The menu changes seasonally, but you might start with a Yucatán chicken tamale or griddled buttermilk corn cakes served with chipotle shrimp and salsa fresca. For entrees, try the red snapper Veracruz (on cilantro rice with roasted peppers, red onion, olives, and capers in a spicy seafood broth) or Sonoran-style Black Angus beef (seared beef chilaquile, grilled tenderloin, and coffee jerky quesadilla on blackened tomatillo-chiltepin sauce). A vegetarian special is offered daily. Andrew Maclauchlan, the pastry chef, has added a wonderfully creative variety of treats to the menu; try the peach star anise upside-down cake with iced gingered peach sauce and vanilla bean ice cream or the apricot and black mission fig linzer tart with brown sugar cream and blueberry ice cream. You can order drinks from the full bar or wine by the glass. The Coyote Cafe has two adjunct establishments. The Rooftop Cantina serves light Mexican fare and cocktails upon a festively painted terrace (try the Yucatán taquitos). On the ground floor is the Coyote Cafe General Store, a retail gourmet Southwestern food market, featuring the Coyote Cafe's own food line called Coyote Cocina (try the salsa), as well as hot sauces and salsas from all over the world.

Geronimo

724 Canyon Rd. ☎ **505/982-1500.** Reservations recommended. Lunch $8–$13; dinner $18–$30. AE, MC, V. Tues–Sun 11:30am–2:15pm; daily 6–10pm. CREATIVE CONTINENTAL.

When Geronimo opened in 1991, no one was sure if it would succeed; so many previous restaurants at this site had failed. But Geronimo has done more than just survive—it has flourished. The restaurant occupies an old adobe structure known as the Borrego House, built by Geronimo Lopez in 1756 and now completely restored. Numerous small dining rooms help it retain the comfortable feel of an old Santa Fe home.

I especially recommend lunch here, because you can get a taste of this often complex food for a fraction of the dinner price. Reserve a spot on the porch and watch the action on Canyon Road. My favorite at lunch is the lobster and vegetable spring rolls served with poblano jicama salad and mango dipping sauce. The Yucatán pork and vegetable tacos with guacamole and tomato salsa are also good. And if you've never tried one, go for the buffalo burger—it's more flavorful than beef. For dinner try the mesquite-grilled Black Angus rib eye with a spicy smoked corn and tomato salsa or the chicken mole relleño served with jalapeño peach salsa and crème fraîche.

For dessert, you won't be disappointed by the mocha pot de crème—especially if you're a chocoholic. The menu changes seasonally, and there is an excellent wine list. Outdoor dining is available in warm weather.

Jack's

135 W. Palace. ☎ **505/983-7220.** Reservations recommended. Main courses $9–$12 at lunch; $16–$25 at dinner. AE, MC, V. Lunch Mon–Sat 11:30am–3pm; dinner Sun–Thur 5:30–9pm, Fri and Sat 5:30–10pm. NEW AMERICAN.

Within 1 week, two friends told me their favorite new restaurant in Santa Fe was Jack's. My opinion? It's well worth a visit. Jack Shaab, the proprietor who was previously a partner in Bistro 315 and Il Piato, both successful Santa Fe eateries, has covered major ground in a short while. He already has a strong following, so expect a crowd. You won't find the stunning Southwestern decor of Santacafé here; this is an edgier, more citified ambience. It's decorated in black and brown, with butcher paper on the tables and crayons you can use for coloring while waiting to be served. The wait won't be long, however; service is prompt and precise.

If you don't feel like trying out this upstart for an expensive dinner, come for lunch. For an appetizer, I suggest the sautéed salmon cakes with wasabi aioli, found on both the lunch and dinner menus. For lunch, I enjoyed the vegetable tart with goat cheese and roasted tomato sauce. If you like hearty flavors, order the cornmeal crusted chicken breast, stuffed with feta, ricotta, and sundried tomatoes and served with a mushroom cream sauce. For dinner, you might try the sautéed Chilean sea bass with roasted sweet-pepper rice. You'll also find duck breast, pork tenderloin, and a mixed grill of lamb. The chocolate mousse I ordered for dessert at first seemed too frozen, but was in fact just extremely rich, almost chewy. It had its own identity, much like the entire dining experience here.

La Casa Sena

125 E. Palace Ave. ☎ **505/988-9232.** Reservations recommended. Lunch $7.75–$10; fixed-price dinner $18–$23. La Cantina, main courses $8.50–$17. AE, CB, DC, DISC, MC, V. Mon–Sat 11:30am–3pm; daily 5:30–10pm. Brunch Sun 11am–3pm. CREATIVE SOUTHWESTERN.

Though this restaurant suffered a fire in 1996, it's been restored and reopened, and many believe the food is even better now. It sits within the Sena compound, a beautiful example of a Spanish hacienda, Territorial-style adobe house built in 1867 by Civil War hero Maj. José Sena for his wife and 23 children. The house, which surrounds a garden courtyard, is today a veritable art gallery, with museum-quality landscapes on the walls and Taos-style handcrafted furniture. The cuisine in the main dining room might be described as Northern New Mexican with a continental flair. Lunches include chicken enchiladas on blue-corn tortillas and almond-encrusted salmon with gazpacho salsa. In the evening, diners might start with a salad of mixed organic greens, goat cheese, and a fresh herb vinaigrette, then move to American corn-fed lamb chops with habañero-papaya sauce, tropical fruit *ensalada,* and crispy root vegetables.

In the adjacent **La Cantina,** waiters and waitresses sing Broadway show tunes as they carry platters from the kitchen to the table. The more moderately priced Cantina menu offers the likes of country-fried rib eye and carne adovada burrito (prime pork roasted with red chile sauce, served with Hatch green chile and cheeses). Both restaurants have exquisite desserts. The award-winning wine list features more than 850 wines. There's patio dining in summer.

Nellie's

211 Old Santa Fe Trail (at Inn at Loretto). ☎ **505/984-7915.** Reservations recommended. Breakfast and lunch $6.75–$9.25; dinner $16–$22. AE, DC, DISC, MC, V. Daily 7am–9:15pm. NEW AMERICAN.

Named for a locally carved wooden snake that hangs over the bar, Nellie's is an interesting new addition to Santa Fe's restaurant scene. With the recent acquisition of sous chef Scott Shampine from the Anasazi, many dishes have the same creative Southwestern flair as those of his former employer. For breakfast, I recommend the pecan brioche French toast, served with Chantilly cream and apricot-infused Vermont maple syrup. Lunch can be especially nice on the patio in summer. Try the Vietnamese–style spaghetti and meat balls (flash-fried tuna and jicama tartare on peanut-dressed Asian noodle salad). I enjoyed the marinated tenderloin of pork for dinner. It came with a delicious blue-corn crepe filled with spinach and New Mexican goat cheese. For dessert the Mexican chocolate pot (shared by three of us) left me dreaming of chocolate for days afterward. The service could have been faster and more attentive. A variety of fine wines is available by the glass or bottle.

Pink Adobe

406 Old Santa Fe Trail. ☎ **505/983-7712.** Reservations recommended. Lunch $4.75–$8.75; dinner $10.75–$23.25. AE, CB, DC, DISC, MC, V. Mon–Fri 11:30am–2:30pm; daily 5:30–10pm. CONTINENTAL/SOUTHWESTERN.

San Pasqual, patron saint of the kitchen, keeps a close eye on this popular restaurant. It's found in the center of the 17th-century Barrio de Analco and across the street from the San Miguel mission. A Santa Fe institution since 1946, it occupies an adobe home believed to be at least 350 years old. Guests enter through a narrow side door into a series of quaint, informal dining rooms with tile or hardwood floors. Stuccoed walls display original modern art or Priscilla Hoback pottery on built-in shelves.

At the dinner hour the Pink Adobe offers the likes of escargot and shrimp rémoulade as appetizers. My favorite main course is the lamb curry or the poulet marengo (chicken baked with brandy wine and mushrooms). For lunch I'll always have a chicken enchilada topped with an egg. The gypsy stew (chicken, green chile, tomatoes, and onions in sherry broth) is also good. You can't leave without trying the hot French apple pie. While dining you can watch an interesting social scene swirl around you—this is the place for Santa Fe old-timers to see and be seen.

Smoking is allowed only in the Dragon Room, the lounge across the alleyway from the restaurant. Under the same ownership, the charming Dragon Room (a real local bar scene) has its own menu offering traditional New Mexican food. The full bar is open daily from 11:30am to 2am, until midnight on Sunday, and there's live entertainment Sunday through Thursday.

✪ Santacafé

231 Washington Ave. ☎ **505/984-1788.** Reservations recommended. Lunch $7–$11 at lunch; dinner $16–$26. AE, MC, V. Mon–Sat 11:30am–2pm; nightly 6–10pm. NEW AMERICAN.

When you eat at this fine restaurant, be prepared for spectacular bursts of flavor. The food is Southwestern with an Asian flair, surrounded by a minimalist decor that accentuates the beautiful architecture of the 18th-century Padre Gallegos House, 2 blocks from the plaza. The white walls are decorated only with deer antlers, and each room contains a fireplace. In warm months you can sit under elm trees in the charming courtyard. Tables are graced by a pot of sunflowers, resplendent against the white tablecloths. Beware that on busy nights the rooms are noisy, despite the precision of the waitstaff's efforts.

The dishes change to take advantage of seasonal specialties. A simple starter such as miso soup is enriched with a lobster-mushroom roulade. One of my favorites is the seared chile-garlic prawns served with fresh pea (or lima) and mushroom risotto. A more adventurous eater might try the pan-seared achiote duck breast with raisin couscous, spiced pecans, and pineapple pasilla puree. There's an extensive wine list, with wine by the glass as well. Desserts are made in-house and are artistically presented. Some have criticized chef Ming Tsai for overdoing his recipes; others claim the prices are too steep. I just find the experience inventively delicious in all respects.

MODERATE

Andiamo!
322 Garfield St. ☎ **505/995-9595.** Reservations recommended. Main courses $7–$16. AE, DISC, MC, V. Wed–Mon 5:30–9:30pm. ITALIAN.

Quite a few new restaurants have sprung up in Santa Fe over the past few years, several of which were created by defectors of some of the city's most popular eateries. Andiamo! is one of those making a successful go of it. Chris Galvin, once the sous chef at Cafe Escalera, has joined forces with business partner Joan Gillcrist at this fine restaurant. They have created an authentically Tuscan atmosphere in which a daily changing menu features antipasto, pasta, and excellent desserts. Still, as with many of these little spinoffs, you tend to feel a bit cramped into the space, and noise levels can get moderately out of hand. I enjoyed the Caesar salad and the penne with merguez, with a bit of musky flavor from the lamb sausage. For dessert I'd recommend the polenta pound cake with lemon crème anglaise. Beer and wine are served at this nonsmoking restaurant.

Babbo Ganzo Trattoria
130 Lincoln Ave. ☎ **505/986-3835.** Reservations recommended. Lunch $4–$9; dinner $10–$22. AE, DC, DISC, MC, V. Mon–Sat 11:30am–2pm; Mon–Sun 5:30–9pm. NORTHERN ITALIAN.

Located atop the only escalator in Santa Fe, the Babbo Ganzo is a great little Italian place that serves lunch and dinner. As you walk into the restaurant, you'll see on your right a wall of pasta for sale and on your left a large bar. There's a wonderful mural, as well as a fireplace and a wine rack. For lunch you could start with bruschetta (grilled bread topped with garlic, tomato, and basil) and then follow with an allo zenzero pizza with red chile, tomato, and mozzarella cheese. Pizzas are baked in the wood-burning oven. At dinner, start with insalata e funghi fritti (shiitake and oyster mushrooms sautéed in balsamic vinegar and served over a bed of lettuce). Move on to pappardelle sui cinghiale (homemade pasta with wild boar in a light tomato sauce), or if you prefer fish, Babbo Ganzo offers a nice selection plus a fresh fish of the day. This is one of the more authentic Italian restaurants in the area.

Bistro 315
315 Old Santa Fe Trail. ☎ **505/986-9190.** Reservations recommended. Main courses $16–$24 at dinner. AE, MC, V. Daily 11:30am–2pm and 5:30–9pm. CONTINENTAL.

Bistro 315 has enjoyed instant success since it opened in 1995, and no wonder—heading it up are Matt Yohalem, a graduate of Johnson and Wales, and Chef Poissonier, formerly with Le Cirque under Chef Daniel Boulud. The restaurant is tiny, with only 27 tables (though an expansion may be completed by press time), but always packed. The food is simply excellent. The menu changes seasonally; on my last visit there I started with croquettes of goat cheese and bell pepper coulis and moved on to piñon-crusted halibut served with spinach and scallion beignets (Yohalem's previous experience in the kitchen of Commander's Palace in New Orleans is evident in his preparation of dishes such as this). The grilled tomato soup was also excellent, and I was fortunate to be there on a night when grilled smoked chicken was on the menu. My favorite dessert here is the warm tarte tatin served with crème fraîche. Because the restaurant is so small and popular, reservations are an absolute must.

Bull Ring

150 Washington Ave. ☎ **505/983-3328.** Reservations recommended. Main courses $10.95–$25. AE, MC, V. Mon–Fri 11:30am–3pm; daily 5–10pm. STEAKS/SEAFOOD.

Legislators and lobbyists from the State Capitol still make a habit of dining and drinking at the Bull Ring even though it recently moved to a new location. Steaks are the dinner specialty. Much ballyhooed is the Bull Ring steak, a 12-ounce charcoal-broiled New York cut with sautéed mushrooms or onions, red or green chile, or au poivre sauce. Seafood dishes include grilled shrimp skewer served on a bed of rice with garlic-herb butter. Enchiladas Nuevo Mexico (blue-corn tortillas with red or green chile, cheese, beef, or chicken) highlight the small regional portion of the menu. Midday diners can also choose half-pound hamburgers or other sandwiches.

Cafe Pasqual's

121 Don Gaspar Ave. ☎ **505/983-9340.** Reservations recommended for dinner. Breakfast $4.75–$8.95; lunch $7.95–$9.95; dinner $15.95–$24.75. AE, MC, V. Mon–Sat 7am–3pm; Mon–Sun 6–10pm (until 10:30pm in summer). Brunch Sun 8am–2pm. INNOVATIVE NEW MEXICAN.

With vivid murals of magical parties, lots of plants, and plenty of Mexican touches, restaurateur Katharine Kagel has created a wonderful escape in this intimate restaurant across from the Hotel St. Francis. The fare is also magical and the service is friendly. My favorite dish for breakfast is the blue-corn pancakes, and for any time of day, the huevos motuleños (two eggs over easy on blue-corn tortillas and black beans topped with sautéed bananas, feta cheese, salsa, and green chile). Soups and salads are also served, and there's a delectable grilled-salmon burrito with herbed goat cheese and cucumber salsa. The frequently changing dinner menu also offers grilled meats and seafoods, plus vegetarian specials. Recently Kagel has introduced even more offerings from interior Mexico, including cuitlacoche quesadillas and cactus fruit sorbet for dessert. There's a communal table for those who would like to meet new people over a meal. Pasqual's offers imported beers and wine by the bottle or glass. Try to go at an odd hour—late morning or afternoon; otherwise you'll have to wait.

Celebrations Restaurant and Bar

613 Canyon Rd. ☎ **505/989-8904.** Reservations recommended. Breakfast/lunch up to $8.95, with daily specials; dinner main courses up to $12.95, with nightly specials. MC, V. Daily 7:30am–2pm; Wed–Sat 5:30–9pm; call for seasonal hours. AMERICAN BISTRO.

Housed in a former art gallery with beautiful stained-glass windows and a kiva fireplace, Celebrations boasts "the ambience of a bistro and the simple charm of another era." In summer, guests can dine on a brick patio facing Canyon Road. The meals here are a delicious melding of flavors, particularly those that cook a while such as the étouffée and stew. Three meals are served daily, starting with breakfast—an omelet with black beans or French toast with orange syrup, for example. Lunch choices include soup, salad, and pasta specials, Swiss raclette, sandwiches such as the oyster poor boy, and the ploughman's lunch. Casseroles, pot pies, and hearty soups are always available in winter. The dinner menu changes with the chef's whims, but recent specialties have included sautéed pecan trout, roasted rack of lamb, and my favorite, crawfish étouffée. All desserts (including the red chile piñon ice cream) are homemade. There is a full bar, as well as a choice of beers and wines by the bottle or glass.

El Farol

808 Canyon Rd. ☎ **505/983-9912.** Reservations recommended. Tapas $3.95–$8.95; main courses $10.95–$22.50. DC, DISC, MC, V. Daily 11:30am–4pm and 6–10pm. SPANISH.

This is the place to head for local ambience and old-fashioned flavor. El Farol (The Lantern) is the Canyon Road artists' quarter's original neighborhood bar. Its low ceilings and dark brown walls have now become the home of one of Santa Fe's largest and most unusual assortments of tapas. Thirty-five varieties are offered, including such delicacies as pulpo à la Gallega (octopus with Spanish paprika sauce) and grilled cactus with ramesco sauce. I prefer simply to order two or three tapas and have my companion do the same so that we can share. However, if you prefer a full dinner, try the paella or the marinated lamb chops with huckleberry, cranberry, or rosemary sauce. Jazz, folk, and ethnic musicians play almost every night beginning at 9:30pm. In summer, an outdoor patio seating 50 is open to diners.

Il Piatto Cucina Italiano

96 West Marcy St. ☎ **505/984-1091.** Reservations recommended. Main courses $7–$15. AE, MC, V. Lunch Mon–Fri 11:30am–2pm; dinner Mon–Sat 5:30–9pm. TRADITIONAL NORTHERN ITALIAN.

This is a spinoff from a more expensive restaurant. A child of Bistro 315, Il Piatto brings executive chef Matt Yohalem's expertise to thinner wallets. It's an Italian cafe, simple and informal, with an incredible collection of cooking utensils, old and new, adorning the walls. The menu changes seasonally, complemented by a few perennial standards. For a starter, try the fresh arugula with pine nuts, raisins, shaved onions, and parmigiano. Among entrees, my favorite is the fettuccine carbonara (rich cream-and-egg sauce and proscuitto), though you can't go wrong with the roast chicken with Italian sausage. A full wine and beer bar is available.

Julian's Restaurant

221 Shelby St. ☎ **505/988-2355.** Reservations recommended. Main courses $14–$21. AE, CB, DC, DISC, MC, V. Daily 5:30pm until the last diners leave. ITALIAN.

Devotees of Julian's in Telluride, Colorado, may be pleased to discover that their favorite northern Italian restaurant resurfaced in Santa Fe in 1989. Elizabeth (Lou) McLeod and Wayne Gustafson have moved to an unpretentious setting just off Alameda Street. Meals are served in an atmosphere of simple elegance around a central kiva fireplace to the strains of light jazz. The menu is primarily Tuscan, though other regional dishes have been known to pop up on occasion. Start with prosciutto di Parma con melone (fresh melon with Italian ham), ostriche alla Genovese (oysters baked with Parmesan cheese, pesto, and seasoned bread crumbs), or melanzana alla griglia con pepperoni (grilled eggplant with roasted peppers and balsamic vinegar). Then move on to the generous main courses—for instance, spigola in guazzetto alla romana (sea bass sautéed with olive oil, garlic, tomatoes, mustard, raisins, and pine nuts), pollo con porcini (chicken grilled with a sauce of porcini mushrooms and dark chicken stock), or veal piccata alla limone. All main courses include bread and a side of pasta. You can order a salad à la carte. There is a full bar with an extensive list of Italian and Californian wines.

La Tertulia

416 Agua Fría St. ☎ **505/988-2769.** Reservations recommended. Lunch $5.50–$7.50; dinner $7.35–$18.50. AE, DC, DISC, MC, V. Tues–Sun 11:30am–2pm and 5–9pm. NEW MEXICAN.

Housed in a former 18th-century convent, La Tertulia's thick adobe walls separate six dining rooms, including the old chapel and a restored sala (living room) with a valuable Spanish colonial art collection. There's also an outside garden patio for summer dining. Dim lighting and viga-beamed ceilings, shuttered windows and wrought-iron chandeliers, lace tablecloths, and hand-carved santos in wall niches lend a feeling of historic authenticity. *La tertulia* means "gathering place" or "discussion." Gourmet regional dishes will give you something to talk about. They include filet y relleños (cheese stuffed chiles and a filet mignon) and camarones con pimientos y tomates (shrimp with peppers and tomatoes). My favorite, however, is the burrito plate, served with rice and posole. If you feel like dessert, try the chocolate piñon-nut truffle torte or natillas (custard). The bar features homemade sangría.

Ore House on the Plaza

50 Lincoln Ave. ☎ **505/983-8687.** Reservations recommended. Main courses $12–$23. AE, MC, V. Mon–Sat 11:30am–2:30pm, Sun noon–2:30pm; daily 5:30–10pm. STEAKS/SEAFOOD.

The Ore House's second-story balcony, at the southwest corner of the plaza, is an ideal spot from which to watch the passing scene while you enjoy cocktails and hors d'oeuvres. In fact, it is *the* place to be between 4 and 6pm every afternoon. The decor is Southwestern, with plants and lanterns hanging amid white walls and booths. The menu, currently presided over by Chef Isaac Modivah, is heavy on fresh seafood and steaks. Daily fresh fish specials include salmon and swordfish (poached, blackened, teriyaki, or lemon), rainbow trout, lobster, and shellfish. Steak Ore House (wrapped in bacon and topped with crabmeat and béarnaise sauce) and chicken Ore House (a grilled breast stuffed with ham, Swiss cheese, green chile, and béarnaise) are local favorites. The Ore House also caters to noncarnivores with vegetable platters.

The bar, with solo music Wednesday through Saturday nights, is proud of its 66 different "custom margaritas." It offers a selection of domestic and imported beers and an excellent wine list. An appetizer menu is served from 2:30 to 5pm daily, and the bar stays open until midnight or later (on Sunday it closes at midnight).

The Palace

142 W. Palace Ave. ☎ **505/982-9891.** Reservations recommended. Lunch $5.25–$10.50; dinner $9.50–$21. AE, DISC, MC, V. Mon–Sat 11:30am–4pm; daily 5:45–10pm. ITALIAN/CONTINENTAL.

When the Burro Alley site of Doña Tules's 19th-century gambling hall was excavated in 1959, an unusual artifact was discovered: a brass door knocker, half shaped like a horseshoe, and the other half like a saloon girl's stockinged leg. That knocker has become the logo of The Palace, which maintains a Victorian flavor as well as a bit of a bordello flair, with lots of plush red upholstery. The brothers Lino, Pietro, and Bruno Pertusini brought a long family tradition into the restaurant business: Their father was chef at the Villa d'Este on Lake Como, Italy. You'll find an older crowd here and fairly small portions. The Caesar salad—prepared tableside—is always good, as are the meat dishes such as the grilled Black Angus New York strip. Fish dishes are inventive and tasty. Try crab cakes with a sauce aurore for lunch or grilled cornmeal-dusted trout with crab filling for dinner. For a light meal, my father likes to order the spaghetti pomodoro; the cannelloni is also nice but much richer. They serve a variety of vegetarian dishes, and there are usually daily specials. The wine list is long and well considered. There is outdoor dining, and the bar is open Monday through Saturday from 11:30am to 2am and Sunday from 5:45pm to midnight, with nightly entertainment including dancing on Saturday after 9pm.

✪ Paul's

72 W. Marcy St. ☎ **505/982-8738.** Reservations recommended for dinner. Lunch $4.95–$7.50; dinner $12.95–$19.95. AE, DISC, MC, V. Mon–Sat 11:30am–2pm; Sun–Thurs 5:30–9pm, Fri–Sat 6–9pm. INTERNATIONAL.

Once just a home-style deli, then a little gourmet restaurant called Santa Fe Gourmet, Paul's (which opened in 1990) is a wonderful place for lunch or dinner. The lunch menu presents a nice selection of main courses, such as salad Niçoise, dill salmon cakes, and an incredible pumpkin bread stuffed with pine nuts, corn, green chile, red chile sauce, queso blanco, and caramelized apples. Sandwiches are also available at lunch. At dinner, the lights are dimmed and the bright Santa Fe interior (with folk art on the walls and colorfully painted screens that divide the restaurant into smaller, more intimate areas) becomes a great place for a romantic dinner. The menu might include red chile duck wontons in a soy ginger cream to start and pecan-herb-crusted baked salmon with sorrel sauce as an entree. The grilled ahi with roasted pepper, artichoke hearts, and green olive salsa is excellent. Every dish is artistically exquisite in its presentation. Paul's won the "Taste of Santa Fe" award for best main course in 1992 (the salmon) and best dessert in 1994. The chocolate ganache is exquisite. A wine list is available. Smoking is not permitted in the restaurant.

✪ Pranzo Italian Grill

540 Montezuma St., Sanbusco Center. ☎ **505/984-2645.** Reservations recommended. Lunch $5.95–9.95 at lunch; dinner $5.95–$17.50. AE, DC, DISC, MC, V. Mon–Sat 11:30am–3pm; Sun–Thurs 5–10pm, Fri–Sat 5–11pm. REGIONAL ITALIAN.

Housed in a renovated warehouse and freshly redecorated in warm Tuscan colors, this sister of Albuquerque's redoubtable Scalo restaurant caters to local Santa Feans with a contemporary atmosphere of modern abstract art and food prepared on an open grill. Homemade soups, salads, creative pizzas, and fresh pastas are among the less expensive menu items. Bianchi e nere al capesante (black-and-white linguine with bay scallops in a light seafood sauce) and pizza pollo affumicato (with smoked chicken, pesto, and roasted peppers) are consistent favorites. Steak, chicken, veal, and fresh seafood grills—heavy on the garlic—dominate the dinner menu. The bar offers the Southwest's largest collection of grappas, as well as a wide selection of wines and champagnes by the glass. The upstairs rooftop terrace is lovely for seasonal moon-watching over a glass of wine. **Portare Via Cafe,** adjacent to the restaurant, is a great place for a light breakfast or lunch. The cinnamon rolls and scones are particularly good, and cappuccino and pastries are served throughout the day. Sandwiches are available at lunch.

Shohko-Cafe and Hiro Sushi

321 Johnson St. ☎ **505/983-7288.** Reservations recommended. Lunch $4.25–$12; dinner $4.25–$18. AE, DISC, MC, V. Mon–Fri 11:30am–2pm; Mon–Thurs 5:30–9pm, Fri–Sat 5:30–9:30pm. JAPANESE SUSHI.

Opened in 1976 as Santa Fe's first Japanese restaurant, in a 150-year-old adobe building that had been a bordello in the 19th century, Shohko celebrated its 20th anniversary in 1996. Its current owners have done a terrific job of blending New Mexican decor (such as ceiling vigas and Mexican tile floors) with traditional Japanese decorative touches (rice paper screens, for instance). Up to 30 fresh varieties of raw seafood, including sushi and sashimi, are served at the 36-foot sushi bar. Here is the place to indulge in teriyaki dishes, sukiyaki, yakitori (skewered chicken), and yakisoba (fried noodles), and two uniquely Southwestern Japanese treats: green chile tempura and, at the sushi bar, the Santa Fe Roll (with green chile, shrimp tempura, and masago). A new menu item—chilled seafood noodle salad—is quite good. There are also some Chinese dishes, a bento box lunch special, and many vegetarian specialties. "Food for Health" is the motto of Shohko. Wine, imported beers, and hot sake are available.

Vanessie of Santa Fe

434 W. San Francisco St. ☎ **505/982-9966.** Reservations strongly recommended. Dinner $12.95–$17.95. AE, CB, DC, DISC, MC, V. Daily 5:30–10:30pm. STEAKS/SEAFOOD.

Vanessie is as much a piano bar as it is a restaurant. The talented musicians Doug Montgomery and Charles Tichenor hold forth at the keyboard, caressing the ivories with a repertoire that ranges from Bach to Barry Manilow. A 10-item menu, served at large, round, wooden tables beneath hanging plants in the main dining room or on a covered patio, never varies: roast chicken, fresh fish, New York sirloin, filet mignon, Australian rock lobster, grilled shrimp, and rack of lamb. I especially like the rotisserie chicken. Portions are large and come with baked potatoes or onion loaf. Fresh vegetables or sautéed mushrooms are available at an extra charge. For dessert, the slice of cheesecake served is large enough for three diners. There's a short wine list.

INEXPENSIVE

Atalaya Restaurant-Bakery

320 S. Guadalupe. ☎ **505/982-2709.** Reservations not required. Main courses $6–$12. AE, DC, DISC, MC, V. Daily 7am–10pm. ECLECTIC.

Atalaya, a trendy bistro, is one of Santa Fe's newer restaurants. The hip decor (exposed brick walls and geometric wall partitions), low prices, and an interesting menu attract a cross section of the city's population to this restaurant day and night. The shrimp and grits done "low-country style" with mushrooms, bacon, green onions, and fried eggs is a great choice for breakfast. My favorite for an early lunch is migas (Tex-Mex scrambled eggs with salsa, Monterrey Jack cheese, and tortilla chips. At lunch try a muffuletta sandwich (a New Orleans staple), filled with ham, salami, provolone, and olive salad. Other menu items include grilled lamb brisket with an orange serrano pepper sauce and mussels served with Thai green curry. Of course, if you'd rather just have a burger, go for the green chile cheeseburger topped with roasted poblanos and Monterrey Jack. The breads here are quite good. Desserts are as varied as the entrees and equally good, and there's an excellent selection of coffees and milkshakes.

Atomic Grill

103 E. Water St. ☎ **505/820-2866.** Most items under $8.50. Summer: daily 7am–3am. Rest of year: Mon–Fri 11am–midnight, Sat–Sun 8am–midnight. AE, DISC, MC, V. CREATIVE MEXICAN/AMERICAN.

A block south of the plaza, this cafe offers decent patio dining at reasonable prices. Of course, there's indoor dining as well. The whole place has a hip and comfortable feel, and the food is prepared imaginatively. For breakfast try the raspberry French toast made with home-baked challah bread, served with apricot butter and maple syrup. For lunch, the green chile stew is tasty, although made with not quite enough chicken habañero sausage. The fish tacos are also nice, if a little bland (ask for extra salsa). For dessert, try the carrot cake. Wine by the glass and 100 different beers are available. They also deliver to the downtown area from 11am to midnight.

Blue Corn Cafe

133 W. Water St. ☎ **505/984-1800.** Reservations accepted for parties of eight or more. Main courses $4.95–$8.75. AE, DC, DISC, MC, V. Daily 11am–11pm. The bar stays open until 2am. NEW MEXICAN.

This lively, attractively decorated Southwestern-style restaurant that opened in 1992 appeals greatly to some and not at all to others. For me the atmosphere is a little like Denny's, but the food is decent and the prices are reasonable. The bar at the entrance lends a party atmosphere to the whole place. While you peruse the menu your waiter will bring freshly made tortilla chips to your table. The Blue Corn Cafe is known for its chile rellenos, and I have enjoyed the achiote grilled half chicken with epazote cream sauce as well as the carne adovada quesadilla (red chile marinated pork, wrapped in a flour tortilla with green chile and Jack and cheddar cheeses, served with guacamole and sour cream). For the more adventurous, the tortilla burger (beef patty covered with cheeses and red or green chile, wrapped in a flour tortilla and served with chile fries) is a good bet. For those who can't decide, the combination plates offer

a nice variety. For dessert, try the Mexican brownies,
ice cream. The bar features 34 varieties of tequila an
margaritas. Live bands play on Friday and Saturday n

The Burrito Co.

111 Washington Ave. ☎ **505/982-4453.** Menu items $1.25–
Sun 10am–5pm. NEW MEXICAN.

If you're in a hurry and find yourself downtown, this is ~p. You
won't enjoy the best food in Santa Fe, but you'll get filled ...ue money. You
can people-watch while you dine on the outdoor patio or enjoy the murals inside,
painted by a local artist. Order and pick up your food at the counter. Breakfast
burritos are popular in the morning; after 11am you can get traditional New Mexican meals with lots of chile. They've recently added pastries and specialty coffees to
the menu.

Carlos' Gosp'l Cafe

125 Lincoln Ave. ☎ **505/983-1841.** Menu items $2.75–$7. No credit cards. Mon–Sat 11am–
3pm. DELI.

A visit may have you singing the praises of the "Say Amen" desserts at this cafe in
the inner courtyard of the First Interstate Bank Building. First, though, try the tortilla or hangover (potato, corn, and green chile chowder with Monterey Jack cheese)
soups or the deli sandwiches. The Gertrude Stein sandwich, with Swiss cheese, tomato, red onion, sprouts, and mayonnaise, is good (although it's hard to say why this
combination is named after Ms. Stein). I prefer the unusual Miles Standish (fresh
turkey breast with cranberries, cream cheese, and mayonnaise). Carlos's has outdoor
tables, but many diners prefer to sit indoors, reading newspapers or chatting around
the large common table. Gospel and soul music play continually; paintings of
churches and performers cover the walls.

✪ Cowgirl Hall of Fame

319 S. Guadalupe St. ☎ **505/982-2565.** Fax 505/982-4047. Reservations recommended.
Lunch $2.95–$7.50; Dinner $4.25–$12.95. AE, DC, DISC, MC, V. Mon–Thurs 11am–11pm, Fri
11am–midnight, Sat 8am–midnight, Sun 8am–10:30pm. The bar is open Mon–Sat until 2am,
Sun until midnight. REGIONAL AMERICAN/BARBECUE.

The Cowgirl Hall of Fame has earned a reputation among Santa Feans as unbridled
fun. Everything at this restaurant has been done in a playful spirit—from the cowgirl paraphernalia decorating the walls to the menu featuring such items as "chicken
wang dangs" (eight chicken wings in a "dandy" citrus-Tabasco marinade). The
restaurant is also known for its good food at reasonable prices. I recommend the
seasonal dish, crawfish étoufée, and the Jamaican jerk pork tenderloin (careful, it can
be *hot*). The bunkhouse smoked brisket with potato salad, barbecue beans, and cole
slaw is excellent, as is the cracker-fried catfish with jalapeño-tartar sauce. Recent specialties of the house included butternut squash casserole and grilled-salmon soft tacos.
There's even a special "kid's corral" that has horseshoes, a rocking horse, a horse-
shaped rubber tire swing, hay bales, and a beanbag toss to keep children entertained
during dinner. The dessert specialty of the house is the original ice-cream baked potato (ice cream molded into a potato shape, rolled in spices, and topped with green
pecans and whipped cream). Happy hour is from 4 to 6pm. There is live music or
comedy almost every night.

✪ Guadalupe Cafe

422 Old Santa Fe Trail. ☎ **505/982-9762.** Breakfast $4.50–$8.75; lunch $6–$12; dinner
$6.95–$15.95. DISC, MC, V. Mon–Fri 7am–2pm, Sat–Sun 8am–2pm; Mon–Sat 5:30–9pm. NEW
MEXICAN.

New Mexican food, this is where I go, and like many Santa Feans, I often. This casually elegant cafe, recently featured in *Bon Appétit* magazine, white stucco building that's warm and friendly and has a nice-sized patio for ining in warmer months. For breakfast try the spinach-mushroom burritos and huevos rancheros, and for lunch the chalupas or stuffed sopaipillas. Any other time, I'd start with fresh roasted ancho chiles (filled with a combination of Montrachet and Monterey Jack cheeses, piñon nuts, and topped with your choice of chile) and move on to the sour-cream chicken enchilada or any of their other Southwestern dishes. Order both red and green chile (Christmas) so that you can sample some of the best sauces in town. To the menu they've recently added some delicious salads, such as a Caesar with chicken. For those who don't enjoy Mexican food, there are also "hamburguesas" (hamburgers) and a selection of traditional favorites such as chicken-fried steak, turkey piñon meat loaf, and chicken salad. Daily specials are available; don't miss the famous chocolate-amaretto adobe pie for dessert. Beer, wine, and margaritas are served.

La Choza

905 Alarid St. ☎ **505/982-0909.** Lunch or dinner $6.75–$8.50. DISC, MC, V. Winter: Mon–Thurs 11am–8pm, Fri–Sat 11am–9pm. Summer: Mon–Sat 11am–9pm. NEW MEXICAN.

The sister restaurant of The Shed (see below) offers some of the best New Mexican food in town at a convenient location near the intersection of Cerrillos Road and St. Francis Drive. When other restaurants are packed, you may have to wait a little while here—although often you'll be seated quickly. It's a warm, casual eatery, the walls vividly painted with magical images, especially popular on cold days when diners gather around the wood-burning stove and fireplace. The menu offers enchiladas, tacos, and burritos on blue-corn tortillas, as well as green chile stew, chile con carne, and carne adovada. Unfortunately, the portions aren't very large. Still, I recommend the chicken enchilada or the bean burrito, and for dessert, you can't leave without trying the mocha cake (chocolate cake with a mocha pudding filling, served with whipped cream). Vegetarians and children have their own menus. Beer and wine are available.

On Lok Yuen

3242 Cerrillos Rd. ☎ **505/473-4133.** Reservations not accepted for Fri evening. Lunch buffet $4.95; dinner $4.95–$8.25. MC, V. Mon–Sat 11am–2:30pm and 4:30–9pm. CHINESE/AMERICAN.

Behind adobe walls you'll find typical Chinese-American fare—longtime favorites such as beef and chicken chow mein, chop suey, egg foo yung, sweet-and-sour pork, and shrimp fried rice. I recommend the deep-fried butterful shrimp. Luncheon specials include wonton soup, rice, and fortune cookie. There's a choice of several dinner combinations.

Osteria d' Assisi

58 S. Federal Place. ☎ **505/986-5858.** Lunch $6.95–$9.50; dinner $9.95–$12.95, with specials up to 16.50. AE, DC, MC, V. Mon–Sat 11am–5pm; Mon–Thurs 5–9pm, Fri–Sat 5–10pm. NORTHERN ITALIAN.

Here again we have a less expensive spinoff of a more expensive restaurant. This one was opened by one of the brothers from The Palace restaurant; expect to get good food at a fraction of the price. However, you sacrifice fine service and consistency here. Located just a few blocks from the Plaza, this restaurant has a quaint country Italian atmosphere with sponge-painted walls, simple wooden furniture, and the sound of Italian from chef Bruno Pertusini, of Lago di Como, Italy, punctuating the air. For antipasto, I enjoyed the caprese (fresh mozzarella with tomatoes and basil, garnished with baby greens in a vinaigrette). For pasta I recommend the paglia e fieno dell ortolano (white and green linguine with portobello mushrooms, artichokes, eggplant, and sun-dried tomatoes). There are daily specials such as the lamb osso bucco served with mashed potatoes, which I enjoyed. All meals are served with homemade Italian bread, and there are a number of special desserts. Beer and wine are by the bottle or glass. There's also a small deli in front from which you can take home meats and cheeses.

San Francisco Street Bar and Grill

114 W. San Francisco St. ☎ 505/982-2044. Fax 505/982-6750. No reservations. Lunch $5.25–$6.75; dinner $5.25–$12.50. AE, DISC, MC, V. Daily 11am–11pm. AMERICAN.

This easygoing eatery offers casual dining amid simple decor in two seating areas: the main restaurant and an indoor courtyard beneath the 3-story Plaza Mercado atrium. Locals come here for thick, juicy hamburgers. There is also a variety of daily specials. The lunch menu consists mainly of soups, sandwiches, and salads; dinner dishes include a tasty chicken breast with roasted red pepper aioli, grilled pork tenderloin with apple and green chile salsa, and New York strip steak. There are also nightly pasta specials such as spinach linguine al Greque (homemade spinach linguini with sun-dried tomatoes, kalamata olives, feta cheese, olive oil, and garlic). The full bar service includes draft beers and daily wine specials.

✪ The Shed

113¹/₂ E. Palace Ave. ☎ 505/982-9030. Reservations accepted at dinner. Lunch $4.75–$8; dinner $6.75–$13.95. DISC, MC, V. Mon–Sat 11am–2:30pm; Thurs–Sat 5:30–9pm. NEW MEXICAN.

Queues often form outside The Shed, half a block east of the Palace of the Governors. A luncheon institution since 1953, it occupies several rooms and the patio of a rambling hacienda that was built in 1692. Festive folk art adorns the doorways and walls. The food is delicious, some of the best in the state, a compliment to traditional Hispanic Pueblo cooking. The chicken or cheese enchilada is renowned in Santa Fe. Tacos and burritos are good, too, all served on blue-corn tortillas with pinto beans and posole. The green chile soup is a local favorite. The Shed's Joshua Carswell has added vegetarian and low-fat Mexican foods to the menu, as well as a wider variety of soups and salads. Don't leave without trying the mocha cake. Beer and wine are available.

Tía Sophia's

210 W. San Francisco St. ☎ 505/983-9880. Breakfast $1.35–$7.50; lunch $3.50–$8.50. MC, V. Mon–Sat 7am–2pm. NEW MEXICAN.

If you want to see how real Santa Fe locals look and eat, go to this friendly downtown restaurant (now in its 23rd year). You'll sit at big wooden booths and sip dinner coffee. Daily breakfast specials include eggs with blue-corn enchiladas (Tuesday) and burritos with chorizo, potatoes, chile, and cheese (Saturday). My favorite is the breakfast burrito or huevos rancheros (eggs over corn tortillas, smothered with chile). Some like the Atrisco plate: two eggs, green chile stew, a cheese enchilada, beans, posole, and a sopaipilla. Beware of what you order because, as the menu states, Tía Sophia's is "not responsible for too hot chile." Because this is a popular place, be prepared to wait for a table.

Tomasita's Cafe

500 S. Guadalupe St. ☎ 505/983-5721. No reservations. Lunch $4.25–$9.50; dinner $4.75–$9.95. MC, V. Mon–Sat 11am–10pm. NEW MEXICAN.

This restaurant may be the one most often recommended by Santa Feans. Why? Some point to the atmosphere; others cite the food and prices. There's often a wait for a table, though once you're seated, the food is usually prepared and served very quickly. Hanging plants and wood decor accent this spacious brick building, adjacent to the old Santa Fe railroad station. It's a bit raucous, with lots of families and large crowds, and a festive atmosphere that spills over from the bar, where many come to drink margaritas. The menu features such traditional New Mexican dishes as chile rellenos and enchiladas (house specialties), as well as stuffed sopaipillas, chalupas, and tacos. Vegetarian dishes, burgers, steaks, and daily specials are also offered. There's full bar service.

Upper Crust Pizza

329 Old Santa Fe Trail. ☎ **505/983-4140.** No reservations. Pizzas $4.95–$16.05. MC, V. Summer: daily 11am–11pm. Winter: Mon–Thurs and Sun 11am–10pm, Fri–Sat 11am–11pm. PIZZA.

Santa Fe's best pizzas may be found here, in an adobe house near the old San Miguel mission. Meals-in-a-dish include the Grecian gourmet pizza (feta and olives) and the whole-wheat vegetarian pizza (topped with sesame seeds). You can either eat here or request free delivery (it takes about 30 minutes) to your downtown hotel. Beer and wine are available, as are salads, calzones, and stromboli.

Zia Diner

326 S. Guadalupe St. ☎ **505/988-7008.** Reservations accepted only for parties of six or more. Lunch $3.50–$8.95; dinner $5.25–$15.95. MC, V. Daily 11:30am–10pm. INTERNATIONAL/ AMERICAN.

In a renovated 1880 coal warehouse, this art deco diner with a turquoise-and-mauve color scheme boasts a stainless-steel soda fountain and a shaded patio. Its business hours make it a convenient stopover after a movie or late outing; however, during key meals on weekends it can get crowded and the wait can be long. The varied menu features homemade soups, salads, fish-and-chips, meat loaf, and of course, enchiladas. Specials range from East Indian curry to spanakopita, Thai-style trout to three-cheese calzone. I like their corn, green chile, and asiago pie, as well as their soup specials. There are fine wines, a full bar, great desserts (for example, tapioca pudding, apple pie, and strawberry rhubarb pie), an espresso bar, and of course, you can get malts, floats, and shakes anytime.

4 Northside

EXPENSIVE

El Nido

NM 22, Tesuque. ☎ **505/988-4340.** Reservations recommended. Main courses $10.95– $30.95. MC, V. Tues–Thurs and Sun 5:30–9:30pm, Fri–Sat 5:30–10pm. STEAK/SEAFOOD.

Life has never been dull at El Nido (The Nest). This 1920s adobe home had been a dance hall and later Ma Nelson's brothel, before it became a restaurant in 1939. Since then it has attracted Santa Feans for the food as well as the atmosphere, lively bar, and occasional flamenco dance performances. Scandinavian wood tables and chairs are surrounded by kiva fireplaces, and the stuccoed walls are covered with local artwork. Longtime favorite appetizers include ceviche and deep-fried oysters. Main courses I've enjoyed include steak with a spicy anise, black-and-pink peppercorn crust and a deliciously moist reddened halibut steak served with a light tequila lime butter. The best desserts here are the profiteroles and the chocolate piñon torte.

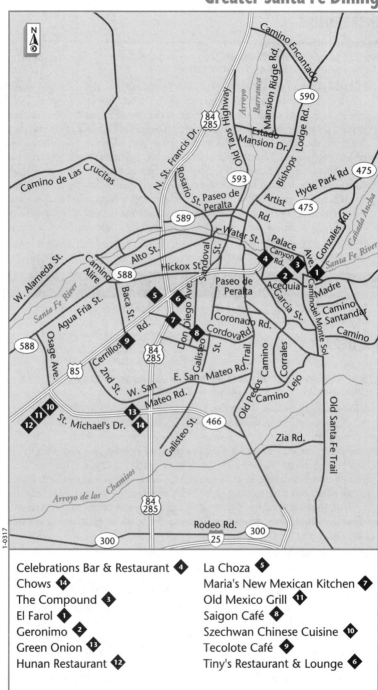

Celebrations Bar & Restaurant **4**

Chows **14**

The Compound **3**

El Farol **1**

Geronimo **2**

Green Onion **13**

Hunan Restaurant **12**

La Choza **5**

Maria's New Mexican Kitchen **7**

Old Mexico Grill **11**

Saigon Café **8**

Szechwan Chinese Cuisine **10**

Tecolote Café **9**

Tiny's Restaurant & Lounge **6**

INEXPENSIVE

Tesuque Village Market

NM 22, Tesuque. ☎ **505/988-8848.** Main courses $4.95–$12. MC, V. Summer: 6am–10pm. Winter: 7am–9pm. AMERICAN/SOUTHWESTERN.

Parked in front of this charming market and restaurant you'll see Range Rovers and beat-up ranch trucks, an indication that the food here has broad appeal. Located under a canopy of cottonwoods at the center of this quaint village, the restaurant is so good it's worth the trip 15 minutes north of Santa Fe. During warmer months you can sit on the porch; in other seasons the interior is comfortable, with plain wooden tables next to a deli counter and upscale market. For me this is a breakfast place, where blue-corn pancakes rule. Friends of mine like the breakfast burritos and huevos rancheros. Lunch and dinner are also popular, and there's always a crowd (though, if you have to wait for a table, the wait is usually brief). For lunch, I recommend the burgers, and for dinner, one of the hearty specials such as lasagna, or my favorite, pork chops verde (boneless pork chops with a green chile sauce). For dessert, there's a variety of pastries and cakes at the deli counter, as well as fancy granola bars and oversized cookies in the market.

5 Southside

Santa Fe's motel strip and other streets south of the Paseo de Peralta have their share of good, reasonably priced restaurants.

MODERATE

Chows

720 St. Michaels Dr. ☎ **505/471-7120.** Main courses $6.50–$14.95. MC, V. Mon–Fri 11:30am–2pm, Sat noon–3pm; daily 5–9pm. CREATIVE CHINESE.

This upscale but casual restaurant, near the intersection of St. Francis and St. Michaels, is Chinese with a touch of health-conscious Santa Fe. The decor is tasteful, with lots of wood and earth tones. The food is unconventional, cooked without MSG. You can get standard pot stickers and fried rice, but you may want to investigate imaginatively named dishes such as firecracker dumplings (carrots, onions, ground turkey, and chile in a Chinese pesto spinach sauce); nuts and birds (chicken, water chestnuts, and zucchini in a Szechwan sauce); or my favorite, Pearl River Splash (whole steamed boneless trout in a ginger onion sauce). For dessert, try the chocolate-dipped fortune cookies. Wine by the bottle or glass as well as beer are available.

Hunan Restaurant

2440 Cerrillos Rd., College Plaza South. ☎ **505/471-6688.** Reservations recommended. Lunch buffet $5.89; dinner $6.95–$25.95. DISC, MC, V. Mon–Thurs 11am–9:30pm, Fri 11am–10:30pm, Sat–Sun 11:30am–9:30pm. NORTHERN AND CENTRAL CHINESE.

A pair of stone lions guards the ostentatious dragon-gate entrance to this Asian delight. The red-and-white decor, together with a large aquarium and antique Oriental furniture, recall a Chinese palace. The hot and spicy Hunan- and Peking-style recipes prepared by chef/owner Alex Lee are equally fascinating. Family dinners of Hunan shredded pork, sha-cha beef, and spiced chicken and shrimp are served with egg roll, wonton soup, fried rice, tea, and fortune cookies. Or you can order such à la carte dishes as whole fish with hot bean sauce or Peking duck. There's an 18-dish luncheon buffet (children's prices available), and food can be taken out.

✪ Old Mexico Grill

2434 Cerrillos Rd., College Plaza South. ☎ **505/473-0338.** Reservations recommended for large parties. Lunch $5.95–$9.95; dinner $8.75–$17.50. DISC, MC, V. Mon–Fri 11:30am–2:30pm; Sun–Thurs 5:30–9pm, Fri–Sat 5:30–9:30pm. MEXICAN.

Here's something unique in Santa Fe: a restaurant that specializes not in Northern New Mexico food but in authentic Mexico City and regional Mexican cuisine. The restaurant's focal point is an exhibition cooking area with an open mesquite grill and French rotisserie, where a tempting array of fajitas, tacos al carbon, and other specialties is prepared. Popular dishes include turkey mole poblano, costillas de puerro en barbacoa de Oaxaca (hickory-smoked baby-back ribs baked in a chipotle and mulato chile, honey-mustard barbecue sauce), shrimp in orange-lime-tequila sauce, and paella Mexicana. There is a nice selection of soups and salads at lunch and a variety of homemade desserts. Servers are attentive, and a full bar serves Mexican beers (10 in all) and margaritas.

✪ Steaksmith at El Gancho

Old Las Vegas Hwy. ☎ **505/988-3333.** Reservations recommended. Main courses $8.95–$24.95. AE, CB, DC, MC, V. Mon–Sat 5:30–10pm, Sun 5–9pm. STEAKS/SEAFOOD.

Santa Fe's most highly regarded steak house is a 15-minute drive up the Old Pecos Trail in the direction of Las Vegas. Guests enjoy attentive service in a pioneer atmosphere of brick walls and viga ceilings. A creative appetizer menu ranges from ceviche Acapulco to grilled pasilla peppers and beef chupadero. For a main course, try the New York sirloin or filet mignon—although all of the meat dishes are excellent. And they're talking up their seafood these days. Try the broiled Canadian scallops or Australian lobster tail. In addition, the chef has added a few vegetarian entrees to the menu. There is also a choice of salads, homemade desserts, and bread, plus a full bar and lounge (serving a tapas menu from 4pm) that even caters to cappuccino lovers.

Szechwan Chinese Cuisine

1965 Cerrillos Rd. ☎ **505/983-1558.** Reservations recommended. Lunch $5–$7; dinner $5.95–$11.95. DISC, MC, V. Daily 11am–9:30pm. NORTHERN CHINESE.

Spicy northern Chinese cuisine at this restaurant pleases the palate. The service is efficient and friendly, and the seafood platter of shrimp, scallops, crab, fish, and vegetables stir-fried in a wine sauce is excellent. Other specialties include Lake Tung Ting shrimp, sesame beef, and General Chung's chicken. Wine and Tsingtao beer from China are available.

Toushie's

4220 Airport Rd. ☎ **505/473-4159.** Reservations recommended on weekends. Lunch $3.95–$8.50; dinner $6.95–$14.95. AE, CB, DC, MC, V. Mon–Fri 11am–midnight; Sat–Sun 4pm–midnight. MEXICAN/AMERICAN.

Located just one block from the Villa Linda Mall, this often-overlooked place on the south side of town is a welcome surprise. Its spacious, semicircular seating area is an ideal place for an intimate area for enjoying a relaxing meal or for listening to mellow dance music on Saturday nights. Appetizers include escargots and sautéed mushrooms. The wide choice of main dishes includes a 16-ounce T-bone steak, prime rib and baked shrimp, and coquille of sea scallops and shrimp. Regional dishes highlight the lunch menu, from menudo (Spanish-style tripe stew) to a combination plate of a taco, a tamale, and a rolled enchilada. A favorite dessert is Toushie's Delight—a flaming sopaipilla.

INEXPENSIVE

Bobcat Bite

Old Las Vegas Hwy. ☎ **505/983-5319.** No reservations. Menu items $3.50–$11.95. No credit cards. Wed–Sat 11am–7:50pm. STEAKS/BURGERS.

This local classic (in business for more than 40 years), located about 5 miles southeast of Santa Fe, is famed for its high-quality steaks—such as the 13-ounce rib eye—and huge hamburgers, including a remarkable green chile cheeseburger. The ranch-style atmosphere appeals to families.

Green Onion

1851 St. Michael's Dr. ☎ **505/983-5198.** Lunch $3–$8; dinner $4.50–$9.50. AE, DISC, MC, V. Daily 11am–10pm. NEW MEXICAN, BURGERS AND SANDWICHES.

The Onion offers up some of the hottest chiles and one of the liveliest local bars in Santa Fe. Roast-beef burritos and chicken enchiladas highlight an established menu, which also features a choice of sandwiches and a great many daily specials.

Maria's New Mexican Kitchen

555 W. Cordova Rd. near St. Francis Dr. ☎ **505/983-7929.** Lunch $4.95–$8.95; dinner $5.75–$15.95. AE, CB, DC, DISC, MC, V. Mon–Fri 11am–10pm, Sat–Sun noon–10pm. NEW MEXICAN.

Built in 1949 by Maria Lopez and her politician husband, Gilbert (the present owners are Al and Laurie Lucero), this restaurant is a prime example of how charm can be created through judicious scavenging. The bricks used for construction came from the old New Mexico State Penitentiary, and most of the furniture once belonged to La Fonda Hotel. The five frescoes in the cantina were painted by master muralist Alfred Morang (1901–58) in exchange for food. Maria's boasts an open tortilla grill, where cooks can be seen making flour tortillas by hand. The food here isn't my favorite in Santa Fe, but the atmosphere appeals to many people. The blue-corn enchiladas are pretty tasty, and some regulars always order Maria's spare rib dinner (pork ribs, cut for easy eating, marinated and baked according to Maria's original recipe from the '40s). If you're a margarita fan, this is the place to sample a variety of them—Maria's features over 70 "real margaritas." In addition, Maria's now offers a tasting sampler of more than 50 tequilas. Strolling mariachi troubadours perform nightly. The restaurant provides patio dining in the summer, and two fireplaces warm the dining room in winter.

Tecolote Cafe

1203 Cerrillos Rd. ☎ **505/988-1362.** Main dishes $2.95–$9.25. AE, CB, DC, DISC, MC, V. Tues–Sun 7am–2pm. NEW MEXICAN/AMERICAN.

This is a local breakfast lovers' favorite. The decor is simple, but the food is elaborate: eggs any style, omelets, huevos rancheros—all served with fresh-baked muffins or biscuits and maple syrup. Give the atole piñon hotcakes (made with blue cornmeal) a try. Luncheon specials include carne adovada burritos and green chile stew, served with beer or wine.

Tiny's Restaurant & Lounge

In the Penn Road Shopping Center, 1015 Penn Rd. (at Cerrillos and St. Francis). ☎ **505/983-9817.** Reservations recommended. Lunch $5.25–$8; dinner $6.95–$15.50. AE, DC, DISC, MC, V. Mon–Fri 11:30am–2pm; Mon–Sat 6–10pm. Bar: Mon–Sat 11:30am–2am. STEAKS/NEW MEXICAN.

A longtime favorite of Santa Feans, Tiny's first opened in 1948 and is decked out in 1950s style. This is where you'll see a neighborhood Santa Fe crowd, rather than the more highbrow locals you'd find at the Pink Adobe. Steaks and shrimp

🏛 Family-Friendly Restaurants

Bobcat Bite *(see page 94)*   The name and the ranch-style atmosphere will appeal to families that are looking for great steaks and huge hamburgers at low prices.

Cowgirl Hall of Fame *(see page 87)*   Kids love the Kid's Corral where, among other things, they can play a game of horseshoes.

Upper Crust Pizza *(see page 90)*   Many people feel they have the best pizza in town, and they'll deliver it to tired tots and their families at downtown hotels.

complement a menu that features fajitas and Northern New Mexico main courses—not the best food in town, but reliable. I order the chicken and guacamole tacos. You might try the house specialty—baked chicken flautas. There's also an indoor/outdoor patio and garden room with full food and cocktail service. A very popular lounge has live jazz entertainment Wednesday and Thursday nights (call for times) and country/Spanish music Friday and Saturday from 9pm to 2am. There's plenty of dancing and good margaritas.

Tortilla Flats

3139 Cerrillos Rd. ☎ **505/471-8685.** Breakfast $1.75–$6.25; lunch $5.25–$9; dinner $6.25–$11. DISC, MC, V. Sun–Thurs 7am–9pm, Fri–Sat 7am–10pm. NEW MEXICAN.

This casual restaurant takes pride in its all-natural ingredients and vegetarian menu offerings (its vegetarian burrito is famous around town). The atmosphere is a bit like Denny's, but don't be fooled: The food is authentic. The blueberry pancakes are delicious, as are the fajitas and eggs with a side of black beans. I also like the blue-corn enchiladas and the chimichangas. The Santa Fe Trail steak (8 ounces of prime rib eye smothered with red or green chile and topped with grilled onions) will satisfy a big appetite. Above all, note the freshness of the tortillas and sopaipillas, made on the spot. You can even peek through a window into the kitchen and watch them being made. Beer and wine are served. A children's menu and take-out service are available.

6 Out of Town

It takes a little extra effort to reach this restaurant, but it's worth it.

Legal Tender

Lamy. ☎ **505/466-1223.** Reservations recommended. Main courses $8.75–$18.95. DISC, MC, V. Summer: Sunday brunch 11am–2pm; regular menu 2pm–9pm; Tues–Thurs 11am–9pm; Fri and Sat 11am–10pm. Winter (Nov 1–Apr 30): Wed 5–9pm; Thurs–Sun 11am–9pm. STEAKS/SEAFOOD.

Just across the road from the old Atchison, Topeka & Santa Fe Railway station in Lamy, the Legal Tender is spectacularly faithful to the Wild West theme of the late 19th century. Originally a general store dating to 1881, it has gone through many incarnations—but the Victorian decor remains. The hand-carved cherrywood bar is the same one its first owner imported from Germany. The tin ceiling in the Americana Room came from the original Hilton Hotel in San Francisco, and its drapes and chandeliers are from the presidential suite of Chicago's Sherman Hotel. Suitably, the building is listed in the National Register of Historic Places.

The food is equally worthy of recognition. I recommend the prime rib or the barbecue. Finish with Lamy lemon cake or the double trouble chocolate brownie. The main dining room now has a hardwood dance floor, and country-and-western bands play on Friday and Saturday, while jazz musicians perform on Sunday and some days during the week. To reach Legal Tender, take I-25 north (actually southeast) toward Las Vegas, get off at exit 290, and follow US 285 south until you see the signs for the Lamy turnoff.

What to See & Do in Santa Fe

7

One of the oldest cities in the United States, Santa Fe has long been a center for the creative and performing arts, so it's not surprising that most of the city's major sights are related to history and the arts. The city's Museum of New Mexico, art galleries and studios, historic churches, and cultural sights associated with local Native American and Hispanic communities all merit a visit. It would be easy to spend a full week sightseeing in the city without ever heading out to any nearby attractions. Of special note is the brand-new Georgia O'Keeffe museum now open to visitors.

SUGGESTED ITINERARIES

If You Have 2 Days

For an overview, start your first day at the Palace of the Governors, and as you leave you might want to pick up a souvenir from the Native Americans selling crafts and jewelry beneath the portal facing the plaza. After lunch, take a self-guided walking tour of old Santa Fe, starting at the Plaza.

On Day 2, spend the morning at the Museum of Fine Arts and the afternoon browsing in the galleries on Canyon Road.

If You Have 3 Days

On the third day, visit the cluster of museums on Camino Lejo—the Museum of International Folk Art, the Museum of Indian Arts and Crafts, and the Wheelwright Museum of the American Indian. Then wander through the historic Barrio de Analco and spend the rest of the afternoon shopping.

If You Have 4 Days or More

Devote your fourth day to exploring the pueblos, including San Juan Pueblo, headquarters of the Eight Northern Indian Pueblos Council, and Santa Clara Pueblo, with its Puye Cliff Dwellings.

On Day 5, go out along the High Road to Taos, with a stop at El Santuario de Chimayo, returning down the Rio Grande Valley.

If you have more time, take a trip to Los Alamos, birthplace of the atomic bomb and home of the Bradbury Science Museum, and Bandelier National Monument.

1 The Top Attractions

✪ Palace of the Governors

North Plaza. ☎ **505/827-6483.** Admission $5 adults, free for children under 17. 4-day passes good at all 4 branches of the Museum of New Mexico cost $8 for adults. Tues–Sun 10am–5pm. Closed Jan 1, Thanksgiving, Dec 25.

The Palace, built in 1610 as the original capital of New Mexico, has been in continuous public use longer than any other structure in the United States. Designated the Museum of New Mexico in 1909, it has become the state history museum, with an adjoining library and photo archives. Several cutaways of doors and windows reveal the early architecture.

Outside the museum, many visitors to Santa Fe are impressed with the sight of Native American artisans sitting shoulder-to-shoulder beneath the long covered portal facing the Plaza. Here on the shaded sidewalk several dozen colorfully dressed members of local Pueblo tribes, plus an occasional Navajo, Apache, or Hopi, spread out their handcrafts: mainly jewelry and pottery, but also woven carpets, beadwork, and paintings. The museum's Portal Program restricts selling space to Native Americans only.

Inside, a series of exhibits chronicles four centuries of New Mexico's Hispanic and American history, from the 16th-century Spanish explorations through the frontier era to modern times. Among Hispanic artifacts are maps dating from the early 1700s. There's even a mid-19th-century chapel, with a simple, bright-colored altarpiece made in 1830 for a Taos church by folk artist José Rafael Aragón.

Governors' offices from the Mexican and 19th-century U.S. eras have been restored and preserved. Displayed artifacts from early New Mexican life include a stagecoach, an early working printing press, and a collection of *mestizajes,* portraits of early Spanish colonists depicting typical costumes of the time. Also on display are pieces from the silver service used aboard the battleship U.S.S. *New Mexico* from 1918 to 1939 and a tiny New Mexico state flag (3 by 4 inches) that went to the moon on one of the Apollo missions.

Most Native American artifacts previously housed here have been moved to the Museum of Indian Arts and Culture. Those that remain include several pieces of ancient pottery from the Puye Plateau culture and photographs depicting a museum-sponsored study of Mayan sites in Mexico's Yucatán.

The bookstore has one of the finest selections of art, history, and anthropology books in the Southwest. There is also a fine print shop and bindery, where limited-edition works are produced on hand-operated presses.

The Palace is the flagship of the Museum of New Mexico system; the main office is at 113 Lincoln Ave. (☎ 505/982-6366). The system comprises five state monuments and four Santa Fe museums—the Palace of the Governors, the Museum of Fine Arts, the Museum of International Folk Art, and the Museum of Indian Arts and Culture.

✪ Museum of Fine Arts

107 W. Palace (at Lincoln Ave.). ☎ **505/827-4455.** Admission $5 adults, free for seniors on Wed, free for children under 17. 4-day passes are available ($8 for 4 museums). Tues–Sun 10am–5pm; Fri evening the museum is open 5–8pm. Closed Jan 1, Easter, Thanksgiving, Dec 25.

Located catercorner from the plaza and just opposite the Palace of the Governors, this was one of the first Pueblo Revival–style buildings constructed in Santa Fe (in 1917). As such, it was a major stimulus in Santa Fe's development as an art colony earlier in this century.

Downtown Santa Fe Attractions

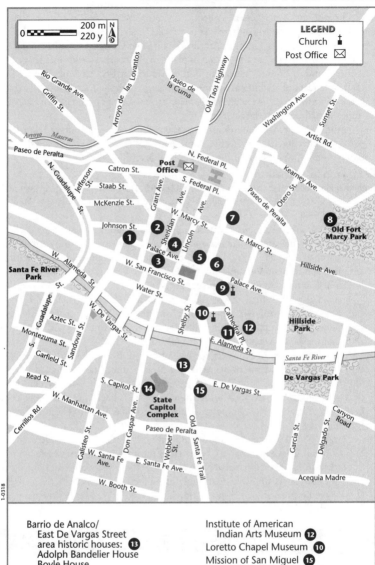

Barrio de Analco/
  East De Vargas Street
  area historic houses: **13**
  Adolph Bandelier House
  Boyle House
  Gregoria Crespin House
  José Alarid House
  Oldest House
  Tudesqui House
Bergere House **2**
Catholic Museum & Lamy Garden **11**
Delgado House **3**
Georgia O'Keefe Museum **1**

Institute of American
  Indian Arts Museum **12**
Loretto Chapel Museum **10**
Mission of San Miguel **15**
Museum of Fine Arts **4**
Old Fort Marcy Park **8**
Padre de Gallegos House **7**
Palace of the Governors **5**
Prince Plaza **6**
Roundhouse (State Capitol) **14**
St. Francis Cathedral **9**

The museum's permanent collection of more than 8,000 works emphasizes regional art and includes landscapes and portraits by all the Taos masters and the contemporary artists R. C. Gorman, Amado Peña Jr., and Georgia O'Keeffe, among others. The museum also has a collection of photographic works by such masters as Ansel Adams, Edward Weston, and Elliot Porter. Modern artists, many of them far from the mainstream of traditional Southwestern art, are featured in temporary exhibits throughout the year. Two sculpture gardens present a range of three-dimensional art from the traditional to the abstract.

Beautiful St. Francis Auditorium, patterned after the interiors of traditional Hispanic mission churches, adjoins the art museum (see chapter 9). A museum shop sells books on Southwestern art, prints, and postcards of the collection.

✪ St. Francis Cathedral

Cathedral Place at San Francisco St. ☎ **505/982-5619.** Donations appreciated. Open daily. Visitors may attend mass Mon–Fri at 7am, 8:15am, and 5:15pm;, Sat at 7am and 5:15pm; Sun at 6, 8, and 10am, noon, and 7pm.

Santa Fe's grandest religious structure is just a block east of the Plaza. An architectural anomaly in Santa Fe, it was built between 1869 and 1886 by Archbishop Jean-Baptiste Lamy in the style of the great cathedrals of Europe. French architects designed the Romanesque building—named after Santa Fe's patron saint—and Italian masons assisted with its construction.

The small adobe Our Lady of the Rosary chapel on the northeast side of the cathedral has a Spanish look. Built in 1807, it's the only portion that remains from Our Lady of the Assumption Church, founded along with Santa Fe in 1610. The new cathedral was built over and around the old church.

A wooden icon set in a niche in the wall of the north chapel, Our Lady of Peace, is the oldest representation of the Madonna in the United States. Rescued from the old church during the 1680 Pueblo Rebellion, it was brought back by Don Diego de Vargas on his (mostly peaceful) reconquest 12 years later; thus the name. Today Our Lady of Peace plays an important part in the annual Feast of Corpus Christi in June and July.

During a $600,000 renovation in 1986, an early 18th-century wooden statue of St. Francis of Assisi was moved to the center of the altar screen. The cathedral's front doors feature 16 carved panels of historic note and a plaque memorializing the 38 Franciscan friars who were martyred during New Mexico's early years. There's also a large bronze statue of Bishop Lamy himself; his grave is under the main altar of the cathedral.

2 More Attractions

MUSEUMS

Catholic Museum and the Archbishop Lamy Commemorative Garden

223 Cathedral Place. ☎ **505/983-3811.** Donations appreciated. Mon–Fri 8:30am–4:30pm; Sat hours vary (call ahead).

Housed in a complex of buildings that date from 1832, the Catholic Museum is one of New Mexico's newest, where visitors have the opportunity to learn about the development and significance of Catholicism in New Mexico. Opened in 1994 by the Archdiocese of Santa Fe, the museum houses a collection of religious relics, including a book printed by Padre Antonio José Martínez of Taos, the Lamy chalice that was given to Archbishop Lamy by Pope Pius IX in 1854, and the document that formally reestablished royal possession of the villa and capital of Santa Fe and was signed by Diego de Vargas in 1692. There are also photographs on display, documenting events pertaining to the 11 archbishops of Santa Fe. The museum shop is stocked with books on New Mexico's history and religious items.

✪ Georgia O'Keeffe Museum

217 Johnson St. ☎ **505/995-0785.** Admission $5 (4-day 5-museum passes to the Museum of New Mexico available). Tues–Sun 10am–5pm (Friday until 8pm).

For years, anxious visitors to Santa Fe have asked, "Where are the O'Keeffes?" Locals flush and are forced to answer: "The Metropolitan Museum of Art in New York and the National Gallery of Art in Washington, D.C." Although this East Coast artist is known world over for her haunting depictions of the shapes and colors of Northern New Mexico, particularly the Abuquiu area, until now little of her work hung in the state. The local Museum of Fine Arts owns just 15 of her works.

The new museum, inaugurated in July 1997, contains the largest collection of O'Keeffes in the world: more than 80 oil paintings, drawings, watercolors, pastels, and sculptures. You can see such "Killer O'Keeffes" as *Jimson Weed,* painted in 1932, and *Evening Star No. VII,* painted in 1917. The rich and varied body of art adorns the walls of a cathedral-like, 10,000-square-foot space—a former Baptist church with adobe walls—downtown. It is the only museum in the United States dedicated solely to one woman's work.

O'Keeffe first visited New Mexico in 1917 and came here for extended periods from the '20s through the '40s. With the artist's images and fame tied inextricably to local desert landscapes, the idea of bringing O'Keeffe's works together came from private collector Anne Windfohr Marion, a Texas heiress who heads up the Burnett Foundation, which, along with the Georgia O'Keeffe Foundation, donated the initial 33 works. Ms. Marion and her husband, John L. Marion, a former chairman of Sotheby's North America, elected architect Richard Gluckman, who also designed the Andy Warhol Museum in Pittsburgh, for the project.

Indian Arts Research Center

School of American Research, 660 Garcia St. (off Canyon Rd.). ☎ **505/982-3584.** Free admission for Native Americans and SAR members; $15 per person for others. Open by appointment only.

With 10,000 objects, the center houses one of the world's finest collections of Southwest Indian art. Admission, however, is restricted. The School of American Research, of which this is a division, was established in 1907 as a center for advanced studies in anthropology and related fields. It sponsors scholarship, academic research, publications, and educational programs. Public tours are offered on Friday and group tours can be scheduled at other times. Call for times and reservations. E-mail: iarc@sarsf.org.

Institute of American Indian Arts Museum

108 Cathedral Place. ☎ **505/988-6211.** Admission (2-day pass) $4 adults, $2 seniors and students, free for children 16 and under. Mon–Sat 10am–5pm, Sun noon–5pm.

The Institute of American Indian Arts (IAIA) is the nation's only congressionally chartered institute of higher education devoted solely to the study and practice of the artistic and cultural traditions of all American Indian and Alaska native peoples. The IAIA museum is the official repository of the most comprehensive collection of contemporary Native American art in the world and has loaned items from its collection to museums all over the world. Many of the best Native American artists of the last 3 decades have visited the IAIA. Their works can often be seen in the many exhibitions offered at the museum throughout the year. The museum's National Collection of Contemporary Indian Art contains painting, sculpture, ceramics, textiles, jewelry, beadwork, basketry, and graphic arts. The museum presents the artistic achievements of IAIA alumni, current students, and other nationally recognized Native American and Alaskan artists.

✪ Museum of Indian Arts and Culture

710 Camino Lejo. ☎ **505/827-6344.** Admission $5 adults, free for children under 17. Tues–Sun 10am–5pm.

Next door to the Museum of International Folk Art, this museum opened in 1987 as the showcase for the adjoining Laboratory of Anthropology. Interpretive displays detail tribal history and contemporary lifestyles of New Mexico's Pueblo, Navajo, and Apache cultures. More than 50,000 pieces of basketry, pottery, clothing, carpets, and jewelry—much of it quite old—are on continual rotating display.

There are frequent demonstrations of traditional skills by tribal artisans and regular programs in a 70-seat multimedia theater. Native American educators run a year-round workshop that encourages visitors to try such activities as weaving and corn grinding. Regular performances of Native American music and dancing by tribal groups are also held. Concession booths give a taste of Native American foods during the summer months.

The laboratory, founded in 1931 by John D. Rockefeller Jr., is a point of interest in itself. Designed by the well-known Santa Fe architect John Gaw Meem, it is an exquisite example of Pueblo Revival architecture. Since the museum opened, the lab has expanded its research and library facilities into its former display wing.

✪ Museum of International Folk Art

706 Camino Lejo. ☎ **505/827-6350.** Admission $5 adults, free for children under 17. Tues–Sun 10am–5pm. The museum is located about 2 miles south of the plaza, in the Sangre de Cristo foothills. Drive southeast on Old Santa Fe Trail, which becomes Old Pecos Trail, and look for signs pointing left onto Camino Lejo.

This branch of the Museum of New Mexico may not seem quite as typically Southwestern as other Santa Fe museums, but it's the largest of its kind in the world. With a collection of some 130,000 objects from more than 100 countries, it is my personal favorite of the city museums.

It was founded in 1953 by the Chicago collector Florence Dibell Bartlett, who said: "If peoples of different countries could have the opportunity to study each other's cultures, it would be one avenue for a closer understanding between men." That's the basis on which the museum operates today.

The special collections include Spanish colonial silver, traditional and contemporary New Mexican religious art, Mexican tribal costumes, Mexican majolica ceramics, Brazilian folk art, European glass, African sculptures, East Indian textiles, and the marvelous Morris Miniature Circus. Particularly delightful are numerous dioramas—all done with colorful miniatures—of people around the world at work and play in typical town, village, and home settings. Recent acquisitions include American weather vanes and quilts, Palestinian costume jewelry and amulets, and Bhutanese and Indonesian textiles.

Children love to look at the hundreds of toys on display throughout the museum. Many of them are housed in a wing built especially to hold part of a collection donated in 1982 by Alexander and Susan Girard. Alexander Girard, a notable architect and interior designer, and his wife, Susan, spent their lives traveling the world collecting dolls, animals, fabrics, masks, and dioramas. They had a home in Santa Fe, where they spent many years before they died. Their donation included over 100,000 pieces, 10,000 of which are exhibited at the museum.

In 1989 the museum opened a new Hispanic Heritage Wing, which houses the country's finest collection of Spanish colonial and Hispanic folk art. Folk art

Fetishes: Gifts of Power

According to Zuni lore, in the early years of man's existence, the Sun sent down his two children to assist humans, who were under siege from earthly predators. The Sun's sons, as it were, shot lightning bolts from their shields and destroyed the predators. For generations, Zunis, traveling across their lands in western New Mexico, have found stones shaped like particular animals. The Zunis believe the stones to be the remains of those long-lost predators, still containing their soul or last breath.

Today in many shops in Santa Fe you too can pick up a carved animal figure, called a fetish. According to belief, the owner of the fetish is able to absorb the power of that creature, whatever it may be. Many fetishes were long ago used for protection and for might in the hunt. Today people own fetishes for many reasons. One might carry a bear for health and strength or an eagle for keen perspective. A mole might be placed in a home's foundation for protection from elements underground, a frog buried with crops for fertility and rain, and a ram carried in the purse for prosperity. For love, some local recommend pairs of fetishes—often foxes or coyotes carved from a single piece of stone. Many fetishes, arranged with bundles on top and attached with sinew, serve as an offering to the animal spirit that resides within the stone.

Fetishes are still carved by many of the pueblos. When selecting, shop around for a little while until you begin to appreciate the difference between clumsily carved ones and more gracefully executed ones. A good fetish is not necessarily one that is meticulously carved. Some fetishes are barely carved at all, since the original shape of the stone already contains the form of the animal. Once you have a sense of the quality and elegance available, decide which animal (and power) suits you best. Native Americans caution, however, that the fetish cannot be expected to impart an attribute you don't already possess. Instead it will help elicit the power that already resides within you. Good sources for fetishes are **Dewey Galleries Limited,** 76 E. San Francisco St. (upstairs; ☎ **505/982-8632**); **Keshi,** 227 Don Gaspar (☎ **505/989-8728**); and **Morning Star Gallery,** 513 Canyon Rd. (☎ **505/982-8187**).

demonstrations, performances, and workshops are often presented here. The 80,000-square-foot museum also has a lecture room, a research library, and a gift shop where a variety of folk art is available for purchase.

Wheelwright Museum of the American Indian

704 Camino Lejo. ☎ **505/982-4636.** Donations appreciated. Mon–Sat 10am–5pm, Sun 1–5pm. Closed Jan 1, Thanksgiving, Dec 25.

Though not part of the state museum system, the Wheelwright is often attached to a trip to the Folk Art and Indian Arts museums, because it's next door. Once known as the Museum of Navajo Ceremonial Art, the Wheelwright was founded in 1937 by Boston scholar Mary Cabot Wheelwright in collaboration with a Navajo medicine man, Hastiin Klah, to preserve and document Navajo ritual beliefs and practices. Klah took the designs of sand paintings used in healing ceremonies and adapted them into the woven pictographs that are a major part of the museum's treasure.

In 1976 the museum's focus was altered to include the living arts of all Native American cultures. Built in the shape of a Navajo hogan, with its doorway facing east (toward the rising sun) and its ceiling formed in the interlocking "whirling log" style, the museum exhibits rotating shows of silverwork, jewelry, tapestry, pottery, basketry, and paintings. In 1997, 600 square feet of new space was added. There's a permanent collection (although it is not always on display), plus an outdoor sculpture garden with works by Allan Houser and other noted artisans. In the basement is the Case Trading Post, an arts-and-crafts shop built to resemble the typical turn-of-the-century trading post found on Navajo reservations. Storyteller Joe Hayes holds the attention of listeners outside a tepee at dusk on certain days in July and August.

CHURCHES

Cristo Rey

Upper Canyon Rd. at Camino Cabra. ☎ **505/983-8528.** Free admission. Open most days, but call for hours.

This Catholic church ("Christ the King" in Spanish), a huge adobe structure, was built in 1940 to commemorate the 400th anniversary of Coronado's exploration of the Southwest. Parishioners did most of the construction work, even making adobe bricks from the earth where the church stands. Architect John Gaw Meem designed the building, in missionary style, as a place to keep some magnificent stone *reredos* (altar screens) created by the Spanish during the colonial era and recovered and restored in the 20th century.

○ Loretto Chapel Museum

207 Old Santa Fe Trail (between Alameda and Water sts.). ☎ **505/984-7971.** Admission $1 adults, free for children 6 and under. Mon–Sat 9:30am–4:30pm, Sun 10:30am–4:30pm.

Though no longer consecrated for worship, the Loretto Chapel is an important site in Santa Fe. Patterned after the famous Sainte-Chapelle church in Paris, it was constructed in 1873—by the same French architects and Italian masons who were building Archbishop Lamy's cathedral—as a chapel for the Sisters of Loretto, who had established a school for young ladies in Santa Fe in 1852.

The chapel is especially notable for its remarkable spiral staircase: It makes two complete 360° turns with no central or other visible support! (A railing was added later.) The structure is steeped in legend. The building was nearly finished in 1878 when workers realized the stairs to the choir loft wouldn't fit. Hoping for a solution more attractive than a ladder, the sisters made a novena to St. Joseph—and were rewarded when a mysterious carpenter appeared astride a donkey and offered to build a staircase. Armed with only a saw, a hammer, and a T-square, the master constructed this work of genius by soaking slats of wood in tubs of water to curve them and holding them together with wooden pegs. Then he disappeared without bothering to collect his fee.

Mission of San Miguel

401 Old Santa Fe Trail (at East de Vargas St.). ☎ **505/983-3974.** Donations appreciated. Mon–Sat 9am–4:30pm, Sun 2–4:30pm. Summer hours start earlier. Mass is said daily at 5pm.

This is one of the oldest churches in America, erected within a couple of years of the 1610 founding of Santa Fe. Tlaxcala tribe members—servants of early Spanish soldiers and missionaries—may have used fragments of a 12th-century pueblo that had been on this site in its construction. Severely damaged during the 1680 Pueblo Rebellion, the church was almost completely rebuilt in 1710 and has been altered numerous times since.

Because of its design, with high windows and thick walls, the structure was occasionally used as a temporary fortress during times of attack by raiding tribes. One painting in the sanctuary has holes that, according to legend, were made by arrows.

The mission and a nearby house—today a gift shop billed as "The Oldest House," though there's no way of knowing for sure—were bought by the Christian Brothers from Archbishop Lamy for $3,000 in 1881, and the order still operates both structures. Among the treasures in the mission are the San José Bell, reportedly cast in Spain in 1356 and brought to Santa Fe via Mexico several centuries later, and a series of buffalo hides and deerskins decorated with Bible stories for Native American converts.

Santuario de Nuestra Señora de Guadalupe
100 S. Guadalupe St. ☎ **505/988-2027.** Donations appreciated. Mon–Sat 9am–4pm. Closed weekends Nov–Apr.

Built between 1795 and 1800 at the end of El Camino Real by Franciscan missionaries, this is believed to be the oldest shrine in the United States honoring the Virgin of Guadalupe, patron saint of Mexico. Better known as Santuario de Guadalupe, the shrine's adobe walls are almost 3 feet thick, and the deep-red plaster wall behind the altar was dyed with oxblood in traditional fashion when the church was restored earlier in this century.

On one wall is a famous oil painting, *Our Lady of Guadalupe,* created in 1783 by the renowned Mexican artist José de Alzibar. Painted expressly for this church, it was brought from Mexico City by mule caravan.

Administered today as a museum by the nonprofit Guadalupe Historic Foundation, the sanctuary is frequently used for chamber music concerts, flamenco dance programs, dramas, lectures, and religious art shows.

PARKS & REFUGES

Old Fort Marcy Park
617 Paseo de Peralta (also access it by traveling 3 blocks up Artist Road and turning right).

Marking the 1846 site of the first U.S. military reservation in the Southwest, this park overlooks the northeast corner of downtown. Only a few mounds remain from the fort, but the Cross of the Martyrs, at the top of a winding brick walkway from Paseo de Peralta near Otero Street, is a popular spot for bird's-eye photographs. The cross was erected in 1920 by the Knights of Columbus and the Historical Society of New Mexico to commemorate the Franciscans killed during the Pueblo Rebellion of 1680. It has since played a role in numerous religious processions. Open daily 24 hours.

Randall Davey Audubon Center
Upper Canyon Rd. ☎ **505/983-4609.** $1 is charged for trail admission. Daily 9am–5pm. House tours conducted sporadically during the summer; call for hours.

Named for the late Santa Fe artist who willed his home to the National Audubon Society, this wildlife refuge occupies 135 acres at the mouth of Santa Fe Canyon. More than 100 species of birds and 120 types of plants live here, and a variety of mammals has been spotted—including black bears, mule deer, mountain lions, bobcats, raccoons, and coyotes. Trails winding through more than 100 acres of the nature sanctuary are open to day hikers.

Santa Fe River State Park
Alameda St. ☎ **505/827-7465.**

This is a lovely spot for an early morning jog, a midday walk beneath the trees, or perhaps a sack lunch at a picnic table. The green strip, which does not close, follows the midtown stream for about 4 miles as it meanders along the Alameda from St. Francis Drive upstream beyond Camino Cabra, near its source.

Regional Santa Fe Attractions & Accommodations

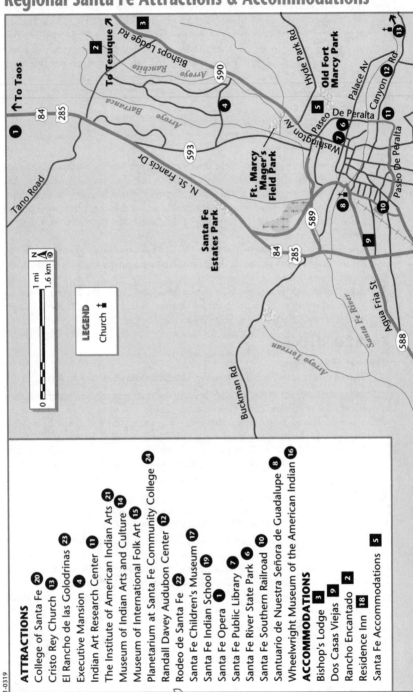

ATTRACTIONS

College of Santa Fe 20
Cristo Rey Church 13
El Rancho de las Golodrinas 23
Executive Mansion 4
Indian Art Research Center 11
The Institute of American Indian Arts 21
Museum of Indian Arts and Culture 14
Museum of International Folk Art 15
Planetarium at Santa Fe Community College 24
Randall Davey Audubon Center 12
Rodeo de Santa Fe 22
Santa Fe Children's Museum 17
Santa Fe Indian School 19
Santa Fe Opera 1
Santa Fe Public Library 7
Santa Fe River State Park 6
Santa Fe Southern Railroad 10
Santuario de Nuestra Señora de Guadalupe 8
Wheelwright Museum of the American Indian 16

ACCOMMODATIONS

Bishop's Lodge 3
Dos Casas Viejas 9
Rancho Encantado 2
Residence Inn 18
Santa Fe Accommodations 5

1-0319

106

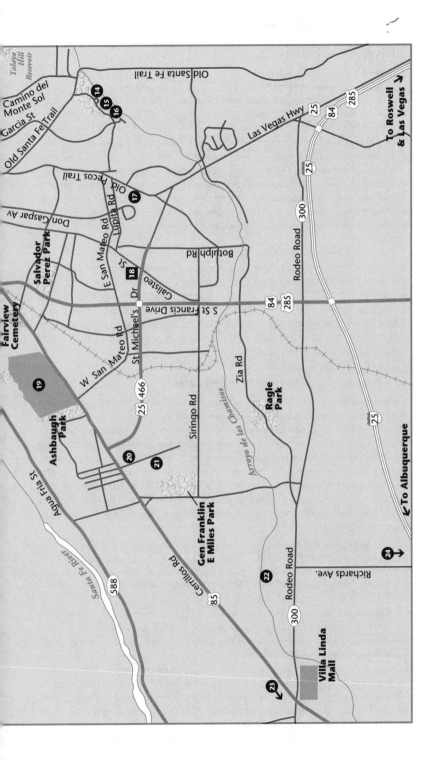

OTHER ATTRACTIONS

✪ El Rancho de las Golondrinas

334 Los Pinos Rd. ☎ 505/471-2261. Admission $4 adults, $3 seniors and teens, $1.50 children 5–12, free for children under 5. Festival weekends, $6 adults, $4 seniors and teens, $2.50 children 5–12. June–Sept, Wed–Sun 10am–4pm; Apr–May and Oct, open by advance arrangement. Closed Nov–Mar.

This 200-acre ranch, about 15 miles south of the Santa Fe Plaza via I-25 (take exit 276), was once the last stopping place on the 1,000-mile El Camino Real from Mexico City to Santa Fe. Today it's a living 18th- and 19th-century Spanish village, comprising a hacienda, a village store, a schoolhouse, and several chapels and kitchens. There's also a working molasses mill, wheelwright and blacksmith shops, shearing and weaving rooms, a threshing ground, a winery and vineyard, and four water mills, as well as dozens of farm animals. A walk around the entire property is 1³/₄ miles in length.

The Spring Festival (the first weekend of June) and the Harvest Festival (the first weekend of October) are the year's highlights at Las Golondrinas ("The Swallows"). On these festival Sundays the museum opens with a procession and mass dedicated to San Ysidro, patron saint of farmers. Other festivals and theme weekends are held throughout the year. Volunteers in authentic costume demonstrate shearing, spinning, weaving, embroidery, wood carving, grain milling, blacksmithing, tinsmithing, soap making, and other activities. There's an exciting atmosphere of Spanish folk dancing, music, theater, and food.

Roundhouse (State Capitol)

Paseo de Peralta and Old Santa Fe Trail. ☎ 505/986-4589.

Some are surprised to learn that this is the only round capitol building in America. It's also the newest, built in 1966. Designed in the shape of a Zia Pueblo emblem (or sun sign, which is also the state symbol), it symbolizes the Circle of Life: four winds, four seasons, four directions, and four sacred obligations. Surrounding the capitol is a lush 6¹/₂-acre garden boasting more than 100 varieties of plants, including roses, plums, almonds, nectarines, Russian olive trees, and sequoias. Benches and sculptures (by local artists) have been placed around the grounds for the enjoyment of visitors. Inside you'll find standard functional offices. The walls are hung with New Mexican art. If you're interested in taking a tour, call the number above for information.

Santa Fe Southern Railroad

410 S. Guadalupe St. ☎ 505/989-8600. Tickets range from $5 for children to $21 for adults; $35 for Friday sunset ride includes dinner (May through October). Trains depart the Santa Fe Depot on Tues, Thurs, Sat, and Sun at 10:30am and return by 3:30pm. The sunset ride departs at about 7:15pm and returns at 10:30pm.

"Riding the old Santa Fe" always referred to riding the Atchison, Topeka & Santa Fe Railroad. Ironically, the main route of the AT&SF bypassed Santa Fe, which probably forestalled some development for the capital city. Still, a spur was run off the main line to Santa Fe in 1880, and today an 18-mile ride along that spur offers views of some of New Mexico's most spectacular scenery and a glimpse of railroad history.

The Santa Fe Depot is a well-preserved tribute to the Mission architecture that the railroad brought to the West in the early 1900s. Characterized by light-colored stuccoed walls, arched openings, and tile roofs, this style was part of an architectural revolution in Santa Fe at a time that builders snubbed the traditional Pueblo style. Standing on the brick platform, you'll hear a bell ring and a whistle blow; the train is ready to roll.

Inside the restored coach, passengers are surrounded by aged mahogany and faded velvet seats. The train snakes through crowded Santa Fe intersections onto the New Mexico plains, broad landscapes spotted with piñon and chamisa, with views of the Sandia and Ortiz mountains. Arriving in the small track town of Lamy, you get another glimpse of a Mission-style station, this one surrounded by spacious lawns where passengers picnic. Others choose to eat hearty burgers, steaks, and lobster at the historic Legal Tender Restaurant. Friday night, May through September, passengers can take an evening train and watch bold New Mexico sunsets while eating a buffet prepared by the excellent local caterer Edible. You'll taste dishes such as chicken skewers in a mole sauce, black bean and squash tamales, vegetables with chipotle chile dip, and chocolate pecan cookies, as well as coffee, tea, and soft drinks. There's also a cash bar selling beer, wine, and margaritas.

COOKING & ART CLASSES

If you have the time and are looking for something to do that's a little off the beaten tourist path, you might consider a cooking or art class.

You can master the flavors of Santa Fe with an entertaining 3-hour demonstration cooking class at the ✪ **Santa Fe School of Cooking and Market,** on the upper level of the Plaza Mercado, 116 W. San Francisco St. (☎ **505/983-4511;** fax 505/983-7540). The class teaches about the flavors and history of traditional New Mexican and contemporary Southwestern cuisines. Cooking light classes are offered for those who prefer to cook with less fat. Prices range from $35 to $55 and include a meal; call for a class schedule. The adjoining market offers a variety of regional foods and cookbooks, with gift baskets available.

If Southwestern art has you hooked, you can take a drawing and painting class led by Santa Fe artist Jane Shoenfeld. Students sketch such outdoor subjects as the Santa Fe landscape and adobe architecture. In case of inclement weather, classes are held in the studio. Each class lasts for 3 hours, and art materials are included in the $60 fee. All levels of experience are welcome. Children's classes can be arranged. Contact Jane at **Sketching Santa Fe,** P.O. Box 5912, Santa Fe, NM 87502 (☎ **505/ 986-1108**).

WINE TASTINGS

If you enjoy sampling regional wines, consider visiting the wineries within easy driving distance of Santa Fe: **Balagna Winery/Il Santo Cellars,** 223 Rio Bravo Dr., in Los Alamos (☎ **505/672-3678**), north on US 84/285 and then west on NM 502; **Santa Fe Vineyards** (P.O. Box 216A), about 20 miles north of Santa Fe on US 84/ 285 (☎ **505/753-8100**); **Madison Vineyards & Winery,** in Ribera (☎ **505/ 421-8028**), about 45 miles east of Santa Fe on I-25 North; and the **Black Mesa Winery,** 1502 NM 68, in Velarde (☎ **800/852-6372**), north on US 84/285 to NM 68.

Be sure to call in advance to find out when the wineries are open for tastings and to get specific directions.

3 Especially for Kids

Don't miss taking the kids to the **Museum of International Folk Art,** where they'll love the international dioramas and the toys (discussed earlier in this chapter). Also visit the tepee at the **Wheelwright Museum of the American Indian** (discussed earlier in this chapter), where storyteller Joe Hayes spins traditional Spanish *cuentos,* Native American folk tales, and Wild West tall tales on weekend evenings. **Bishop's Lodge** and **Rancho Encantado** both have extensive children's programs during the summer. These include horseback riding, swimming, arts and crafts programs, as well as special activities such as archery and tennis. Kids are sure to enjoy ❂ **El Rancho de las Golondrinas** (discussed above), a 200-acre ranch 15 miles south of Santa Fe, today a living 18th- and 19th-century Spanish village comprising a hacienda, a village store, a schoolhouse, and several chapels and kitchens.

Planetarium at Santa Fe Community College

6401 Richards Ave. (South of Rodeo Rd.). ☎ **505/471-8200,** or 505/438-1777 for the information line. Admission $3.50 adults, $2 seniors and children 12 and under. Friday adult and family shows 6:30–7:30pm and 8–9pm; Saturday children's show 10:30–11:30am; Celestial Highlights, a live program mapping the night sky for that particular month, is on the first Thursday of the month from 7–8pm.

The Planetarium offers imaginative programs, combining star shows with storytelling and other interactive techniques. Some of the titles reveal the inventiveness of the programs: Rusty Rocket's Last Blast, in which kids launch a model rocket; Planet Patrol; and the Solar System Stakeout, in which kids build a solar system. There's also a 10-minute segment on the current night sky. Programs vary, from those designed for preschoolers to ones for high school kids.

Rodeo de Santa Fe

2801 Rodeo Rd. ☎ **505/471-4300.**

The rodeo is held annually the weekend following the Fourth of July. (See "Northern New Mexico Calendar of Events," in chapter 2, for details.)

Santa Fe Children's Museum

1050 Old Pecos Trail. ☎ **505/989-8359.** Admission $3 adults, $2 children under 12. Thurs and Sat 10am–5pm, Fri 9am–5pm, Sun noon–5pm.

Designed for whole families to experience, this museum offers interactive exhibits and hands-on activities in the arts, humanities, science, and technology. Special performances and hands-on sessions with artists and scientists are regularly scheduled.

Santa Fe Public Library

145 Washington Ave. ☎ **505/984-6780.**

Children's story hours and other special programs can be found here in the center of town at various times throughout the week. Call for additional information.

4 Santa Fe Strolls

Santa Fe, with its intricate streets and resonant historical architecture, lends itself to walking. The city's downtown core extends only a few blocks in any direction from the plaza, and the ancient Barrio de Analco and the Canyon Road artists' colony are a mere stone's throw away.

WALKING TOUR 1
The Plaza Area

Start: The Plaza.
Finish: Loretto Chapel.
Time: 1 to 5 hours, depending on the length of visits to the museums and churches.
Best Times: Any morning after breakfast (before the afternoon heat), but after the Native American traders have spread out their wares.

1. The Plaza has been the heart and soul of Santa Fe, as well as its literal center, since its concurrent establishment with the city in 1610. Originally designed as a

Walking Tour—The Plaza Area

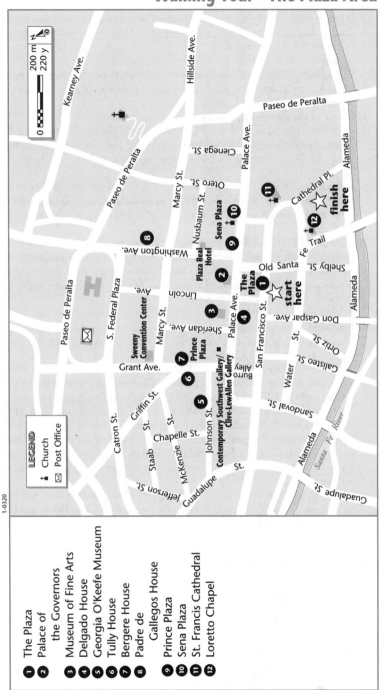

LEGEND
- ✝ Church
- ⊠ Post Office

1. The Plaza
2. Palace of the Governors
3. Museum of Fine Arts
4. Delgado House
5. Georgia O'Keefe Museum
6. Tully House
7. Bergere House
8. Padre de Gallegos House
9. Prince Plaza
10. Sena Plaza
11. St. Francis Cathedral
12. Loretto Chapel

1-0320

meeting place, it has been the site of innumerable festivals and other historical, cultural, and social events. Long ago the plaza was a dusty hive of activity as the staging ground and terminus of the Santa Fe Trail. Today those who congregate around the central fountain enjoy the best people-watching in New Mexico. Santa Feans understandably feel nostalgic for the days when the plaza, now the hub of the tourist trade, still belonged to natives and not outside commercial interests.

Facing the Plaza on its north side is the:

2. **Palace of the Governors,** which has functioned continually as a public building since it was erected in 1610 as the capitol of Nuevo Mexico. Today it's the flagship of the New Mexico State Museum system (see "The Top Attractions," above). Every day Native American artisans spread out their crafts for sale beneath its portico.

Even though you are just two stops into it, you might want to fortify your strength for the rest of the walk. I recommend the carnitas or tamales from the street vendor immediately opposite the Palace, at Lincoln and Palace avenues, in front of the:

3. **Museum of Fine Arts,** with its renowned St. Francis Auditorium (see "The Top Attractions," above, and "Major Performing Arts Companies," in chapter 9). The works of Georgia O'Keeffe and other famed 20th-century Taos and Santa Fe artists are some of the highlights of a visit here. The building is a fine example of Pueblo Revival–style architecture.

Virtually across the street is the:

4. **Delgado House,** 124 W. Palace Ave., a Victorian mansion that's an excellent example of local adobe construction modified by late 19th-century architectural detail. It was built in 1890 by Felipe B. Delgado, a merchant most known for his business of running mule and ox freight trains over the Santa Fe Trail to Independence and the Camino Real to Chihuahua. The home remained in the Delgado family until 1970. It now belongs to the Historic Santa Fe Foundation.

If you continue west on Palace Avenue you'll come to two galleries worth perusing. The first is the **Contemporary Southwest Gallery,** where you'll find lots of vivid landscapes, but nothing that extends the limits of the genre. Next door, however, you'll come to **Cline-LewAllen Gallery,** where you'll find an array of paintings, sculpture, ceramics, and work in other media that may indeed make you stop and ponder.

Nearby, you'll see a narrow lane—Burro Alley—jutting south toward San Francisco Street. You may want to head down the lane and peek into **Down and Outdoors in Santa Fe,** where you'll find a selection of comforters, pillows, outerwear, and fine sweaters. Head back to Palace and make your way north on Grant Street, turning left on Johnson Street to the:

5. **Georgia O'Keeffe Museum,** 217 Johnson St. Just opened in 1997, it houses the largest collection of O'Keeffe works in the world. The 10,000-square-foot space is the only museum in the United States dedicated solely to one woman's work.

Head back to Grant Street where you'll find the:

6. **Tully House,** 136 Grant Ave., built in 1851 in Territorial style. This is the headquarters of the Historic Santa Fe Foundation. (A publication of the foundation, *Old Santa Fe Today,* gives detailed descriptions, with a map and photos, of 50 sites within walking distance of the Plaza.)

Across the street is the:

7. **Bergere House,** 135 Grant Ave., built around 1870. U.S. president Ulysses S. Grant and his wife, Julia, stayed here during an 1880 visit to Santa Fe.

Proceed north on Grant, turning right on Marcy. On the north side of this corner is the **Sweeney Convention Center,** host of major exhibitions and home of the Santa Fe Convention and Visitors Bureau.

Three blocks farther east, through a residential, office, and restaurant district, turn left on Washington Avenue. A short distance along on your right, note the:

8. Padre de Gallegos House, 227–237 Washington Ave., built in 1857 in the Territorial style. Padre de Gallegos was a priest, who in the eyes of newly arrived Archbishop Jean-Baptiste Lamy, kept too high a social profile and was therefore defrocked in 1852. Gallegos later represented the territory in Congress and eventually became the federal superintendent of Native American affairs.

Reverse course and turn south again on Washington Avenue, passing en route the public library and some handsomely renovated accommodations such as the Territorial Inn. This is a good time to stop for refreshments at the **Plaza Real Hotel;** during summer you'll find a variety of drinks served on the veranda, and in winter the small bar can be quite cozy. When you leave there, you'll notice across the street the entrance to the Palace of the Governors' archives. As you approach the Plaza, turn left (east) on Palace Avenue. A short distance farther on your left is:

9. Prince Plaza, 113 E. Palace Ave., a former governor's home. This Territorial-style structure, which now houses The Shed restaurant, had huge wooden gates to keep out tribal attacks.

Next door is:

10. Sena Plaza, 125 E. Palace Ave. This city landmark offers a quiet respite from the busy streets with its parklike patio. **La Casa Sena** restaurant (a great place to stop for lunch or dinner) is the primary occupant of what was once the 31-room Sena family adobe hacienda, built in 1831. The Territorial legislature met in the upper rooms of the hacienda in the 1890s.

Turn right (south) on Cathedral Place to enter the doors of:

11. St. Francis Cathedral, built in Romanesque style between 1869 and 1886 by Archbishop Lamy. Santa Fe's grandest religious edifice, it has a famous 17th-century wooden Madonna known as *Our Lady of Peace* (see "The Top Attractions," above).

After leaving the cathedral, walk around the back side of the illustrious La Fonda Hotel—south on Cathedral Place and west on Water Street—to the intersection of the Old Santa Fe Trail. Here, in the northwest corner of the Best Western Inn at Loretto, you'll find the:

12. Loretto Chapel, more formally known as the Chapel of Our Lady of Light. Lamy was also behind the construction of this chapel, built for the Sisters of Loretto. It is remarkable for its spiral staircase, which has no central or other visible support (see "More Attractions," above). By now you may be tired and hungry. I recommend heading straight back to the Plaza to a small general store where **F. W. Woolworth's,** the now-defunct, legendary five-and-dime store, once stood. Like Woolworth's, it serves up a cherished local, er, delicacy called Frito pie: a bag of Fritos smothered in chile con carne, in a plastic bag with a spoon and a napkin.

WALKING TOUR 2
Barrio de Analco/Canyon Road

Start: Don Gaspar Avenue and East de Vargas Street.
Finish: Any of the quaint restaurants on Canyon Road.

Time: 1 to 3 hours, depending on how long you spend in the art galleries.
Best Times: Anytime.

The Barrio de Analco, now called East de Vargas Street, is beyond question one of the oldest continuously inhabited avenues in the United States. Spanish colonists, with their Mexican and Native American servants, built homes here in the early 1600s, when Santa Fe was founded; some of them survive to this day.

Most of the houses you'll see as you walk east on de Vargas are private residences, not open for viewing. But they are well worth looking at because of the feeling they evoke of Santa Fe life now relegated to bygone days. Most have interpretive historical plaques on their outer walls. The first you'll see is:

1. **Tudesqui House,** 129 E. de Vargas St., dating from the early 19th century, now recognizable for the wisteria growing over its adobe walls.
   Across the street is the:
2. **Gregoria Crespin House,** 132 E. de Vargas St., whose records date back at least to 1747 (when it was sold for 50 pesos). Originally of pueblo design, it later had Territorial embellishments added in the trim of bricks along its roofline.
   Just down the road is the:
3. **Santa Fe Playhouse,** 142 E. de Vargas St., home of the oldest existing thespian group in New Mexico. Actors still perform in this original adobe theater (see "The Performing Arts," in chapter 9).
   In the next block, just east of the **Old Santa Fe Trail,** is the:
4. **Mission of San Miguel,** built around 1612 and ranking as one of the oldest churches in America (see "More Attractions," above). Today it's maintained and operated by the Christian Brothers. This might be a good time to stop in at the **Pink Adobe's Dragon Room** (across Old Santa Fe Trail from the Mission); there won't be another refreshment stop until many blocks up Canyon Road.
   Across de Vargas Street from the Mission is the so-called:
5. **Oldest House in the United States.** Whether or not this is true is anybody's guess, but it's among the last of the poured-mud adobe houses and may have been built by Pueblo people. Sadly, modern graffiti has defiled some interior walls. The entrance is through a gift shop.
   A few doors down you'll come to **Maslak Mcleod,** Inuit Native Canadian Art, a gallery of rich artifacts you'll want to check out.
   There are more homes at the east end of de Vargas, before its junction with Canyon Road. Among them is the:
6. **Boyle House,** 327 E. de Vargas St., built in the mid-18th century as a hacienda.
   Nearby is the:
7. **José Alarid House,** 338 E. de Vargas St., built in the 1830s and now an art gallery.
   A few houses farther down the road is the:
8. **Adolph Bandelier House,** 352 E. de Vargas St., home of the famous archaeologist who unearthed the prehistoric ruins at Bandelier National Monument.
   De Vargas intersects narrow, winding Canyon Road after it crosses the Paseo de Peralta. Extending about 2 miles from the Paseo to Camino Cabra, Canyon Road today is lined with art galleries, shops, and restaurants. But it was once a Native American trail used by the Pueblo tribes when they came to launch their 1680 insurrection against the Spanish colonists. Historic buildings include the:
9. **Juan José Prada House,** 519 Canyon Rd., which dates from about 1760. A few doors before this house, you'll come to **Morning Star Gallery,** a great place to glimpse museum-quality Native American arts and artifacts.

Walking Tour—Barrio de Analco/Canyon Road

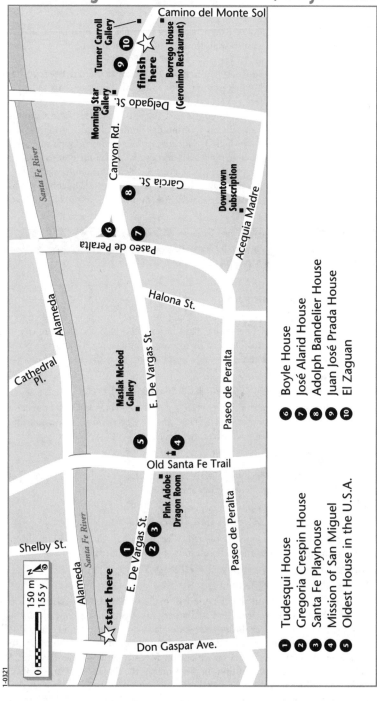

Camino del Monte Sol

Turner Carroll Gallery

Borrego House (Geronimo Restaurant)

finish here

Morning Star Gallery

Delgado St.

Canyon Rd.

Santa Fe River

Garcia St.

Downtown Subscription

Acequia Madre

Paseo de Peralta

Halona St.

Alameda

Cathedral Pl.

E. De Vargas St.

Maslak Mcleod Gallery

Paseo de Peralta

Santa Fe River

Old Santa Fe Trail

Pink Adobe Dragon Room

E. De Vargas St.

Paseo de Peralta

Shelby St.

Alameda

start here

150 m
155 y

Don Gaspar Ave.

1. Tudesqui House
2. Gregoria Crespin House
3. Santa Fe Playhouse
4. Mission of San Miguel
5. Oldest House in the U.S.A.

6. Boyle House
7. José Alarid House
8. Adolph Bandelier House
9. Juan José Prada House
10. El Zaguan

1-0321

Farther up the road is:

10. **El Zaguan,** 545 Canyon Rd., a beautiful example of a Spanish hacienda. In warmer months be sure to walk through the garden, subject of many Santa Fe paintings.

Farther up Canyon Road you may want to stop at the old **Borrego house,** which is now Geronimo Restaurant, for lunch or dinner. Across the street is **Turner Carroll Gallery,** where you'll find very eclectic art.

If you have the stamina to continue, I suggest turning right on Camino del Monte Sol, then right again on Camino del Poniente, and bearing right onto Acequia Madre ("mother ditch"). This narrow lane winds through one of Santa Fe's oldest and most notable residential districts. It follows the mother ditch, used for centuries to irrigate gardens in the area. A 30- to 45-minute walk will bring you to **Downtown Subscription,** where can have some excellent coffee and baked goods. From there you can take Garcia Street north back to the base of Canyon Road.

5 Organized Tours

BUS, CAR & TRAM TOURS

Gray Line Tours
1330 Hickox St. ☎ **505/983-9491.**

The trolley-like Roadrunner departs several times daily in summer (less often in winter) from the La Fonda Hotel on the Plaza, beginning at 9am, for 1 1/2-hour city tours. Buy tickets as you board. Daily tours to Taos, Chimayo, and Bandelier National Monument are also offered.

LorettoLine
At the Inn at Loretto, 211 Old Santa Fe Trail. ☎ **505/983-3701.**

For an open-air tour of the city, contact LorettoLine. Tours last 1 1/2 hours and are offered daily from May to October. Tour times are every hour on the hour during the day from 10am to 3pm. Tickets are $9 for adults, $4 for children.

WALKING TOURS

As with the above independent walking tours, these are the best way to get an appreciable feel for Santa Fe's history and culture.

Afoot in Santa Fe
At the Inn at Loretto, 211 Old Santa Fe Trail. ☎ **505/983-3701.**

Personalized 2 1/2-hour tours are offered twice daily (9:30am and 1:30pm) from the Inn at Loretto at a cost of $12. Reservations are not required.

Storytellers and the Southwest: A Literary Tour
941 Calle Mejia. ☎ **505/989-4561.**

This is a new operation that I hope will make it. Barbara Harrelson, a former Smithsonian museum docent and avid reader, takes you on a 2-hour literary walking tour of downtown, exploring the history, legends, characters, and authors of the region through its landmarks and historic sites. It's a great way to absorb the unique character of Santa Fe. Tours take place on Monday and Wednesday from May through September; meet at 6pm in the center of the Plaza. The cost is $10. I recommend calling ahead.

Walking Tour of Santa Fe
107 Washington Ave. ☎ **800/338-6877** or 505/983-6565.

One of Santa Fe's best walking tours begins under the T-shirt tree at Tees & Skis, 107 Washington Ave., near the northeast corner of the Plaza (at 9:30am and 1:30pm) and lasts about 2¹/₂ hours. The tour costs $10 for adults. Children are free.

MISCELLANEOUS TOURS

Pathways Customized Tours
161-F Calle Ojo Feliz. ☎ **505/982-5382.**

Don Dietz offers several planned tours, including a downtown Santa Fe walking tour, a full city tour, a trip to the cliff dwellings and native pueblos, a "Taos adventure," and a trip to Georgia O'Keeffe country (with a focus on the landscape that inspired the art now possible to view in the new O'Keeffe Museum). He will try to accommodate any special requests you might have. These tours last anywhere from 1¹/₂ to 9 hours, depending on the one you choose. Don has extensive knowledge of the area's culture, history, geology, and flora and fauna and will help you make the most of your precious vacation time.

Rain Parrish
535 Cordova Rd., Suite 250. ☎ **505/984-8236.**

A Navajo anthropologist, artist, and freelance curator offers custom guide services focusing on cultural anthropology, Native American arts, and the history of the Native Americans of the Southwest. Ms. Parrish includes visits to local Pueblo villages.

Recursos de Santa Fe
826 Camino de Monte Rey. ☎ **505/982-9301.**

This organization is a full-service destination management company, emphasizing custom-designed itineraries to meet the interests of any group. They specialize in the archaeology, art, literature, spirituality, architecture, environment, food, and history of the Southwest.

Rojo Tours & Services, Inc.
P.O. Box 15744. ☎ **505/474-8333.** Fax 505/474-2992.

Customized private tours are arranged to pueblos, cliff dwellings, and ruins, as well as art tours with studio visits.

Santa Fe Detours
107 Washington Ave. ☎ **800/DETOURS** or 505/983-6565.

Santa Fe's most extensive tour-booking agency accommodates almost all travelers' tastes, from bus and rail tours to river rafting, backpacking, and cross-country skiing.

✪ Southwest Safaris
P.O. Box 945, Santa Fe, NM 87504. ☎ **800/842-4246** or 505/988-4246.

This tour is one of the most interesting Southwest experiences I've had. We flew in a small plane at 500 feet off the ground from Santa Fe to the Grand Canyon while experienced pilot Bruce Adams explained 300 million years of geologic history. We passed over the ancient ruins of Chaco Canyon and the vivid colors of the Painted Desert, as well as over many land formations on Navajo Nation land so remote they remain nameless. Then, of course, there was the spectacular Grand Canyon, where we landed for a jeep tour and lunch on a canyon-side bench. Trips to many Southwest destinations are available, including Monument Valley, Mesa Verde, Canyon de Chelly, Arches/Canyonlands, as well as the ruins at Aztec, NM. Prices range from $199 to $449.

6 Outdoor Activities

Note: In addition to all the activities and recreation centers listed below, there will be a new full-service family recreation center in Santa Fe by early- to mid-1998. The complex will include a 25-meter pool, leisure and therapy pools, an ice-skating rink, three gyms, a workout room, racquetball courts, and an indoor running track. Contact the Santa Fe Convention and Visitors Bureau for more information.

BALLOONING New Mexico is renowned for its spectacular Balloon Fiesta, which takes place annually in Albuquerque. If you want to take a ride, you'll probably have to go to Albuquerque or Taos, but you can book your trip in Santa Fe through **Santa Fe Detours,** 107 Washington Ave. ☎ **800/338-6877** or 505/983-6565. Flights take place early in the day. Rates begin at around $135 a flight. If you've got your heart set on a balloon flight, I would suggest that you make your reservations early because flights are often canceled due to bad weather. That way, if you have to reschedule, you'll have enough time to do so.

BIKING You can cycle along main roadways and paved country roads year-round in Santa Fe, but be aware that traffic is particularly heavy around the Plaza, and all over town motorists are not especially attentive to bicyclists, so you need to be especially alert. Mountain biking has exploded here and is especially popular in the spring, summer, and fall; the high-desert terrain is rugged and challenging, but mountain bikers of all levels can find exhilarating rides. The Santa Fe Convention and Visitors Bureau can supply you with bike maps. I recommend the following trails: West of Santa Fe, the Caja del Rio area has nice dirt roads and some light technical biking; Bland Canyon in the Jemez Mountains to the south near Cochiti Pueblo is an exciting ride with views and stream crossings; the railroad tracks south of Santa Fe provide wide-open biking on beginner to intermediate technical trails; and the Borrego Trail up toward the Santa Fe Ski Area is a very challenging technical ride.

In Santa Fe bookstores, look for *Mountain Biking in Northern New Mexico: Historical and Natural History Rides* by Craig Martin and *The New Mexican Mountain Bike Guide* by Brant Hayengand and Chris Shaw. Both are excellent guides to trails in Santa Fe, Taos, and Albuquerque. The books outline tours for beginner, intermediate, and advanced riders. **Palace Bike Rentals,** 409 E. Palace Ave. (☎ **505/986-0455**), rents high-quality mountain bikes for $15/half day and $20/full day (until sunset). They also have multiple-day rates. Children's bikes are available, as are child seats and child trailers. **Sun Mountain Bike Rentals,** 121 Sandoval St. (☎ **505/820-2902**), rents quality regular and front-suspension mountain bikes for $12–$15/half day and $18–$30/full day. They also rent street cruisers for $5/hour, $12/half day and $20/full day. Weekly rentals can be arranged. Both shops supply accessories such as helmets, locks, water, maps, and trail information.

FISHING In the lakes and waterways around Santa Fe, anglers typically catch trout (there are five varieties in the area). Other local fish include bass, perch, and kokanee salmon. The most popular fishing holes are Cochiti and Abiquiu lakes as well as the Rio Chama and Pecos streams. Fly-fishing is popular in the Rio Grande. Check with the **New Mexico Game and Fish Department** (☎ **505/827-7911**) for information and licenses. **High Desert Angler,** 435 S. Guadalupe St. (☎ **505/988-7688**), specializes in fly-fishing gear and guide service.

GOLF There are two public courses in the Santa Fe area: the 18-hole **Santa Fe Country Club,** on Airport Road (☎ **505/471-2626**); and the often-praised, 18-hole **Cochiti Lake Golf Course,** 5200 Cochiti Hwy., Cochiti Lake, about 35 miles southwest of Santa Fe via I-25 and NM 16 and 22 (☎ **505/465-2239**). **Santa Fe Golf**

In case you want to be welcomed there.

We're here to see that you're always welcomed at establishments everywhere. That's why millions of people carry the American Express® Card – for peace of mind, confidence, and security, around the world or just around the corner.

In case you're running low.

We're here to help with more than 118,000 Express Cash locations around the world. In order to enroll, just call American Express before you start your vacation.

do more

Express Cash

And just in case.

We're here with American Express® Travelers Cheques and Cheques *for Two.*® They're the safest way to carry money on your vacation and the surest way to get a refund, practically anywhere, anytime.

Another way we help you...

do more

Travelers Cheques

and Driving Range, 4680 Wagon Rd. (☎ **505/474-4680**), is also open to the public throughout the year. They have 42 practice tees, golf merchandise, and rental clubs, and will provide instruction.

HIKING   It's hard to decide which of the 1,000 miles of nearby national forest trails to challenge. Four wilderness areas are especially attractive: **Pecos Wilderness,** with 223,000 acres east of Santa Fe; **Chama River Canyon Wilderness,** 50,300 acres west of Ghost Ranch Museum; **Dome Wilderness,** 5,200 acres of rugged canyonland adjacent to Bandelier National Monument; and **San Pedro Parks Wilderness,** 41,000 acres west of Los Alamos. Also visit the 58,000-acre **Jemez Mountain National Recreation Area.** Information on these and other wilderness areas is available from the **Santa Fe National Forest,** 1220 St. Francis Dr. (P.O. Box 1689), Santa Fe, NM 87504 (☎ **505/438-7840**). If you're looking for company on your trek, contact the Santa Fe branch of the **Sierra Club** (☎ **505/983-2703**), or **Tracks,** 417 San Pasqual, Santa Fe, NM 87501 (☎ **505/982-2586**). I enjoy taking a chairlift ride to the summit of the **Santa Fe Ski Area** (☎ **505/983-9155**) and hiking around up there in the spring and summer months. You might also consider purchasing *The Hiker's Guide to New Mexico* (Falcon Press Publishing Co., Inc.) by Laurence Parent; it outlines 70 hikes throughout the state.

HORSEBACK RIDING   Trips ranging in length from a few hours to overnight can be arranged by **Santa Fe Detours,** 107 Washington Ave. (☎ **800/338-6877** or 505/983-6565). You'll ride with "experienced wranglers" and can even arrange a trip that includes a cookout or brunch. Rides are also major activities at two local guest ranches: **The Bishop's Lodge** and **Rancho Encantado** (see "Where to Stay in Santa Fe," in chapter 5). In addition, **Rocky Mountain Tours,** 217 W. Manhattan St., Santa Fe, NM 87501 (☎ **505/984-1684**), arranges escorted rides for individuals of all ability levels as well as families. Trips can run from 90 minutes to a full day. Special packages such as "Raft and Ride," and "Design Your Own Ride" are also available.

HUNTING   Mule deer and elk are taken by hunters in the **Pecos Wilderness** and **Jemez Mountains,** as are occasional black bears and bighorn sheep. Wild turkeys and grouse are frequently bagged in the uplands, geese and ducks at lower elevations. Check with the **New Mexico Game and Fish Department** (☎ **505/827-7911**) for information and licenses.

JOGGING   Despite its elevation, Santa Fe is popular with runners and hosts numerous competitions, including the annual Old Santa Fe Trail Run on Labor Day. "Fun Runs," sponsored by **Santa Fe Striders** (☎ **505/983-2144**), begin at the Plaza on Wednesday at 6pm year-round (5:30pm in winter).

RIVER RAFTING   Although Taos is the real rafting center of New Mexico, several companies serve Santa Fe during the April to October white-water season. They include the **Southwest Wilderness Adventures,** P.O. Box 9380, Santa Fe, NM 87501 (☎ **800/869-7238** or 505/983-7262); **New Wave Rafting,** 103 E Water St. (upstairs; ☎ **505/984-1444**); and the **Santa Fe Rafting Co.,** 1000 Cerrillos Rd. (☎ **505/988-4914**).

SKIING   There's something available for every ability level at **Ski Santa Fe,** about 16 miles northeast of Santa Fe via Hyde Park (Ski Basin) Road. Lots of locals ski here, particularly on weekends; if you can, go on weekdays. It's a good family area and fairly small, so it's easy to split off from and later reconnect with your party.

Built on the upper reaches of 12,000-foot Tesuque Peak, the area has an average annual snowfall of 225 inches and a vertical drop of 1,650 feet. Seven lifts, including a 5,000-foot triple chair and a new quad chair, serve 39 runs and 590 acres of terrain, with a total capacity of 7,800 riders an hour. Base facilities, at 10,350 feet, center around La Casa Mall, with a cafeteria, lounge, ski shop, and boutique. Another restaurant, Totemoff's, has a midmountain patio.

The ski area is open daily from 9am to 4pm; the season often runs from Thanksgiving to early April, depending on snow conditions. Rates for all lifts are $39 for adults, $24 for children and seniors, free for kids less than 46 inches tall (in their ski boots), and free for seniors 73 and older. For more information, contact **Ski Santa Fe,** 1210 Luisa St., Suite 5, Santa Fe, NM 87505 (☎ **505/982-4429**). For 24-hour reports on snow conditions, call ☎ **505/983-9155.** The New Mexico Snow Phone (☎ **505/984-0606**) gives statewide reports. Ski packages are available through **Santa Fe Central Reservations** (☎ **800/776-7669** outside New Mexico or 505/983-8200 within New Mexico).

Cross-country skiers find seemingly endless miles of snow to track in the **Santa Fe National Forest** (☎ **505/438-7840**). A favorite place to start is at the Black Canyon campground, about 9 miles from downtown en route to the Santa Fe Ski Area. In the same area are the Borrego Trail (high intermediate) and the Norski Trail, 7 miles up from Black Canyon. Basic Nordic lessons and backcountry tours are offered by **Bill Neuwirth's Tracks,** 417 San Pasqual, Santa Fe, NM 87501 (☎ **505/ 982-2586**).

Other popular activities at the ski area in winter include snowshoeing, snowboarding, sledding, and inner tubing. Snowshoe and snowboard rentals are available at a number of downtown shops and the ski area.

SOARING   Soaring is available for those who don't believe the sky is the limit. There are two types of soaring. In one, you and the pilot are in a propless and motorless plane that generally seats only two people. A powered plane tows you up and you glide about catching updrafts to stay afloat for hours. In the other, the plane is equipped with a retractable prop. This allows you to take off without use of a tow plane. Once at altitude, the prop retracts, leaving you to glide. Either type allows for plenty of scenic viewing and the thrill of free birdlike flight. For information and rates, call **Santa Fe Soaring** (☎ **505/424-1928**).

SPAS   If traveling, skiing, or other activities have left you weary, a great place to treat your body and mind is **Ten Thousand Waves,** a Japanese-style health spa about 3 miles northeast of Santa Fe on Hyde Park Road (☎ **505/982-9304**). This serene retreat, nestled in a grove of piñon, offers hot tubs, saunas, and cold plunges, plus a variety of massage and other bodywork techniques.

Bathing suits are optional in the 10-foot communal hot tub, where you can stay as long as you want for $13. Nine private hot tubs cost $18 to $25 an hour, with discounts for seniors and children. You can also arrange therapeutic massage, hot-oil massage, in-water watsu massage, herbal wraps, salt glows, and facials. New in 1996 were four treatment rooms that feature dry brush aromatherapy treatments and Ayurvedic treatments; a women's communal tub; and lodging at the Houses of the Moon, a six-room Japanese-style inn. The spa is open on Sunday, Monday, Wednesday, and Thursday from 10am to 9:30pm; Tuesday from 4:30 to 9:30pm; and Friday and Saturday from 10am to 11pm. Reservations are recommended, especially on weekends.

SWIMMING   The City of Santa Fe operates four indoor pools and one outdoor pool. The pool closest to downtown is found at the **Fort Marcy Complex** (☎ **505/ 984-6725**) on Camino Santiago, off Bishop's Lodge Road. Admission to the pool is $1.25 for adults, $1 for students, and 50¢ for children 8 to 13. Call the Santa Fe Convention and Visitors Bureau for information about the other area pools.

TENNIS   Santa Fe has 44 public tennis courts and 4 major private facilities. The **City Recreation Department** (☎ **505/438-1485**) can locate all indoor, outdoor, and lighted public courts.

7 Spectator Sports

HORSE RACING   The ponies run from Memorial Day through September at **The Downs at Santa Fe** (☎ **505/471-3311**), about 11 miles south of Santa Fe off US 85, near La Cienega. Post time for eight-race cards is 3:15pm on Friday; for nine-race cards, 3:15pm on Saturday and Sunday. General admission is free. A closed-circuit TV system shows instant replays of each race's final-stretch run and transmits out-of-state races for legal betting.

RODEO   The **Rodeo de Santa Fe,** 2801 Rodeo Rd. (☎ **505/471-4300**), is held annually the weekend following the Fourth of July. (See "Northern New Mexico Calendar of Events," in chapter 2, for details.)

8 Santa Fe Shopping

Shopping in Northern New Mexico means handcrafts, art, and artifacts. For traditional Native American crafts as well as Hispanic folk art and abstract contemporary works, Santa Fe is the place to shop. Galleries speckle the downtown area, and as an artists' thoroughfare Canyon Road is preeminent. Still, the greatest concentration of Native American crafts is displayed beneath the portal of the Palace of the Governors. Any serious arts aficionado should try to attend one or more of the city's great arts festivals—the Spring Festival of the Arts in May, the Spanish Market in July, the Indian Market in August, and the Fall Festival of the Arts in October.

1 The Shopping Scene

Few visitors to Santa Fe leave the city without acquiring at least one item from the Native American artisans at the Palace of the Governors. When you are thinking of making such a purchase, keep the following pointers in mind:

Silver jewelry should have a harmony of design, clean lines, and neatly executed soldering. Navajo jewelry typically features large stones, with designs shaped around the stone. Zuni jewelry usually has patterns of small or inlaid stones. Hopi jewelry rarely uses stones; it usually has a motif incised into the top layer of silver and darkened.

Turquoise of a deeper color is usually higher quality, so long as it hasn't been color treated. Heishi bead necklaces usually use stabilized turquoise.

Pottery is traditionally hand-coiled and of natural clay, not thrown on a potter's wheel using commercial clay. It is hand-polished with a stone, hand-painted, and fired in an outdoor oven (usually an open firepit) rather than an electric kiln. Look for an even shape; clean, accurate painting; a high polish (if it is a polished piece); and an artist's signature.

Navajo rugs are appraised according to tightness and evenness of weave, symmetry of design, and whether natural (preferred) or commercial dyes have been used.

Kachina dolls are more highly valued according to the detail of their carving: fingers, toes, muscles, rib cages, feathers, etc. Elaborate costumes are also desirable. Oil staining is preferred to the use of bright acrylic paints.

How to Buy a Navajo Rug

After you arrive in New Mexico, you'll quickly notice beautiful hand-loomed Navajo rugs in shops, hotels, and museums. The colors and designs are so striking that, chances are, you'll begin looking for one to take home.

There are several things to keep in mind when shopping for a Navajo rug. First of all, be sure that the shop where you're browsing has a solid reputation (ask for recommendations at your hotel). Although all Navajo rugs are authentic, they may not all be good quality (none is completely perfect, however). Of course, you want to buy a rug that pleases you aesthetically, but it is unwise to buy on impulse because you will be spending quite a lot of money. Take your time looking. Spread the rugs out completely to make sure there are no obvious flaws (holes or fraying wool). Check to make sure the design and color are uniform throughout. The weave lines should be straight, without any visible loose ends. The rug should be of equal thickness throughout. If the rug you like meets all of the above criteria and the price is right, buy it—you'll regret it later if you don't.

Sand paintings should display clean, narrow lines, even colors, balance, an intricacy of design, and smooth craftsmanship.

Local museums, particularly the Wheelwright Museum and the Institute of American Indian Art, can provide a good orientation to contemporary craftsmanship.

Contemporary artists are mainly painters, sculptors, ceramists, and fiber artists, including weavers. Peruse one of the outstanding **gallery catalogs** for an introduction to local dealers—*The Collector's Guide to Santa Fe and Taos* by Wingspread Incorporated (P.O. Box 13566-M, Albuquerque, NM 87192), *Santa Fe and Taos Arts* by The Book of Santa Fe (535 Cordova Rd., Suite 241, Santa Fe, NM 87501), or *The Santa Fe Catalogue* by Modell Associates (P.O. Box 1007, Aspen, CO 81612). They're widely available at shops or can be ordered directly from the publishers.

A terrific insight into Santa Fe art and artists is made possible by the personalized studio tours offered by ✪ **Studio Entrada,** P.O. Box 4934, Santa Fe, NM 87502 (☎ 505/983-8786). At a cost of $100 for two people, director Linda Morton takes small groups into private studios to meet the artists and learn about their work. Each itinerary lasts about 2½ hours and includes two or three studio gallery visits.

Business hours vary quite a bit among establishments, but most are open at least Monday through Friday from 10am to 5pm, with mall stores open until 9pm. Most shops are open similar hours on Saturday, and many also open on Sunday afternoon during the summer. Winter hours tend to be more limited.

After the high-rolling 1980s, during which art markets around the country soared, came the penny-pinching 1990s. Many galleries in Santa Fe were forced to shut their doors. Those that remained tended to specialize in particular types of art, a refinement process that has improved the gallery scene here. Still, some worry that the lack of serious art buyers in the area leads to fewer good galleries and more T-shirt and trinket stores. The Plaza has its share of those, but still has a good number of serious galleries, appealing to those buyers whose interests run to accessible art—Southwestern landscapes and the like. On Canyon Road, the art is often more experimental and more diverse, from contemporary sculpture to eastern European portraiture.

2 The Top Galleries

CONTEMPORARY ART

Adieb Khadoure Fine Art
610 Canyon Rd. ☎ **505/820-2666.**

This is a working artists' studio featuring contemporary artists Jeff Uffelman and Hal Larsen and Santa Fe artist Phyllis Kapp. Their works are shown in the gallery daily from 10am to 6pm, and Adieb Khadoure also sells beautiful rugs, furniture, and pottery from around the world.

Canyon Road Contemporary Art
403 Canyon Rd. ☎ **505/983-0433.**

This gallery represents some of the finest emerging U.S. contemporary artists as well as internationally known artists. You'll find figurative, landscape, and abstract paintings, as well as raku pottery.

✪ Cline LewAllen Gallery
129 W. Palace Ave. ☎ **505/989-8702.**

This is one of my favorite galleries. You'll find bizarre and beautiful contemporary works in a range of mediums from granite to clay to twigs. There are always exciting works on canvas as well.

Hahn Ross Gallery
409 Canyon Rd. ☎ **505/984-8434.**

Owner Tom Ross, a children's book illustrator, specializes in representing artists who create colorful, fantasy-oriented works. I'm especially fond of the wild party scenes by Susan Contreras.

Leslie Muth Gallery
131 W. Palace Ave. ☎ **505/989-4620.**

Here you'll find "Outsider Art," wild works made by untrained artists in a bizarre variety of media, from sculptures fashioned from pop bottle lids to portraits painted on flattened beer cans. Much of it is extraordinary and affordable.

✪ Shidoni Foundry and Gallery
Bishop's Lodge Rd. Tesuque. ☎ **505/988-8001.**

Shidoni Foundry is one of the area's most exciting spots for sculptors and sculpture enthusiasts. At the foundry visitors may take a tour through the facilities to view casting processes. In addition, there is a 5,000-square-foot contemporary gallery, a bronze gallery, and a wonderful sculpture garden.

NATIVE AMERICAN & OTHER INDIGENOUS ART

Dreamtime Gallery
223¹/₂ Canyon Rd. ☎ **505/986-0344.**

If you're at all interested in Australian Aboriginal artwork, this is the place to visit. There are some very interesting Aboriginal paintings and sculptures, as well as original weavings, bark paintings, and digeridoos. Works start as low as $25.

Frank Howell Gallery
103 Washington Ave. ☎ **505/984-1074.**

If you've never seen the wonderful illustrative hand of Frank Howell, you'll want to visit this gallery. You'll find a variety of contemporary American and American

Indian art. There's sculpture by award-winner Tim Nicola, as well as fine art, jewelry, and graphics.

Glenn Green Galleries

50 E. San Francisco St. ☎ **505/988-4168.**

The gallery, which maintains exclusive representation for Allan Houser's bronze and stone sculptures, also exhibits paintings, prints, photographs, and jewelry by other important artists.

✪ Joshua Baer & Company

116^1/$_2$ E. Palace Ave. ☎ **505/988-8944.**

This is a great place to explore. You'll find 19th-century Navajo blankets, pottery, jewelry, as well as primitive art from around the world.

Maslak-Mcleod

225 E. de Vargas. ☎ **505/820-6389.**

Enter a world of strange creatures from the north in this gallery that specializes in Inuit and other native Canadian art. Here you'll find seals carved from bone and native myths emerging from stone.

✪ Morning Star Gallery

513 Canyon Rd. ☎ **505/982-8187.**

This is one of my favorite places to browse. Throughout the rambling gallery are American Indian art masterpieces elegantly displayed. You'll see a broad range of works, from late-19th-century Navajo blankets to 1920s Zuni needlepoint jewelry.

PHOTOGRAPHY

Andrew Smith Gallery

203 W. San Francisco St. ☎ **505/984-1234.**

I'm always amazed when I enter this gallery and see works I've seen reprinted in major magazines for years. There they are, photographic prints, large and beautiful, hanging on the wall. Here you'll see famous works by Edward Curtis, Eliot Porter, Ansel Adams, Annie Leibovitz, and others.

Photo-eye Gallery

370 Garcia St. ☎ **505/988-5152.**

You're bound to be surprised each time you step into this new gallery a few blocks off Canyon Road. Dealing in contemporary photography, the gallery represents 40 renowned, as well as emerging, artists. They've also taken on the Platinum Gallery collection, from the notable Santa Fe gallery that closed in 1996.

SPANISH AND HISPANIC ART

Santos of New Mexico

2712 Paseo de Tularosa. ☎ **505/473-7941.**

Here you'll find the work of award-winning *Santero* Charles M. Carillo: Traditional New Mexican *santos* crafted out of cottonwood root and decorated with homemade pigments, as well as hand-adzed panels. By appointment only.

TRADITIONAL ART

Alterman & Morris Galleries

225 Canyon Rd. ☎ **505/983-1590.**

This is a well of interesting traditional art, mostly 19th- and 20th-century American paintings and sculpture. The gallery represents Remington and Russell, in addition to Taos founders, Santa Fe artists, and members of the Cowboy Artists of America and National Academy of Western Art.

✪ Gerald Peters Gallery
439 Camino del Monte Sol (P.O. Box 908). ☎ **505/988-8961.**

By April 1998, Gerald Peters plans to be moved into a 2-story pueblo-style building at 1011 Paseo de Peralta. The works displayed here are so fine you'll feel as though you're in a museum. You'll find 19th- and 20th-century American painting and sculpture, featuring art of Georgia O'Keeffe, William Wegman, and the founders of the Santa Fe and Taos artist colonies.

Mayans Gallery Ltd.
601 Canyon Rd. ☎ **505/983-8068.**

Established in 1977, this is one of the oldest galleries in Santa Fe; you'll find 20th-century American and Latin American paintings, photography, prints, and sculpture. E-mail: arte2@aol.com; web: http://www.artnet.com/mayans.html.

✪ Nedra Matteucci's Fenn Galleries
1075 Paseo de Peralta. ☎ **505/982-4631.**

When you enter this gallery, note the beautifully sculpted wall that surrounds it, merely a taste of what's to come. The gallery specializes in 19th- and 20th-century American art. Inside you'll find a lot of high-ticket works such as those of early Taos and Santa Fe painters, as well as classic American impressionism, historical Western modernism, and contemporary Southwestern landscapes and sculpture, including monumental pieces displayed in the sculpture garden.

✪ Owings-Dewey Fine Art
76 E. San Francisco St., upstairs. ☎ **505/982-6244.**

These are treasure-filled rooms. You'll find 19th- and 20th-century American painting and sculpture including works by Georgia O'Keeffe, Robert Henri, Maynard Dixon, Fremont Ellis, and Andrew Dasburg, as well as antique works such as Spanish colonial *retablos, bultos,* and tin works. Don't miss the Day of the Dead exhibition around Halloween.

3 More Shopping A to Z

ANTIQUES

El Paso Import Company
418 Cerrillos Rd. ☎ **505/982-5698.**

Whenever I'm in the vicinity of this shop, I always browse through. It's packed—and I mean packed—with colorful, weathered colonial and ranchero furniture. The home furnishings and folk art here are imported from Mexico and have a primitive feel.

Scarlett's Antique Shop & Gallery
225 Canyon Rd. ☎ **505/983-7092.**

A good place to browse in search of bizarre finds, this shop is packed with Early American antiques, fine crystal, vintage Hollywood jewelry, pre-1920 postcards, collected western books, and jewelry "confections" by international artist Helga Wagner.

BELTS

Caballo
727 Canyon Rd. ☎ **505/984-0971.**

The craftspeople at Caballo fashion "one of a kind, one at a time" custom-made belts. Everything is hand-tooled, hand-carved, and hand-stamped. The remarkable buckles are themselves worthy of special attention. This shop merits a stop.

BOOKS

Collected Works Bookstore

208-B W. San Francisco St. ☎ **505/988-4226.**

This is a good downtown book source, with a carefully chosen selection of books up front, in case you're not sure what you want, and shelves of southwest, travel, nature, and other books.

Horizons—The Discovery Store

328 S. Guadalupe St. ☎ **505/983-1554.**

Here you'll find adult and children's books, science-oriented games and toys, telescopes, binoculars, and a variety of unusual educational items. I always find interesting gifts for my little nieces in this store.

Nicholas Potter, Bookseller

211 E. Palace Ave. ☎ **505/983-5434.**

This bookstore handles rare and used hardcover books, as well as tickets to many local events.

CRAFTS

Davis Mather Folk Art Gallery

141 Lincoln Ave. ☎ **505/983-1660.**

This small shop is a wild animal adventure. You'll find New Mexican animal wood carvings in shapes of lions, tigers, and bears—even chickens, as well as other folk and Hispanic arts.

Gallery 10

225 Canyon Rd. ☎ **505/983-9707.**

"Important art by native peoples" is how this gallery dubs its offerings, And they're right. This is definitely museum-quality Native American pottery, weavings, basketry, and contemporary paintings and photography. My favorite potter, Tammy Garcia, has work here when she's not sold out of it.

Kent Galleries

130 Lincoln Ave. ☎ **505/988-1001.**

One of the city's best showcases for (very) fine crafts and home furnishings, the gallery is, as the local writer John Villani put it, "a great place to find that $7,000 rocking chair you've been dreaming of." There are also ceramics, paintings, and jewelry.

✪ Nambe Mills, Inc.

924 Paseo de Peralta (at Canyon Rd.). ☎ **505/988-5528.**

Here you'll find cooking, serving, and decorating pieces, fashioned from an exquisite sand-cast and handcrafted alloy. Also available at **Plaza Mercado,** 104 W. San Francisco St. (☎ **505/988-3574**), and **216A Paseo del Pueblo Norte** (Yucca Plaza), Taos (☎ **505/758-8221**).

FASHIONS

Dewey & Sons Trading Company

53 Old Santa Fe Trail. ☎ **505/983-5855.**

Look for Native American trade blankets and men's and women's apparel here.

Jane Smith Ltd.
554 Canyon Rd. ☎ **505/988-4775.**

This is the place for flashy western-style clothing. You'll find boots with more colors on them than there are crayolas and jackets made by Elvis's very own designer. There are also flamboyant takes on household goods such as bedding and furniture.

Judy's Unique Apparel
714 Canyon Rd. ☎ **505/988-5746.**

Judy's has eclectic separates made locally or imported from around the globe. You'll find a wide variety of items here, many at surprisingly reasonable prices.

Origins
135 W. San Francisco St. ☎ **505/988-2323.**

A little like a Guatemalan or Turkish marketplace, this store is packed with wearable art, folk art, and work of local designers. Look for good buys on ethnic jewelry. Throughout the summer there are trunk shows, with a chance to meet the artists.

Rancho
554 Canyon Rd. ☎ **505/986-1688.**

Here you'll find authentic, comfortable, functional western wear. This store features clothing by Schaefer Outfitter for both men and women, as well as by the Great American Cowboy and Wild Mustangs.

FOOD

Chile Shop
109 E. Water St. ☎ **505/983-6080.**

This store has too many cheap trinketlike items for me. But many find novelty items to take back home. You'll find everything from salsas to cornmeal and tortilla chips. The shop also stocks cookbooks and pottery items.

Cookworks Gourmet
322 S Guadalupe St. ☎ **505/988-7676.**

For the chef or merely the wannabe, this is a fun place for browsing. You'll find inventive food products and cooking items spread across three shops. There's also gourmet food and cooking classes.

Coyote Cafe General Store
132 Water St. ☎ **505/982-2454.**

This store is an adjunct to one of Santa Fe's most popular restaurants. The big thing here is the enormous selection of hot sauces; however, you can also get fresh fruits and vegetables, a wide variety of Southwestern food items, T-shirts, and aprons.

Señor Murphy Candy Maker
100 E. San Francisco St. (La Fonda Hotel). ☎ **505/982-0461.**

This candy store is unlike any you'll find in other parts of the country—everything here is made with local ingredients. The chile piñon nut brittle is a taste sensation! Señor Murphy has another shop at 223 Canyon Rd. (☎ **505/983-9243**).

FURNITURE

Southwest Spanish Craftsmen
328 S. Guadalupe St. ☎ **505/982-1767.**

The Spanish colonial and Spanish provincial furniture, doors, and home accessories in this store are a bit too elaborate for my tastes, but if you find yourself drooling over carved wood, this is your place.

Taos Furniture
232 Galisteo St. (P.O. Box 5555). ☎ **505/988-1229.**

Classic Southwestern furnishings handcrafted in solid Ponderosa pine—both contemporary and traditional pieces.

GIFTS & SOUVENIRS

El Nicho
227 Don Gaspar Ave. ☎ **505/984-2830.**

For the thrifty art shopper this is the place to be. Inside the funky Santa Fe village, you'll find handcrafted Navajo and Oaxacan folk art as well as carvings, jewelry, and other items by local artisans.

Thea
612A Agua Fría St. ☎ **505/995-9618.**

This new shop is so rich and enticing that you won't want to leave. Owned by Svetlana Britt, a beautiful Russian woman, it features candles, scents, and aromatherapy. These are excellent gift items, colorfully and elaborately packaged. Named for the goddess of light, the whole store has a luminous quality.

JEWELRY

Packards
61 Old Santa Fe Trail. ☎ **505/983-9241.**

Opened by a notable trader, Al Packard, later sold to new owners, this store on the Plaza is worth checking out to see some of the best jewelry available. You'll also find exquisite rugs and pottery.

Tresa Vorenberg Goldsmiths
656 Canyon Rd. ☎ **505/988-7125.**

You'll find some wildly imaginative designs in this jewelry store where more than 30 artisans are represented. All items are handcrafted and custom commissions are welcomed.

MALLS & SHOPPING CENTERS

de Vargas Center Mall
N. Guadalupe St. and Paseo de Peralta. ☎ **505/982-2655.**

There are more than 55 merchants and restaurants in this mall just northwest of downtown. This is Santa Fe's small, struggling mall. Though there are fewer shops than Villa Linda, this is where I shop because I don't tend to get the mall phobia I get in the more massive places. Open Monday through Thursday from 10am to 7pm, Friday from 10am to 9pm, Saturday from 10am to 6pm, and Sunday from noon to 5pm.

Sanbusco Market Center
500 Montezuma St. ☎ **505/989-9390.**

Unique shops and restaurants occupy this remodeled warehouse near the old Santa Fe Railroad Yard. Though most of the shops in this little mall are overpriced, it's a fun place to window-shop. There's a farmers' market in the south parking lot. Open from 7am to noon on Tuesday and Saturday in the summer.

Villa Linda Mall

4250 Cerrillos Rd. (at Rodeo Rd.). ☎ **505/473-4253.**

Santa Fe's largest mall (including department stores) is near the southwestern city limits, not far from the I-25 on-ramp. If you're from a major city, you'll probably find shopping here very provincial. Anchors include JCPenney, Sears, Dillard's, and Mervyn's. Open Monday through Friday from 10am to 9pm, Saturday from 10am to 6pm, and Sunday from noon to 5pm.

MARKETS

Farmers' Market

In the parking lot of Sanbusco Market Center, 500 Montezuma St. No phone.

Every Saturday and Tuesday from 7 to 11:30am, you'll find a farmers' market in the parking lot of Sanbusco Market Center. Everything is here from fruits, vegetables, and flowers to cheeses, cider, and salsas.

Trader Jack's Flea Market

US 84/285 (about 8 miles north of Santa Fe). No phone.

If you're a flea-market hound, you'll be happy to find Trader Jack's. More than 500 vendors here sell everything from used cowboy boots (you might find some real beauties) to clothing, jewelry, books, and furniture. The flea market is open from mid-April to late November on Friday, Saturday, and Sunday.

NATURAL ART

Mineral & Fossil Gallery of Santa Fe

127 W. San Francisco St. ☎ **800/762-9777** or 505/984-1682.

You'll find ancient artwork here, from fossils to geodes in all sizes and shapes. There's also natural mineral jewelry and decorative items for the home, including lamps, wall clocks, furniture, art glass, and carvings.

POTTERY & TILES

Artesanos Imports Company

222 Galisteo St. ☎ **505/983-1743.**

This is like a trip south of the border, with all the scents and colors you'd expect on such a journey. You'll find a wide selection of Talavera tile and pottery, as well as light fixtures and many other accessories for the home. There's even an outdoor market where you can buy fountains and chile ristras.

Santa Fe Pottery

323 S. Guadalupe St. ☎ **505/989-3363.**

The work of more than 50 master potters from New Mexico and the Southwest is on display here. You'll find everything from mugs to lamps.

RUGS

Seret & Sons Rugs and Tapestries

149 E Alameda St. and 232 Galisteo St. ☎ **505/988-9151** or 505/983-5008.

If you're like me and find Middle Eastern decor irresistible, you need to wander through either of these shops. You'll find kilims and Persian and Turkish rugs, as well as some of the Moorish-style ancient doors and furnishings that you see around Santa Fe.

WINES

The Winery

500 Montezuma St. ☎ **505/982-WINE.**

Perhaps the best-stocked wine shop in New Mexico, The Winery also carries gourmet foods, beers, and gift baskets.

9

Santa Fe After Dark

Santa Fe is a city committed to the arts. Its night scene is dominated by highbrow cultural events, beginning with the world-famous Santa Fe Opera; the club and music scene run a distant second.

Complete information on all major cultural events can be obtained from the **Santa Fe Convention and Visitors Bureau** (☎ 800/777-CITY or 505/984-6760) or from the **City of Santa Fe Arts Commission** (☎ 505/984-6707). Current listings are published each Friday in the "Pasatiempo" section of *The New Mexican,* the city's daily newspaper, and in the *Santa Fe Reporter,* published every Wednesday.

Nicholas Potter, Bookseller, 203 E. Palace Ave. (☎ 505/983-5434), carries tickets to select events. You can order by phone from TicketMaster (☎ 505/842-5387 for information, 505/884-0999 to order). Discount tickets may be available on the night of a performance; the opera, for example, offers standing-room tickets at a greatly reduced price just 1 hour before curtain time.

A variety of free concerts, lectures, and other events are presented in the summer, cosponsored by the City of Santa Fe and the Chamber of Commerce. In 1997 the **El Corazón de Santa Fe** ("the heart of the city") program featured Saturday night musical and cultural events on the Plaza, and the city hopes to continue them into 1998.

The **Santa Fe Summer Concert Series,** at the Paolo Soleri Outdoor Amphitheatre on the campus of the Santa Fe Indian School (Cerrillos Road), has brought such name performers as B. B. King, Frank Zappa, and Kenny Loggins to the city. More than two dozen concerts and special events are scheduled each summer.

Note: Many companies noted here perform at locations other than their listed addresses, so check the site of the performance you plan to attend.

1 The Performing Arts

No fewer than 24 performing-arts groups flourish in this city of 600,000. Many of them perform year-round, but others are seasonal. The acclaimed Santa Fe Opera, for instance, has just a 2-month summer season: July and August.

MAJOR PERFORMING ARTS COMPANIES
OPERA & CLASSICAL MUSIC

✪ Santa Fe Opera

P.O. Box 2408, Santa Fe, NM 87504-2408. ☎ **800/280-4654** or 505/986-5900 for tickets. Tickets $20–$106 Mon–Thurs; $26–$112 Fri–Sat. Wheelchair seating $14 Mon–Thurs; $20 Fri–Sat. Standing room (sold on day of performance beginning at 10am) $6 Mon–Thurs; $8 Fri–Sat; $15 Opening Night Gala. Backstage tours: First Mon in July to last Fri in Aug, Mon–Sat at 1pm; $6 adults, free for children 15 and under.

Even if your visit isn't timed to coincide with the opera season, you shouldn't miss seeing the company's open-air amphitheater. Located on a wooded hilltop 7 miles north of the city off US 84/285, the sweeping curves of this serene structure seem perfectly attuned to the contour of the surrounding terrain. At night, the lights of Los Alamos can be seen in the distance under clear skies.

Many rank the Santa Fe Opera second only to the Metropolitan Opera of New York as the finest company in the United States today. Established in 1957 by John Crosby—still the opera's artistic director—it consistently attracts famed conductors, directors, and singers (the list has included Igor Stravinsky). At the height of the season the company is 500 strong, including the skilled craftspeople and designers who work on the sets. The opera company is noted for its performances of the classics, little-known works by classical European composers, and American premieres of 20th-century works.

The 9-week, 40-performance opera season runs from late June through late August. In 1998, you can see such classics as Puccini's *Madama Butterfly* and Mozart's *The Magic Flute.* Usually there is a Strauss opera; in 1998 it will be *Salome.* And for the less traditional-minded, there's *Beatrice and Benedict* by Hector Berlioz and *A Dream Play* by Ingvar Lidholm. All performances begin at 9pm. At press time, the theater was undergoing a major renovation, but will be open for the 1998 season.

ORCHESTRAL & CHAMBER MUSIC

Oncydium Chamber Baroque

210 E. Marcy St., Suite 15, Santa Fe, NM 87501. ☎ **505/988-0703.**

This new chamber ensemble presents Renaissance, classical, and baroque concerts six times during the year at various gallery spaces in Santa Fe. On Sunday in August the group performs at brunch and afternoon teas. Call for information and schedules, but expect to have a hard time getting someone on the phone.

Santa Fe Pro Musica Chamber Orchestra & Ensemble

320 Galisteo, Suite 502 (P.O. Box 2091), Santa Fe, NM 87504-2091. ☎ **505/988-4640.** Tickets $7–$35.

This chamber ensemble performs everything from Bach to Vivaldi and contemporary masters. During Holy Week the Santa Fe Pro Musica presents its annual Baroque Festival Concert. Christmas brings candlelight chamber ensemble concerts. Pro Musica's season runs October through April.

✪ Santa Fe Symphony and Chorus

P.O. Box 9692, Santa Fe, NM 87504. ☎ **505/983-1414.** Tickets $10–$35 (six seating categories).

This 60-piece professional symphony orchestra has grown rapidly in stature since its founding in 1984. Matinee and evening performances of classical and popular works are presented in a subscription series at Sweeney Center (Grant Avenue at Marcy

Street) from August to May. There's a preconcert lecture before each performance. During the spring there are music festivals (call for details).

Serenata of Santa Fe

P.O. Box 8410, Santa Fe, NM 87504. ☎ **505/989-7988.** Tickets $10 general admission, $15 reserved seats.

This professional chamber-music group specializes in bringing lesser-known works of the masters to the concert stage. Concerts are presented from September to May at the Santuario de Nuestra Señora de Guadalupe (100 S. Guadalupe St.). Call the number above for dates and details.

CHORAL GROUPS

Desert Chorale

219 Shelby St. (P.O. Box 2813), Santa Fe, NM 87501. ☎ **800/905-3315** (ProTix) or 505/988-7505. Tickets $18–$34 adults, half price for students.

This 24- to 30-member vocal ensemble, New Mexico's only professional choral group, recruits members from all over the country. It's nationally recognized for its eclectic blend of both Renaissance melodies and modern avant-garde compositions. During summer months the chorale performs classic concerts at both the historic Santuario de Nuestra Señora de Guadalupe and the Loretto Chapel, as well as smaller cameo concerts at more intimate settings throughout Santa Fe and Albuquerque. The chorale also performs a popular series of Christmas concerts during December. Most concerts begin at 8pm (3 or 6pm on Sunday).

Sangre de Cristo Chorale

P.O. Box 4462, Santa Fe, NM 87502. ☎ **505/662-9717.** Tickets $18–$35 Christmas, for catered dinner and concerts; $10 ($8 at the door) in the spring.

This 34-member ensemble has a repertoire ranging from classical, baroque, and Renaissance works to more recent folk music and spirituals. Much of it is presented a cappella. The group gives concerts in Santa Fe, Los Alamos, and Albuquerque. The Christmas dinner concerts are extremely popular.

Santa Fe Women's Ensemble

424 Kathryn Place, Santa Fe, NM 87501. ☎ **505/983-2137.** Tickets $13 and $16, students $8 and $12.

This choral group of 12 semiprofessional singers, sponsored by the Santa Fe Concert Association (see below), offers classical works sung a cappella as well as with varied instrumental accompaniment during the spring and fall season. Both the "Christmas Offering" concerts (in mid-December) and the annual "Spring Offering" concerts are held in the Loretto Chapel (Old Santa Fe Trail at Water Street). Call for tickets.

MUSIC FESTIVALS & CONCERT SERIES

Santa Fe Chamber Music Festival

239 Johnson St. (P.O. Box 853), Santa Fe, NM 87504. ☎ **505/983-2075** or 505/982-1890 for the box office (after June 23). Tickets $32–$36.

The festival brings an extraordinary group of international artists to Santa Fe every summer. Its 6-week season of some 50 concerts runs from mid-July through mid-August and is held in the beautiful St. Francis Auditorium. Each festival season features chamber-music masterpieces, new music by a composer in residence, jazz, free youth concerts, preconcert lectures, and open rehearsals. Performances are Monday and Wednesday through Friday at 8pm and Saturday and Sunday at 6pm. Open rehearsals, youth concerts, and preconcert lectures are free to the public.

Santa Fe Concert Association

P.O. Box 4626, Santa Fe, NM 87502. ☎ **800/905-3315** or 505/984-8759 for tickets. Tickets $15–$65.

Founded in 1938, the oldest musical organization in Northern New Mexico has a September-to-May season that includes approximately 15 annual events. Among them are a distinguished artists' series featuring renowned instrumental and vocal soloists and chamber ensembles, a special Christmas Eve concert, and sponsored performances by local artists. All performances are held at the St. Francis Auditorium; tickets are sold by Nicholas Potter, Bookseller (see "Books" under "More Shopping A to Z" in chapter 8).

THEATER COMPANIES

Greer Garson Theater Center

College of Santa Fe, St. Michael's Dr. ☎ **505/473-6511** or 505/473-6439. Tickets $6–$12 adults, $5–$11 students and seniors ($22–$35 for summer season Santa Fe Stages performances).

In this beautiful, intimate theater, the college's Performing Arts Department produces four plays annually, with five presentations of each, given between October and May. Usually the season consists of a comedy, a drama, a musical, and a classic.

✪ Santa Fe Community Theatre

142 E. de Vargas St., Santa Fe, NM 87501. ☎ **505/988-4262.** Tickets $10 adults, $8 students and seniors; on Sun people are asked to "pay what you wish."

Founded in the 1920s, this is the oldest extant theater group in New Mexico. Still performing in a historic adobe theater in the Barrio de Analco, it attracts thousands for its dramas, avant-garde theater, and musical comedy. Its popular one-act melodramas call on the public to boo the sneering villain and swoon for the damsel in distress.

Santa Fe Stages

105 East Marcy St., Suite 107. ☎ **505/982-6683.** Tickets $22–$35.

I highly recommend these performances staged at two theaters on the College of Santa Fe campus. Artistic director Martin Platt does his best to bring internationally known works, from comedy to dance, to Santa Fe. The theaters are small and intimate, adding to the audience's enjoyment.

Shakespeare in Santa Fe

355 E. Palace Ave. (box office only), Santa Fe, NM 87501. ☎ **505/982-2910.** Tickets free. Reserved seating available for a donation.

Every Friday, Saturday, and Sunday during July and August, in the library courtyard of St. John's College (southeast of downtown—off Camino del Monte Sol), Shakespeare in Santa Fe presents Shakespeare in the Park.

DANCE COMPANIES

✪ María Benitez Teatro Flamenco

Institute for Spanish Arts, P.O. Box 8418, Santa Fe, NM 87501. ☎ **800/905-3315** or 505/982-1237 for tickets. Tickets $16–$27 (subject to change).

This is a performance you won't want to miss. True flamenco is one of the most thrilling of dance forms, displaying the inner spirit and verve of the gypsies of Spanish

Andalusia, and María Benitez, trained in Spain, is a fabulous performer. The Benitez Company's "Estampa Flamenca" summer series is performed nightly except Tuesday from July through early September. At press time, the María Benitez Theater at the Radisson Hotel was being completely redesigned.

MAJOR CONCERT HALLS & ALL-PURPOSE AUDITORIUMS

Center for Contemporary Arts

1050 Old Pecos Trail. ☎ **505/982-1338.** Tickets for films $6.

The Center for Contemporary Arts (CCA) presents the work of internationally, nationally, and regionally known contemporary artists in art exhibitions, dance, new music concerts, poetry readings, performance-art events, theater, and video screenings. The CCA Cinématique screens films from around the world nightly, with special series presented regularly. A permanent outdoor James Turrell Skyspace is located on the CCA grounds. The CCA Warehouse/Teen Project is a unique program designed to encourage creativity, individuality, and free expression by giving teens a safe, free place to create programs and events, including workshops, art exhibitions, a radio show and publication, theater ensemble, cafe (with open mike opportunities), and concerts featuring local teen bands. CCA's galleries are open Monday through Friday from noon to 8pm and Saturday from 1 to 8pm.

Paolo Soleri Amphitheatre

At the Santa Fe Indian School, 1501 Cerrillos Rd. ☎ **505/989-6318.** For tickets, call Santa Fe Indian School at 505/989-6318 or contact Ticketmaster at 505/884-0999.

This outdoor arena is the venue for many warm-weather events, with a large number of concerts presented here each summer. In recent years the facility has attracted such big-name acts as Joan Armatrading, the Grateful Dead, B. B. King, Kenny Loggins, Anne Murray, Suzanne Vega, Ziggy Marley, Lyle Lovett, Dave Matthews, Allan Parsons Project, and the Reggae Sunsplash. For information on scheduled performers while you're in Santa Fe, contact Big River Corporation, P.O. Box 8036, Albuquerque, NM 87198 (☎ 505/256-1777).

✪ St. Francis Auditorium

In the Museum of Fine Arts, Lincoln and Palace aves. ☎ **505/827-4455.** Ticket prices vary; see above for specific performing-arts companies.

This beautiful music hall, patterned after the interiors of traditional Hispanic mission churches, is noted for its excellent acoustics. The hall hosts a wide variety of musical events, including the Santa Fe Chamber Music Festival in July and August. The Santa Fe Symphony Festival Series, the Santa Fe Concert Association, the Santa Fe Women's Ensemble, and various other programs are also held here.

Sweeney Convention Center

201 W. Marcy St. ☎ **800/777-2489** or 505/984-6760. Tickets $10–$30, depending on seating and performances. Tickets are never sold at Sweeney Convention Center; event sponsors handle ticket sales.

Santa Fe's largest indoor arena hosts a wide variety of trade expositions and other events during the year. It's also the home of the Santa Fe Symphony Orchestra and the New Mexico Symphony Orchestra's annual Santa Fe Series.

2 The Club & Music Scene

In addition to the clubs and bars listed below, there are a number of hotels whose bars and lounges feature some type of entertainment. (See chapter 5, "Where to Stay in Santa Fe.")

COUNTRY, JAZZ & FOLK

Cowgirl Hall of Fame

319 S Guadalupe St. ☎ **505/982-2565.** No cover for music in the main restaurant. Other performances $2–$7.

It's difficult to categorize what goes on at night at this bar and restaurant. Some nights there's blues guitar, others there's comedy, and others there's flamenco music and dance. You might also find something called cowboy poetry or live theater. In the summer this is a great place to sit under the stars and listen to music or see some romping fun entertainment inside.

✪ El Farol

808 Canyon Rd. ☎ **505/983-9912.** Cover $3–$5.

The original neighborhood bar of the Canyon Road artists' quarter (its name means "the lantern") is the place to head for local ambience. Its low ceilings and dark brown walls are the home of Santa Fe's largest and most unusual selection of *tapas* (bar snacks and appetizers), from pulpo à la Gallega (octopus with Spanish paprika sauce) to grilled cactus with ramesco sauce. Jazz, folk, and ethnic musicians—some of national note—perform most nights.

Fiesta Lounge

In La Fonda Hotel, 110 E. San Francisco St. ☎ **505/982-5511.** No cover.

This lively lobby bar offers cocktails and live entertainment nightly.

Rodeo Nites

2911 Cerrillos Rd. ☎ **505/473-4138.** Cover $2.50 Mon–Wed, $3 Fri–Sat. Closed Sun.

There's live country dance music nightly at this popular club.

ROCK & DISCO

Chelsea Street Pub & Grill

In the Villa Linda Mall, Rodeo and Cerrillos rds. ☎ **505/473-5105.** No cover.

Burgers and beer are served here during the lunch and dinner hours, but when the shopping mall closes at 9pm, the pub really starts hopping. Recently, it became a sports bar, with a number of televisions on order. Rather noisy college students frequent the place.

The Drama Club

125 N Guadalupe St. ☎ **505/988-4374.**

This place is doing its best to present Santa Fe with a club scene. Some nights work better than others. On nights when there's a deejay, it can be quite dull, but other nights there are live performances by jazz, rock-and-roll, and world music groups, and the place lights up. There are also stage acts ranging from drag queen shows to nationally known comedy acts. You'll find pool tables up front. The crowd here varies greatly, but it's mostly a gay and lesbian club.

3 The Bar Scene

Evangelo's

200 W. San Francisco St. ☎ **505/982-9014.** No cover.

Food is not served at Evangelo's, but the tropical decor and mahogany bar are unique to Santa Fe. More than 250 varieties of imported beer are available, and pool tables are an added attraction. On Saturday night starting at 9pm, live bands play; some nights there's jazz, others rock, and others reggae. Evangelo's is extremely popular with the local crowd. You'll find your share of business people, artists, and even bikers here. Open daily from noon until 1 or 2am.

Dana's After Dark

222 N. Guadalupe. ☎ **505/982-5225.** No cover.

This after-hours club caters to late-night lesbian and gay crowds, but it's such a cool place I'd recommend it for anyone. Opened in the spring of 1997, it's in an old adobe house that has been painted inside with wild colors like summer squash orange and lime green. Open Tuesday through Saturday from 6pm to 4am, it's the place for late-night food, fine coffee, and carefully prepared desserts. You can listen to a variety of types of music from "ultra-lounge music" to cha-cha and mambo. Or you can go into the music room and play piano or conga drums, or into the game room and play retro-board games from the '60s.

Vanessie of Santa Fe

434 W. San Francisco St. ☎ **505/982-9966.** No cover.

This is unquestionably Santa Fe's most popular piano bar. The talented Doug Montgomery and Charles Tichenor have a loyal local following. Their repertoire ranges from Bach to Billy Joel, Gershwin to Barry Manilow. They play Monday through Saturday from 8:30pm to 2am and Sunday from 8pm to midnight. There's an extra microphone, so if you're daring (or drunk), you can stand up and accompany the piano and vocals. National celebrities have even joined in—including Harry Connick, Jr. (and he wasn't even drunk). Vanessie's offers a great bar menu.

10

Excursions from Santa Fe

Native American pueblos and ruins, a national monument and national park, Los Alamos (the A-bomb capital of the United States), and the scenic and fascinating High Road to Taos are all easy day trips from Santa Fe. A longer drive will take you to Chaco Culture National Historic Park (well worth the trip).

1 Exploring the Northern Pueblos

Of the eight northern pueblos, Tesuque, Pojoaque, Nambe, San Ildefonso, San Juan, and Santa Clara are within about 30 miles of Santa Fe. Picuris (San Lorenzo) is on the High Road to Taos (see section 3, below), and Taos Pueblo is just outside the town of Taos.

The six pueblos described in this section can easily be visited in a single day's round-trip from Santa Fe. Plan to focus most of your attention on San Juan and Santa Clara, including the former's arts cooperative and the latter's Puye cliff dwellings.

Certain **rules of etiquette** should be observed in visiting the pueblos. These are personal dwellings and/or important historic sites—and they must be respected as such. Don't climb on the buildings or peek into doors or windows. Don't enter sacred grounds, such as cemeteries and kivas. If you attend a dance or ceremony, remain silent while it is taking place and refrain from applause when it's over. Many pueblos prohibit photography or sketches; others require you to pay a fee for a permit. If you don't respect the privacy of the Native Americans who live at the pueblo, you'll be asked to leave (I've seen it happen more than once).

TESUQUE PUEBLO

Tesuque (Te-*soo*-keh) Pueblo is located about 9 miles north of Santa Fe on US 84/285. You will know that you are approaching the pueblo when you see the unusual Camel Rock and a large roadside casino. Despite this concession to the late 20th century, the 400 pueblo dwellers are faithful to their traditional religion, rituals, and ceremonies. Excavations confirm that a pueblo has existed here at least since the year A.D. 1200; accordingly, this pueblo is now on the National Register of Historic Places. A mission church and adobe houses surround the Plaza, the area in which visitors are asked to remain.

Some Tesuque women are skilled potters; Ignacia Duran's black-and-white and red micaceous pottery and Teresa Tapia's miniatures and pots with animal figures are especially noteworthy. The **San Diego Feast Day,** featuring buffalo, deer, flag, or Comanche dances, is November 12.

The Tesuque Pueblo address is Route 5, Box 360-T, Santa Fe, NM 87501 (☎ **505/983-2667**). Admission to the pueblo is free; however, there is a $20 charge for still cameras; special permission is required for movie cameras, sketching, and painting. The pueblo is open daily from 9am to 5pm. Camel Rock Casino (☎ 505/ 984-8414) is open 24 hours and has a snack bar on the premises. In addition, Tesuque Pueblo provides an RV and campground park (☎ 505/455-2661), which is open year-round.

POJOAQUE PUEBLO

About 6 miles farther north on US 84/285, at the junction of NM 502, is Pojoaque (Po-*hwa*-keh). Though small (population 200) and without a definable village (more modern dwellings exist now), Pojoaque is important as a center for traveler services; in fact, Pojoaque, in its Tewa form, means "water-drinking place." The historical accounts of the Pojoaque people are sketchy, but we do know that in 1890 small-pox took its toll on the Pojoaque population, forcing most of the Pueblo residents to abandon their village. Since the 1930s the population has gradually increased, and in 1990 a war chief and two war captains were appointed. Today visitors won't find much to look at, but the Poeh Center, operated by the pueblo, features a museum and crafts store. Indigenous pottery, embroidery, silverwork, and beadwork are available for sale at the Pojoaque Pueblo Tourist Center.

A modern community center is located near the site of the old pueblo and church. **Our Lady of Guadalupe Day,** the annual feast day celebrated on December 12, features a bow-and-arrow or buffalo dance.

The pueblo's address is Route 11, Box 71, Santa Fe, NM 87501 (☎ **505/ 455-2278**). Admission is free. Contact the Governor's Office for information about sketching and camera fees. The pueblo is open every day during daylight hours.

NAMBE PUEBLO

Drive east about 3 miles from Pojoaque on NM 503, then turn right at the Bureau of Reclamation sign for Nambe Falls. Approximately 2 miles farther is Nambe ("mound of earth in the corner"), a 700-year-old Tewa-speaking pueblo (population 450), with a solar-powered tribal headquarters, at the foot of the Sangre de Cristo Range. Only a few original pueblo buildings remain, including a large round kiva, used today in ceremonies. Pueblo artisans make woven belts, beadwork, and brown micaceous pottery.

Nambe Falls make a stunning three-tier drop through a cleft in a rock face about 4 miles beyond the pueblo, tumbling into Nambe Reservoir. A recreational site at the reservoir offers fishing, boating (nonmotor boats only), hiking, camping, and picnicking. The **Waterfall Dances** on July 4 and the **Saint Francis of Assisi Feast Day** on October 4, which has an elk dance ceremony, are observed at the sacred falls.

The address is Route 1, Box 117, Santa Fe, NM 87501 (☎ **505/455-2036** or 505/455-2304 for the Ranger Station). Admission to the pueblo is free, but there is a $5 charge for still cameras, $10 for movie cameras, and $10 for sketching. At the recreational site, the charge for fishing is $6 per day for adults, $4 per day for children; for camping it is $8 per night. The pueblo is open daily from 8am to 5pm. The recreational site is open March, September, and October from 7am to 7pm, April and May from 7am to 8pm, and June through August from 6am to 8pm.

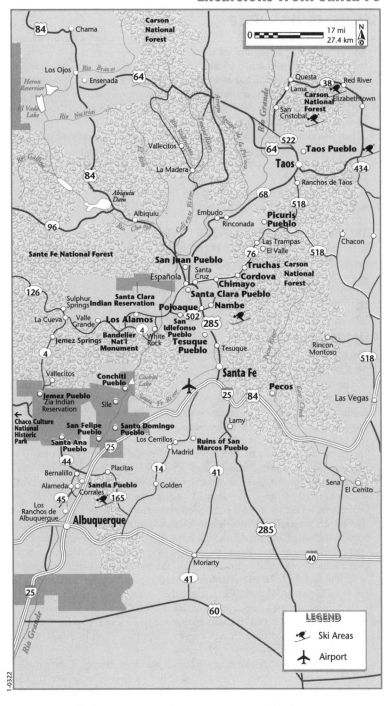

SAN ILDEFONSO PUEBLO

If you turn left on NM 502 at Pojoaque, it's about 6 miles to the turnoff to this pueblo, nationally famous for its matte-finish black-on-black pottery, developed by tribeswoman María Martinez in the 1920s. The pottery-making process is explained at the **San Ildefonso Pueblo Museum,** where exhibits of pueblo history, arts, and crafts are presented Monday through Friday. A couple of Westerns were filmed here in the 1940s. San Ildefonso is one of the most-visited pueblos in Northern New Mexico, attracting more than 20,000 visitors a year.

San Ildefonso Feast Day (January 23) is a good time to observe the social and religious traditions of the pueblo, when buffalo, deer, and Comanche dances are presented. Dances may be scheduled for Easter, the **Harvest Festival** (early September), and Christmas.

The pueblo has a 4½-acre fishing lake that is open April through October. Picnicking is encouraged; camping is not.

The pueblo's address is Route 5, Box 315A, Santa Fe, NM 87501 (☎ **505/455-3549**). The admission charge is $3 for a noncommercial vehicle and $10 for a commercial vehicle, plus 50¢ per passenger. The charge for using a still camera is $5; for a video camera or sketching, it is $15. If you plan to fish, the charge is $8 for adults, $4 for children 6 to 12 years of age, and free for children under 6. The pueblo is open in the summer, daily from 8am to 5pm; call for weekend hours. In the winter, it is open Monday through Friday from 8am to 4:30pm. It is closed for major holidays and tribal events.

SAN JUAN PUEBLO

If you continue north on US 84/285, you will reach the pueblo via NM 74, a mile off NM 68, about 4 miles north of Española.

The largest (population 1,950) and northernmost of the Tewa-speaking pueblos and headquarters of the Eight Northern Indian Pueblos Council, San Juan is located on the east side of the Rio Grande—opposite the 1598 site of San Gabriel, the first Spanish settlement west of the Mississippi River and the first capital of New Spain. In 1598, the Spanish, impressed with the openness and helpfulness of the people of San Juan, decided to establish a capital there (it was moved to Santa Fe 10 years later)—making San Juan Pueblo the first to be subjected to Spanish colonization. The Indians were generous, providing food, clothing, shelter, and fuel—they even helped sustain the settlement when its leader Conquistador Juan de Oñate became preoccupied with his search for gold and neglected the needs of his people. Unfortunately, the Spanish subjugation of the Indians left them virtual slaves, forced to provide the Spanish with corn, venison, cloth, and labor. They were compelled to participate in Spanish religious ceremonies and to abandon their own religious practices. Under no circumstances were Indian ceremonials allowed; those caught participating in them were punished. In 1676 several Indians were accused of "sorcery" and jailed in Santa Fe. Later they were led to the Plaza, where they were flogged or hanged. This despicable incident became a turning point in Indian-Spanish relations, generating an overwhelming feeling of rage in the Indian community. One of the accused, a San Juan Pueblo Indian named Pop, became a leader in the Great Pueblo Revolt, which led to freedom from Spanish rule for 12 years.

The past and present cohabit here. The San Juan tribe, though Roman Catholics, still practice traditional religious rituals; thus, two rectangular kivas flank the church in the main plaza, and *caciques* (pueblo priests) share power with civil authorities. The annual **San Juan Fiesta** is June 23 and 24, with buffalo and Comanche dances. Another annual ceremony is the **turtle dance** on December 26.

The address of the pueblo is P.O. Box 1099, San Juan Pueblo, NM 87566 (☎ **505/852-4400**). Admission is free. Photography or sketching may be allowed with prior permission from the Governor's Office. The charge for fishing is $10 for adults and $5 for children and seniors. The pueblo is open every day during daylight hours.

The **Eight Northern Indian Pueblos Council** (☎ **505/852-4265**) is a sort of chamber of commerce and social-service agency.

A crafts shop, **Oke Oweenge Arts and Crafts Cooperative** (☎ **505/852-2372**), specializes in local wares. This is a fine place to seek out San Juan's distinctive red pottery, a lustrous ceramic incised with traditional geometric symbols. Also displayed for sale are seed, turquoise, and silver jewelry; wood and stone carvings; indigenous clothing and weavings; embroidery; and paintings. Artisans often work on the premises; visitors can watch. The co-op is open Monday through Saturday from 9am to 5pm (but is closed San Juan Feast Day). **Sunrise Crafts,** another crafts shop, is located to the right of the co-op. There you'll find one-of-a-kind handcrafted pipes, beadwork, and burned and painted gourds.

Right on the main road that goes through the pueblo is the **Tewa Indian Restaurant,** serving traditional pueblo chile stews, breads, blue-corn dishes, posole, teas, and desserts. It's open Monday through Friday from 9am to 2:30pm; closed holidays and feast days.

Fishing and picnicking are encouraged at the San Juan Tribal Lakes, open year-round. As with most of the other pueblos, San Juan offers bingo. In summer, the doors open Wednesday through Sunday at 5:30pm; in winter, on Sunday at noon.

SANTA CLARA PUEBLO

Close to Española (on NM 5), Santa Clara, with a population of about 1,600, is one of the largest pueblos. If you contact the pueblo a week in advance, you can take one of the driving and walking tours offered Monday through Friday, including visits to the pueblo's historic church and artists' studios. Visitors can enter specified studios to watch artists making baskets and highly polished red-and-black pottery.

There are corn and harvest dances on **Santa Clara Feast Day** (August 12); other special days include buffalo and deer dances (early February) and children's dances (December 28).

The famed **Puye Cliff Dwellings** (see below) are on the Santa Clara reservation.

The pueblo's address is P.O. Box 580, Española, NM 87532 (☎ **505/753-7326**). Admission is free. The charge for still cameras is $5; for movie cameras and sketching, the charge is $15. The pueblo is open every day during daylight hours; the visitor center is open Monday through Friday from 9am to 4:30pm.

PUYE CLIFF DWELLINGS

In the 13th century the Santa Clara people migrated to their home on the Rio Grande from a former home high on the Pajarito Plateau to the west. Dwellings from their previous settlement have been preserved at this site, an 11-mile climb west of the pueblo. It is believed that this site at the mouth of the Santa Clara Canyon (now a National Historic Landmark) was occupied from about 1250 to 1577.

High on a nearly featureless plateau, the volcanic tuff rises in a soft tan facade 200 feet high. Here the Anasazi found niches to build their homes. Visitors can descend via staircases and ladders from the 7,000-foot mesa top into the 740-room pueblo ruin, which includes a ceremonial chamber and community house. Petroglyphs are evident in many of the rocky cliff walls.

About 6 miles farther west is the **Santa Clara Canyon Recreational Area,** a sylvan setting for camping that is open year-round for picnicking, hiking, and fishing in ponds and Santa Clara Creek.

If you would like to visit the cliff dwellings, call ☎ **505/753-7326.** The admission is $5 for adults, $4 for children and seniors. There is an additional charge of $2 for guided tours (1 week advance notice is required). The dwellings are open daily in the summer from 8am to 8pm; daily in the winter from 9am to 4:30pm.

2   Los Alamos & the Ancient Cliff Dwellings of Bandelier National Monument

Pueblo tribes lived in this rugged area for well over 1,000 years, and an exclusive boys' school operated atop the 7,300-foot plateau from 1928 to 1943. Then the **Los Alamos National Laboratory** was established here in secrecy—Project Y of the Manhattan Engineer District, the hush-hush wartime program that developed the world's first nuclear weapons.

Project director J. Robert Oppenheimer, later succeeded by Norris E. Bradbury, worked with a team of 30 to 100 scientists in research, development, and production of the weapons. Today 3,000 scientists and another 4,800 support staff work at the Los Alamos National Laboratory, making it the largest employer in Northern New Mexico. Still operated by the University of California for the federal Department of Energy, its 32 technical areas occupy 43 square miles of mesa-top land.

The laboratory is known today as one of the world's foremost scientific institutions. It's still oriented primarily toward the defense industry—the Trident and Minuteman strategic warheads were created here, for example—but it has many other research programs, including studies in nuclear fusion and fission, energy conservation, nuclear safety, the environment, and nuclear waste. Its international resources include a genetic-sequence data bank, with wide implications for medicine and agriculture, and an Institute for Geophysics and Planetary Physics.

In 1995 workers at Los Alamos began preparations to build a plutonium core for a U.S. stockpile nuclear warhead—the first of its kind to be built in more than 40 years. It is reported that Los Alamos will build one or two cores a year for the U.S. Navy in order to replace older warheads.

ORIENTATION/USEFUL INFORMATION

Los Alamos is located about 35 miles west of Santa Fe and about 65 miles southwest of Taos. From Santa Fe, take US 84/285 north approximately 16 miles to the Pojoaque junction, then turn west on NM 502. Driving time is only about 50 minutes.

Los Alamos is a town of 18,000 spread over the colorful, fingerlike mesas of the Pajarito Plateau, between the Jemez Mountains and the Rio Grande Valley. As NM 502 enters Los Alamos from Santa Fe, it follows Trinity Drive, where accommodations, restaurants, and other services are located. Central Avenue parallels Trinity Drive and has restaurants, galleries, and shops, as well as the Los Alamos Historical Museum (free) and the Bradbury Science Museum (free).

The **Los Alamos Chamber of Commerce,** P.O. Box 460, Los Alamos, NM 87544 (☎ **505/662-8105;** fax 505/662-8399; e-mail: lacoc@unix.nets.com), runs a visitor center that is open Monday through Friday from 9am to 4pm and Saturday from 10am to 4pm.

EVENTS

The Los Alamos events schedule includes a **Sports Skiesta** in late March or early April; **art-and-crafts fairs** in May, August, and November; a **county fair, rodeo, and arts festival** in August; and a **triathlon** in August/September.

WHAT TO SEE & DO

Aside from the sights described below, Los Alamos offers the **Pajarito Mountain ski area,** Camp May Road (P.O. Box 155), Los Alamos, NM 87544 (☎ 505/662-SNOW), with five chairlifts—it's open only on Saturday, Sunday, Wednesday, and federal holidays; the **Los Alamos Golf Course,** 4250 Diamond Dr. (☎ 505/662-8139), at the edge of town; and the **Larry R. Walkup Aquatic Center,** 2760 Canyon Rd. (☎ 505/662-8170), the highest-altitude indoor Olympic-size swimming pool in the United States. There's even an outdoor ice-skating rink.

IN LOS ALAMOS

✪ Bradbury Science Museum

At the Los Alamos National Laboratory, 15th St. and Central Ave. ☎ **505/667-4444.** Free admission. Tues–Fri 9am–5pm, Sat–Mon 1–5pm. Closed major holidays.

This outstanding museum is the lab's public showcase. Atomic research is emphasized in a wide variety of scientific and historical displays, including more than 35 hands-on exhibits. Visitors can peruse photographs and documents depicting the earliest days of Project Y, including a 1939 letter from Albert Einstein to President Franklin D. Roosevelt suggesting research into uranium as a new and important source of energy. There are exhibits on weapons research, including an overview of the nation's nuclear arsenal; achievements in alternative-energy research, from solar and geothermal to laser and magnetic-fusion energy; biomedical research and development; computer technology; and basic research into the nature of nuclei, atoms, and molecules. Visitors can explore the museum, experiment with lasers, use computers, and view laboratory research in energy, defense, environment, and health. Self-guided exhibits have interesting hands-on features and video monitors. Educational and historical films are shown continuously.

Fuller Lodge Art Center

2132 Central Ave. ☎ **505/662-9331.** Free admission. Mon–Sat 10am–4pm, Sun 1–4pm.

Works of Northern New Mexico artists and traveling exhibitions of regional and national importance are displayed here. Four annual arts-and-crafts fairs are also held here—in May, August, October, and November.

Los Alamos Historical Museum

2132 Central Ave. ☎ **505/662-4493.** Free admission. Summer, Mon–Sat 9:30am–4:30pm, Sun 11am–5pm; winter, Mon–Sat 10am–4pm, Sun 1–4pm.

The massive log building that once housed the dining and recreation hall for the Los Alamos Ranch School for Boys is now a National Historic Landmark known as the Fuller Lodge. Its current occupants include the museum office and research archives, the Fuller Lodge Art Center (see above), and the Los Alamos County Chamber of Commerce, which doubles as a visitor information center. The museum, located in the small log-and-stone building to the north of Fuller Lodge, depicts area history, from prehistoric cliff dwellers to the present. It also has exhibits ranging from Native American artifacts to school memorabilia and an excellent new permanent Manhattan Project exhibit. The museum sponsors guest speakers and operates a tax-free bookstore. Now you can visit the museum on the World Wide Web at **http://www.vla.com/lahistory** or **http://www.losalamos.com/lahistory**.

NEARBY

✪ Bandelier National Monument

NM 4 (HCR 1, Box 1, Suite 15, Los Alamos, NM 87544). ☎ **505/672-3861.** Admission $10 per vehicle. Open every day during daylight hours. Closed Jan 1 and Dec 25.

Less than 15 miles south of Los Alamos along NM 4, this National Park Service area contains both extensive ruins of the ancient cliff-dwelling Anasazi Pueblo culture and 46 square miles of canyon-and-mesa wilderness.

After an orientation stop at the visitor center and museum to learn about the culture that flourished here between A.D. 1100 and 1550, most visitors follow a trail along Frijoles Creek to the principal ruins. The pueblo site, including an underground kiva, has been stabilized. The biggest thrill for most folks, though, is climbing hardy Ponderosa pine ladders to visit an alcove—140 feet above the canyon floor—that was once home to prehistoric people. Tours are self-guided or led by a National Park Service ranger. Be aware that dogs are not allowed on trails.

On summer nights rangers offer campfire talks about the history, culture, and geology of the area. Some summer evenings, the guided night walks reveal a different, spooky aspect of the ruins and cave houses, outlined in the two-dimensional chiaroscuro of the thin cold light from the starry sky. During the day, nature programs are sometimes offered for adults and children. The small museum at the visitor center displays artifacts found in the area.

Elsewhere in the monument area, 70 miles of maintained trails lead to more tribal ruins, waterfalls, and wildlife habitats. However, a recent fire has decimated parts of this area, so periodic closings will take place in order to allow the land to reforest.

The separate **Tsankawi** section, reached by an ancient 2-mile trail close to White Rock, has a large unexcavated ruin on a high mesa overlooking the Rio Grande Valley. The town of **White Rock,** about 10 miles southeast of Los Alamos on NM 4, offers spectacular panoramas of the river valley in the direction of Santa Fe; the White Rock Overlook is a great picnic spot.

Within Bandelier areas have been set aside for picnicking and camping. The national monument is named after the Swiss-American archaeologist Adolph Bandelier, who explored here in the 1880s.

Past Bandelier National Monument on NM4, beginning about 15 miles from Los Alamos, is **Valle Grande,** a vast meadow 16 miles in area—all that remains of a massive volcano that erupted nearly a million years ago. When the mountain spewed ashes and dust as far away as Kansas and Nebraska, its underground magma chambers collapsed, forming this great valley—the largest volcanic caldera in the world. However, lava domes that pushed up after the collapse obstruct a full view across the expanse. Valle Grande is now privately owned land.

3　Along the High Road to Taos

Unless you're in a hurry to get from Santa Fe to Taos, the High Road—also called the Mountain Road or the King's Road—is by far the most fascinating route. It runs through tiny ridgetop villages where Hispanic traditions and way of life continue much as they did a century ago.

CHIMAYO

About 28 miles north of Santa Fe on NM 84/285 is the historic weaving center of Chimayo. It's approximately 16 miles past the Pojoaque junction, at the junction of NM 520 and NM 76 via NM 503. In this small village, families—such as the

Ortegas—maintain the tradition of crafting hand-woven textiles initiated by their ancestors seven generations ago, in the early 1800s. Both **Ortega's Weaving Shop** and **Galeria Ortega** are fine places to take a close look at this ancient craft.

Today, however, many more people come to Chimayo to visit ✪ **El Santuario de Nuestro Señor de Esquipulas** (the Shrine of Our Lord of Esquipulas), better known simply as "El Santuario de Chimayo." Ascribed with miraculous powers of healing, this church has attracted thousands of pilgrims since its construction in 1814–16. Up to 30,000 people participate in the annual Good Friday pilgrimage, many of them walking from as far away as Albuquerque.

Although only the earth in the anteroom beside the altar is presumed to have the gift of healing powers, the entire shrine radiates true serenity. It's quite moving to peruse the written testimonies of rapid recovery from illness or injury on the walls of the anteroom, and equally poignant to read the as-yet-unanswered entreaties made on behalf of loved ones.

A National Historic Landmark, the church has five beautiful *reredos,* or panels of sacred paintings, one behind the main altar and two on each side of the nave. Each year during the fourth weekend in July, the military exploits of the 9th-century Spanish saint Santiago are celebrated in a weekend fiesta, including the historic play **Los Moros y Los Cristianos** (Moors and Christians).

Lovely **Santa Cruz Lake** has a dual purpose: The artificial lake provides water for Chimayo Valley farms and also offers a recreation site for trout fishing and camping at the edge of the Pecos Wilderness. To reach it, turn south 4 miles on NM 503, about 2 miles east of Chimayo.

WHERE TO DINE

✪ Restaurante Rancho de Chimayo

CTR 98. ☎ **505/351-4444.** Reservations recommended. Lunch $6–$10; dinner $10–$15. MC, V. Daily 11:30am–9pm. Closed Mon during winter. NEW MEXICAN.

Many travelers schedule their outings in order to have lunch at this well-known restaurant. The adobe home, built by Hermenegildo Jaramillo in the 1880s, has been in the food business for about three decades. Native New Mexican cuisine, prepared from generations-old Jaramillo family recipes, is served on terraced patios and in cozy dining rooms beneath hand-stripped vigas.

CORDOVA

Just as Chimayo is famous for its weaving, the village of Cordova, about 7 miles east on NM 76, is noted for its wood-carvers. Small shops and studios along the highway display *santos* (carved saints) and various decorative items carved from aspen and cedar.

TRUCHAS

Robert Redford's 1988 movie *The Milagro Beanfield War* featured the town of Truchas (which means "trout"). A former Spanish colonial outpost built on top of an 8,000-foot mesa, 4 miles east of Cordova, it was chosen as the site for the film in part because traditional Hispanic culture is still very much in evidence. Subsistence farming is prevalent here. The scenery is spectacular: 13,101-foot Truchas Peak dominates one side of the mesa, and the broad Rio Grande Valley dominates the other.

About 6 miles east of Truchas on NM 76 is the small town of **Las Trampas,** noted for its San José Church, which some call the most beautiful of all churches built during the Spanish colonial period.

PICURIS (SAN LORENZO) PUEBLO

Near the regional education center of Peñasco, about 24 miles from Chimayo near the intersection of NM 75 and NM 76, is the Picuris (San Lorenzo) Pueblo (☎ **505/ 587-2519**). The 270 citizens of this 15,000-acre mountain pueblo, native Tiwa speakers, consider themselves a sovereign nation: Their forebears never made a treaty with any foreign country, including the United States. Thus, they observe a traditional form of tribal council government. Their **annual feast day** at San Lorenzo Church is August 10.

Still, the people are modern enough to have fully computerized their public showcase operations, Picuris Tribal Enterprises. Besides running the Hotel Santa Fe in the state capital, they own the **Picuris Pueblo Museum,** where weaving, beadwork, and distinctive reddish-brown clay cooking pottery are exhibited Monday through Friday from 9am to 6pm. Self-guided tours through the old village ruins begin at the museum; camera fees start at $5. There's also an information center, crafts shop, grocery and other shops, and a cafe that serves Pueblo and American food at lunchtime. Fishing permits ($6 for adults and children) are available, as are permits to camp at Pu-Na and Tu-Tah Lakes, regularly stocked with trout.

About a mile east of Peñasco on NM 75 is Vadito, the former center for a conservative Catholic brotherhood, the Penitentes, earlier in this century.

DIXON & EMBUDO

Taos is about 24 miles north of Peñasco via NM 518. But day-trippers from Santa Fe can loop back to the capital by taking NM 75 west from Picuris Pueblo. Dixon, approximately 12 miles west of Picuris, and its twin village Embudo, a mile farther on NM 68 at the Rio Grande, are home to many artists and craftspeople who exhibit their works during the annual **autumn show** sponsored by the Dixon Arts Association. For a taste of the local grape, you can follow signs to **La Chiripada Winery** (☎ 505/579-4437), whose product is surprisingly good. Local pottery is also sold in the tasting room. The winery is open Monday through Saturday from 10am to 5pm.

Near Dixon is the **Harding Mine,** a University of New Mexico property where visitors can gather mineral specimens without going underground. If you haven't signed a liability release at the Albuquerque campus, ask at Labeo's Store in Dixon. They'll direct you to the home of a local resident who can get you started on your hunt almost immediately.

Two more small villages lie in the Rio Grande Valley at 6-mile intervals south of Embudo on NM 68. **Velarde** is a fruit-growing center; in season, the road here is lined with stands selling fresh fruit or crimson chile ristras and wreaths of native plants. **Alcalde** is the site of Los Luceros, an early-17th-century home that is to be refurbished as an arts and history center. The unique **Dance of the Matachines,** a Moorish-style ritual brought from Spain by the conquistadors, is performed here on holidays and feast days.

ESPAÑOLA

The commercial center of Española (population 7,000) no longer has the railroad that led to its establishment in the 1880s, but it may have New Mexico's greatest concentration of **"low riders."** Their owners lavish attention on these late-model customized cars, so called because their suspension leaves them sitting quite close to the ground. The cars have inspired a unique auto subculture. You can't miss the cars—they cruise the main streets of town, especially on weekend evenings.

Sights of interest in Española include the **Bond House Museum,** a Victorian-era adobe home that exhibits local history and art; and the **Santa Cruz Church,** built

Georgia O'Keeffe & New Mexico: A Desert Romance

In June 1917, during a short visit to the Southwest, the painter Georgia O'Keeffe (born 1887) visited New Mexico for the first time. She was immediately enchanted by the stark scenery—even after her return to the energy and chaos of New York City, her mind wandered frequently to New Mexico's arid land and undulating mesas. However, not until coaxed by the arts patroness and "collector of people" Mabel Dodge Luhan 12 years later did O'Keeffe return to the multihued desert of her daydreams.

When she arrived in Santa Fe in April 1929, O'Keeffe was reportedly ill, both physically and emotionally. New Mexico seemed to soothe her spirit and heal her physical ailments almost magically. Two days after her arrival, Mabel Dodge persuaded O'Keeffe to move into her home in Taos. There she would be free to paint and socialize as she liked.

In Taos, O'Keeffe began painting what would become some of her best-known canvases—close-ups of desert flowers and objects such as cow and horse skulls. "The color up there is different . . . the blue-green of the sage and the mountains, the wildflowers in bloom," O'Keeffe once said of Taos. "It's a different kind of color from any I've ever seen—there's nothing like that in north Texas or even in Colorado." Taos transformed not only her art, but her personality as well. She bought a car and learned to drive. Sometimes, on warm days, she ran stark naked through the sage fields. That August, a new, rejuvenated O'Keeffe rejoined her husband, photographer Alfred Stieglitz, in New York.

The artist returned to New Mexico year after year, spending time with Mabel Dodge as well as staying at isolated Ghost Ranch. She drove through the countryside in her snappy Ford, stopping to paint in her favorite spots along the way. Up until 1949, O'Keeffe always returned to New York in the fall. Three years after Stieglitz's death, though, she relocated permanently to New Mexico, spending each winter and spring in Abiquiu and each summer and fall at Ghost Ranch. Georgia O'Keeffe died in Santa Fe in 1986.

---

in 1733 and renovated in 1979, which houses many fine examples of Spanish colonial religious art. Major events include the July **Fiesta de Oñate,** commemorating the valley's founding in 1596; the October **Tri-Cultural Art Festival** on the Northern New Mexico Community College campus; and the weeklong **Summer Solstice** celebration staged in June by the nearby Ram Das Puri ashram of the Sikhs (☎ **505/ 753-6341**).

Complete information on Española and the vicinity can be obtained from the **Española Valley Chamber of Commerce,** 417 Big Rock Center, Española, NM 87532 (☎ **505/753-2831**).

If you admire the work of Georgia O'Keeffe, try to plan a short trip to **Abiquiu,** a tiny town at a bend of the Rio Chama, 14 miles south of Ghost Ranch and 22 miles north of Española on US 84. Once you see the surrounding terrain, it will be clear that this was the inspiration for many of her startling landscapes. Since March 1995, **O'Keeffe's adobe home** (where she lived and painted) has been open for public tours. However, a reservation must be made in advance; the charge is $20 for a 1-hour tour. A number of tours are given each week—on Tuesday, Thursday, and Friday—and a limited number of people are accepted per tour. Visitors are not permitted to take pictures. Fortunately, O'Keeffe's home remains as it was when she lived there (until 1986). Call months in advance for reservations (☎ 505/ 685-4539).

4 Pecos National Monument

About 15 miles east of Santa Fe, I-25 meanders through **Glorieta Pass,** site of an important Civil War skirmish. In March 1862, volunteers from Colorado and New Mexico, along with Fort Union regulars, defeated a Confederate force marching on Santa Fe, thereby turning the tide of Southern encroachment in the West.

Take NM 50 east to **Pecos,** a distance of about 7 miles. This quaint town, well off the beaten track since the interstate was constructed, is the site of a noted **Benedictine monastery.** About 26 miles north of here on NM 63 is the village of **Cowles,** gateway to the natural wonderland of the Pecos Wilderness. There are many camping, picnicking, and fishing locales en route.

Pecos National Historical Park (☎ 505/757-6414), about 2 miles south of the town of Pecos off NM 63, contains the ruins of a 15th-century pueblo and 17th- and 18th-century missions. Coronado mentioned Pecos Pueblo in 1540: "It is feared through the land," he wrote. With a population of about 2,000, the Native Americans farmed in irrigated fields and hunted wild game. Their pueblo had 660 rooms and many kivas. By 1620 Franciscan monks had established a church and convent. Military and natural disasters took their toll on the pueblo, and in 1838 the 20 surviving Pecos went to live with relatives at the Jemez Pueblo.

The **E. E. Fogelson Visitor Center** tells the history of the Pecos people in a well-done, chronologically organized exhibit, complete with dioramas. A 1¹/₂-mile loop trail begins at the center and continues through Pecos Pueblo and the **Misión de Nuestra Señora de Los Angeles de Porciuncula** (as the church was formerly called). This excavated structure—170 feet long and 90 feet wide at the transept—was once the most magnificent church north of Mexico City.

Pecos National Historical Park is open Memorial Day to Labor Day, daily from 8am to 6pm; the rest of the year, daily from 8am to 5pm; closed January 1 and December 25. Admission is $2.

5 Chaco Culture National Historic Park

A combination of a stunning setting and well-preserved ruins makes the long drive to Chaco Canyon worth the trip. Whether you come from the north or south, you drive in on a dusty (and sometimes muddy) road that seems to add to the authenticity and adventure of this remote New Mexico experience.

When you finally arrive, you walk through stark desert country that seems perhaps ill-suited as a center of culture. However, the ancient Anasazi people successfully farmed the lowlands and built great masonry towns, which connected with other towns over a wide-ranging network of roads crossing this desolate place.

What's most interesting here is how changes in architecture—beginning in the mid-800s when the Anasazi started building on a larger scale than they had previously—chart the area's cultural progress. They used the same masonry techniques that tribes had used in smaller villages in the region, walls one stone thick with generous use of mud mortar, but they built stone villages of multiple stories with rooms several times larger than in the previous stage of their culture. Within a century, six large pueblos were underway. This pattern of a single large pueblo with oversized rooms, surrounded by conventional villages, caught on throughout the region. New communities built along these lines sprang up. Old villages built similarly large pueblos. Eventually there were more than 75 such towns, most of them closely tied to Chaco by an extensive system of roads.

This progress led to Chaco becoming the economic center of the San Juan Basin by A.D. 1000. As many as 5,000 people may have lived in some 400 settlements in and around Chaco. As masonry techniques advanced through the years, walls rose more than four stories in height. Some of these are still visible today.

Chaco's decline after a century and a half of success coincided with a drought in the San Juan Basin between A.D. 1130 and 1180. Scientists still argue vehemently over why the sight was abandoned and where the Chacoans went. Many believe that an influx of outsiders may have brought new rituals to the region, causing a schism among tribal members. Most agree, however, that the people drifted away to more hospitable places in the region and that their descendants live among the Pueblo people today.

You may want to focus your energy on seeing **Pueblo Bonito,** the largest prehistoric Southwest Native American dwelling ever excavated. It contains giant kivas and 800 rooms covering more than 3 acres. Also, the **Pueblo Alto Trail** is a nice hike that takes you up on the canyon rim so you can see the ruins from above—in the afternoon, with thunderheads building, the views are spectacular.

This trip involves some 380 miles of driving. You may want to call ahead to inquire about road conditions because the roads can get extremely muddy. To get to Chaco from Santa Fe, take I-25 south to Bernalillo, then NM 44 northwest through Cuba to Nageezi. Turn left onto a dirt road that runs almost 30 miles south to the park's boundary. The trip takes about 3¹/₂ to 4 hours. Overnight camping is permitted year-round. Or, a nice stop on the way back is the **Jemez River Bed and Breakfast Inn** off State Highway 44, 16 miles on Highway 4 in Jemez Springs (☎ **800/ 809-3262** or 505/829-3262).

The park's address is Star NM 4, Box 6500, Bloomfield, NM 87413 (☎ **505/ 786-7014**). Admission is $4 per car, campsite extra. The visitor center is open Memorial Day through Labor Day, daily from 8am to 6pm; the rest of the year, daily from 8am to 5pm. Trails are open from sunrise to sunset.

11 Getting to Know Taos

Wedged between the western flank of the Sangre de Cristo Range and the semiarid high desert of the upper Rio Grande Valley, Taos combines nature and culture, history and progress. There's considerably less commercialization here than in the state capital.

Located just 40 miles south of the Colorado border, about 70 miles north of Santa Fe and 135 miles from Albuquerque, Taos is best known for its thriving art colony, its historic Native American Pueblo, and the nearby ski area, one of the most highly regarded in the Rockies. Taos also has several fine museums (including a brand-new one that opened in 1995) and a wide choice of accommodations and restaurants.

About 5,000 people consider themselves Taoseños (permanent residents of Taos) today. This area may have been inhabited for 5,000 years; throughout the Taos valley there are ruins that date back more than 1,000 years.

The Spanish first visited this area in 1540, colonizing it in 1598. In the last 2 decades of the 17th century, they put down three rebellions at the Taos Pueblo. During the 18th and 19th centuries Taos was an important trade center: New Mexico's annual caravan to Chihuahua, Mexico, couldn't leave until after the annual midsummer Taos Fair. French trappers began attending the fair in 1739. Even though the Plains tribes often attacked the pueblos at other times, they would attend the market festival under a temporary annual truce. By the early 1800s Taos had become a meeting place for American "mountain men," the most famous of whom, Kit Carson, made his home in Taos from 1826 to 1868.

Thoroughly Hispanic, Taos remained loyal to Mexico during the Mexican War of 1846. The town rebelled against its new U.S. landlord in 1847, even killing newly appointed Governor Charles Bent in his Taos home. Nevertheless, the town was eventually incorporated into the Territory of New Mexico in 1850. During the Civil War, Taos fell into Confederate hands for 6 weeks; afterward, Carson and two other men raised the Union flag over Taos Plaza and guarded it day and night. Since that time, Taos has had the honor of flying the flag 24 hours a day.

Taos's population declined when the railroad bypassed it in favor of Santa Fe. In 1898 two East Coast artists—Ernest Blumenschein and Bert Phillips—discovered the dramatic, varied effects of sunlight

Taos Downtown & Area

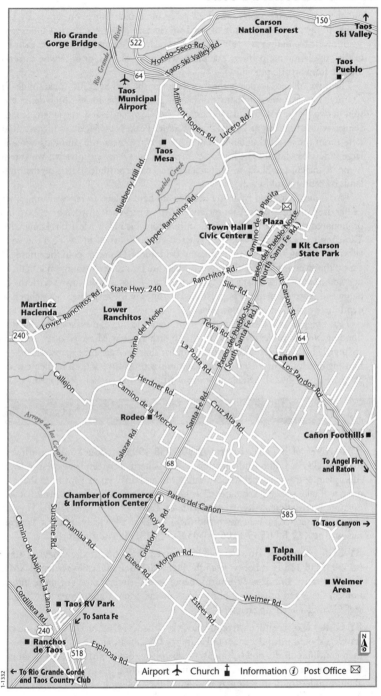

Airport ✈ Church ■ Information ⓘ Post Office ⊠

on the natural environment of the Taos valley and depicted them on canvas. By 1912, thanks to the growing influence of the **Taos Society of Artists,** the town had gained a worldwide reputation as a cultural center. Today it is estimated that more than 15% of the population are painters, sculptors, writers, musicians, or otherwise earn their income from artistic pursuits.

The town of Taos is merely the focal point of the rugged 2,200-square-mile Taos County. Two features dominate this sparsely populated region: the high desert mesa, split in two by the 650-foot-deep chasm of the Rio Grande; and the Sangre de Cristo Range, which tops out at 13,161-foot Wheeler Peak, New Mexico's highest mountain. From the forested uplands to the sage-carpeted mesa, the county is home to a large variety of wildlife. The human element includes Native Americans who are still living in ancient pueblos and Hispanic farmers who continue to irrigate their farmlands by centuries-old methods.

Taos is also inhabited by many people who have chosen to retreat from, or altogether drop out of, mainstream society. There's a laid-back attitude here, even more pronounced than the general *mañana* attitude for which New Mexico is known. Most Taoseños live here to play here—and that means outdoors. Many work at the ski area all winter (skiing whenever they can) and work for raft companies in the summer (to get on the river as much as they can). Others are into rock climbing, mountain biking, and backpacking. That's not to say that Taos is just a resort town. With the Hispanic and Native American populations' histories in the area, there's a richness and depth here that most resort towns lack.

Taos's biggest task these days is to try to stem the tide of overdevelopment that is flooding Northern New Mexico. In the "Northern New Mexico Today" section of chapter 1, I addressed the city's success in battling back airport expansion and some housing developments. A grassroots community program has been implemented recently that will give all neighborhoods a say in how their area is developed.

1 Orientation

ARRIVING

BY PLANE   The **Taos Airport** (☎ 505/758-4995) is about 8 miles northwest of town on US 64. Call for information on local charter services. It's easiest to fly into Albuquerque International Airport, rent a car, and drive up to Taos from there. The drive will take you approximately 2¹/₂ hours. If you'd rather be picked up at Albuquerque International Airport, call **Pride of Taos** (☎ 505/758-8340). They offer charter bus service to Taos town and the Taos Ski Valley daily. **Faust's Transportation, Inc.** (☎ 800/535-1106 or 505/758-3410) offers a similar service.

BY BUS   The Taos Bus Center, Paseo del Pueblo Sur at the Chevron station (☎ 505/758-1144), is not far from the Plaza. **Greyhound/Trailways** and **TNM&O Coaches** arrive and depart from this depot several times a day. For more information on these and other local bus services to and from Albuquerque and Santa Fe, see "Getting There," in chapter 2.

BY CAR   Most visitors arrive in Taos via either NM 68 or US 64. Northbound travelers should exit I-25 at Santa Fe, follow US 285 as far as San Juan Pueblo, and then continue on the divided highway when it becomes NM 68. Taos is about 79 miles from the I-25 junction. Southbound travelers from Denver on I-25 should exit about 6 miles south of Raton at US 64 and then follow it about 95 miles to Taos. Another major route is US 64 from the west (214 miles from Farmington).

VISITOR INFORMATION

The **Taos County Chamber of Commerce,** at the junction of US 64 and NM 68 (P.O. Drawer I), Taos, NM 87571 (☎ **800/732-TAOS** or 505/758-3873), is open year-round, daily from 9am to 5pm; closed major holidays. **Carson National Forest** also has an information center in the same building as the chamber. On the Internet you can access information about Taos at http://taoswebb.com/ nmusa/TAOS. The e-mail address for the Taos County Chamber of Commerce is taos@taoswebb.com.

CITY LAYOUT

The Plaza is a short block west of Taos's major intersection—where US 64 (Kit Carson Road) from the east joins NM 68, **Paseo del Pueblo Sur** (also known as South Pueblo Road or South Santa Fe Road). US 64 proceeds north from the intersection as **Paseo del Pueblo Norte** (North Pueblo Road). **Camino de la Placita** (Placitas Road) circles the west side of downtown, passing within a block of the other side of the Plaza. Many of the streets that join these thoroughfares are winding lanes lined by traditional adobe homes, many of them over 100 years old.

Most of the art galleries are located on or near the Plaza, which was paved over with bricks several years ago, and along neighboring streets. Others are located in the **Ranchos de Taos** area a few miles south of the Plaza.

MAPS   To find your way around town, pick up a free copy of the Taos map from the **Chamber of Commerce at Taos Visitor Center,** 1139 Paseo del Pueblo Sur (☎ 505/758-3873). Good, detailed city maps can be found at area bookstores as well (see "More Shopping A to Z," in chapter 8).

2 Getting Around

BY BUS & TAXI

If you're in Taos without a car, you're in luck because there is now a local bus service, provided by **Taos Transit** (☎ 505/751-2000). It begins at around 6:30am and ends at about 11pm. The route runs from Kachina Lodge on Paseo del Pueblo Norte and ends at the Ranchos Post Office on the south side of town. Bus fares are 50¢ one-way, $1 all day, and $5 for a 7-day pass.

In addition, Taos has two private bus companies. **Pride of Taos** (☎ 505/ 758-8340) has a shuttle bus service linking town hotels and Taos Ski Valley; they run four times a day for $10 round-trip. A night bus ($10 with a 10-person minimum) brings skiers staying at the ski valley into town for dinner and returns them to their lodgings.

Faust's Transportation (☎ 505/758-3410) offers town taxi service daily from 7am to 9pm, with fares of $7 anywhere within the city limits for up to two people ($2 per additional person), $35 to Albuquerque International Airport, and $30 to Taos Ski Valley from Taos town.

BY CAR

With offices at the Taos airport, **Dollar** (☎ 800/369-4226 or 505/758-9501) is reliable and efficient. Other car-rental agencies are available out of Albuquerque. See chapter 15 for details.

PARKING   Parking can be difficult during the summer rush, when the stream of tourists' cars moving north and south through town never ceases. If you can't find parking on the street or in the Plaza, check out some of the nearby roads (Kit Carson Road, for instance) because there are plenty of metered and unmetered lots in Taos town.

WARNING FOR DRIVERS   Reliable paved roads lead to starting points for side trips up poorer forest roads to many recreation sites. Once you get off the main roads, you won't find gas stations or cafes. Four-wheel-drive vehicles are recommended on snow and much of the otherwise unpaved terrain of the region. If you're doing some off-road adventuring, it's wise to go with a full gas tank, extra food and water, and warm clothing—just in case. At the higher-than-10,000-foot elevations of Northern New Mexico, sudden summer snowstorms are not unheard of.

ROAD CONDITIONS   Information on road conditions in the Taos area can be obtained free from the **State Police** (☎ 505/758-8878 within New Mexico). Also, for highway conditions throughout the state, call the **State Highway Department** (☎ 800/432-4269).

BY BICYCLE/ON FOOT

Bicycle rentals are available from **Gearing Up Bicycle Shop,** 129 Paseo del Pueblo Sur (☎ 505/751-0365); daily rentals run $20 for a mountain bike with front suspension. **Hot Tracks Cyclery & Ski Touring Service,** 214 Paseo del Pueblo Sur (☎ 505/751-0949), rents front-suspension mountain bikes and full-suspension bikes for $15/half day and $20/full day. **Native Sons Adventures,** 715 Paseo del Pueblo Sur (☎ 800/753-7559 or 505/758-9342), rents unsuspended bikes for $15/half day and $20/full day, front-suspension bikes for $25/half day and $35/full day, and full-suspension bikes for $35/half day and $45/full day; they also rent car racks for $5. Each shop supplies helmets and water bottles with rentals.

Most of Taos's attractions can easily be reached by foot since they are within a few blocks of the Plaza.

FAST FACTS: Taos

Airport   See "Orientation," above.

Area Code   The telephone area code for all of New Mexico is 505.

Business Hours   Most businesses are open Monday through Friday from 10am to 5pm, though some may open an hour earlier and close an hour later. Many tourist-oriented shops are also open on Saturday morning, and some art galleries are open all day Saturday and Sunday, especially during peak tourist seasons. Banks are generally open Monday through Thursday from 9am to 3pm and Friday from 6am to 6pm. Call establishments for specific hours.

Car Rentals   See "Getting Around," above.

Climate   Taos's climate is similar to that of Santa Fe. Summer days are dry and sunny, except for frequent afternoon thunderstorms. Winter days are often bracing, with snowfalls common but rarely lasting too long. Average summer temperatures range from 50°F to 87°F. Winter temperatures vary between 9°F and 40°F. Annual rainfall is 12 inches; annual snowfall is 35 inches in town, 300 inches at Taos Ski Valley, elevation 9,207 feet. (A foot of snow is equal to an inch of rain.)

Currency Exchange   Foreign currency can be exchanged at the Centinel Bank of Taos, 512 Paseo del Pueblo Sur (☎ 505/758-6700).

Dentists If you need dental work, try Dr. Walter Jakiela, 536 Paseo del Pueblo Norte (☎ **505/758-8654**); Dr. Michael Rivera (☎ **505/758-0531**); or Dr. Tom Simms, 623-B Paseo del Pueblo Sur (☎ **505/758-8303**).

Doctors Members of the Taos Medical Group, on Weimer Road (☎ **505/ 758-2224**), are highly respected. Also recommended are Family Practice Associates of Taos, on Don Fernando Street (☎ **505/758-3005**), a short distance west of the Plaza.

Driving Rules See "Getting Around," above.

Drugstores See "Pharmacies," below.

Embassies/Consulates See chapter 3, "Fast Facts: For the Foreign Traveler."

Emergencies Dial ☎ **911** for police, fire, and ambulance.

Eyeglasses Taos Eyewear, in Cruz Alta Plaza (☎ **505/758-8758**), handles most needs Monday through Friday between 8:30am and 5pm. It also has emergency service.

Hospital Holy Cross Hospital, 1397 Weimer Rd., off Paseo del Canyon (☎ **505/ 758-8883**), has 24-hour emergency service. Serious cases are transferred to Santa Fe or Albuquerque.

Hotlines The crisis hotline (☎ **505/758-9888**) is available for emergency counseling.

Information See "Visitor Information," above.

Library The Taos Public Library, off Camino de la Placita (☎ **505/758-3063**), has a general collection for Taos residents, a children's library, and special collections on the Southwest and Taos art.

Liquor Laws As in Santa Fe, bars must close by 2am Monday through Saturday and can be open only between noon and midnight on Sunday. The legal drinking age is 21.

Lost Property Check with the police (☎ **505/758-2216**).

Newspapers/Magazines The *Taos News* (☎ **505/758-2241**) and the *Sangre de Cristo Chronicle* (☎ **505/377-2358**) are published every Thursday. *Taos Magazine* is also a good source of local information. The *Albuquerque Journal,* the *New Mexican* from Santa Fe, and the *Denver Post* are easily obtained at the Fernandez de Taos Bookstore on the Plaza.

Pharmacies There are several full-service pharmacies in Taos. Furr's Pharmacy (☎ **505/758-1203**), Smith's Pharmacy (☎ **505/758-4824**), and Wal-Mart Pharmacy (☎ **505/758-2743**) are all located on Pueblo Sur and are easily seen from the road.

Photographic Needs Check Plaza Photo, Taos Main Plaza (☎ **505/758-3420**). Minor camera repairs can be done the same day, but major repairs must be sent to Santa Fe and usually require several days. Plaza Photo offers a full line of photo accessories and 1-hour processing. April's 1-Hour Photos, at 613E N. Pueblo Rd. (☎ **505/758-0515**), is another good choice.

Police In case of emergency, dial ☎ **911**. All other inquiries should be directed to Taos Police, Civic Plaza Drive (☎ **505/758-2216**). The Taos County Sheriff, with jurisdiction outside the city limits, is located in the county courthouse on Paseo del Pueblo Sur (☎ **505/758-3361**).

Post Offices   The main Taos Post Office is at 318 Paseo del Pueblo Norte, Taos, NM 87571 (☎ **505/758-2081**), a few blocks north of the Plaza traffic light. There are smaller offices in Ranchos de Taos (☎ **505/758-3944**) and at El Prado (☎ **505/758-4810**). The ZIP code for Taos is 87571.

Radio   Local stations are KRZA-FM (88.7); the National Public Radio Station— KKIT-AM (1340)—for news, sports, weather, and a daily event calendar at 6:30am (☎ **505/758-2231**); and KTAO-FM (101.7), which broadcasts an entertainment calendar daily (☎ **505/758-1017**).

Taxes   Gross receipts tax for Taos town is 6.8125%, and for Taos County it's 6.3125%. There is an additional local bed tax of 3.5% in Taos town and 3% on hotel rooms in Taos County.

Television   Channel 2, the local access station, is available in most hostelries. For a few hours a day there is local programming. Cable networks carry Santa Fe and Albuquerque stations.

Time   As is true throughout New Mexico, Taos is in the Mountain time zone. It's 2 hours earlier than New York, 1 hour earlier than Chicago, and 1 hour later than Los Angeles. Daylight saving time is in effect from early April to late October.

Useful Telephone Numbers   For information on road conditions in the Taos area, call the state police at ☎ **505/758-8878** or dial ☎ **800/432-4269** (within New Mexico) for the state highway department. Taos County offices are at ☎ **505/ 758-8834.**

Where to Stay in Taos

12

A tiny town with a big tourist market, Taos has some 2,100 rooms in 87 hotels, motels, condominiums, and bed-and-breakfasts. Many new properties have recently opened, turning this into a buyer's market. In the slower season, you may even want to try bargaining your room rate down, because competition for travelers is steep. Most of the hotels and motels are located on Paseo del Pueblo Sur and Norte, with a few scattered just east of the town center along Kit Carson Road. The condos and bed-and-breakfasts are generally scattered throughout Taos's back streets.

During peak seasons, visitors without reservations may have difficulty finding a vacant room. **Taos Central Reservations,** P.O. Box 1713, Taos, NM 87571 (☎ **800/821-2437** or 505/758-9767), might be able to help.

One thousand or so of Taos County's beds are in condominiums and lodges at or near the Taos Ski Valley. The **Taos Valley Resort Association,** P.O. Box 85, Taos Ski Valley, NM 87525 (☎ **800/776-1111** or 505/776-2233; fax 505/776-8842), can book these as well as rooms in Taos, and unadvertised condominium vacancies. The World Wide Web address for Taos Valley Resort Association is http://taoswebb.com/nmusa/nmResv. If you'd rather e-mail for a reservation, the address is res@taoswebb.com.

Some three dozen bed-and-breakfasts are listed with the Taos Chamber of Commerce. The **Taos Bed and Breakfast Association** (☎ **800/876-7857** or 505/758-4747) and the **Traditional Taos Bed and Breakfast Association** (☎ **800/525-8267**), both with strict guidelines for membership, will provide information and make reservations for member homes.

Affordable Accommodations and Tours, P.O. Box 1258, Taos, NM 87571 (☎ **800/290-5384** or 505/751-1292), will help you find accommodations from bed-and-breakfasts to home rentals, hotels, and cabins throughout Taos and Northern New Mexico. They'll also help you arrange rental cars and reservations for outdoor activities such as white-water rafting, horseback riding, fishing/hunting trips, and ski packages.

Unlike in Santa Fe, there are two high seasons in Taos: winter (the Christmas-to-Easter ski season) and summer. Spring and fall are shoulder seasons, often with lower rates. The period between Easter and Memorial Day is notoriously slow in the tourist industry here, and many restaurants and other businesses take their annual vacations at this time. Book well ahead for ski holiday periods (especially Christmas) and for the annual arts festivals (late May to mid-June and late September to early October).

In these listings, the following categories have been used to describe peak-season prices for a double: **Expensive** applies to accommodations that charge over $100 per night; **Moderate** refers to $75 to $100; and **Inexpensive** indicates $75 and under. A tax of 11.38% in Taos town and 9.8125% in Taos County will be added to every hotel bill.

1 Best Bets

- **Best Historic Hotel:** The **Historic Taos Inn,** 125 Paseo del Pueblo Norte (☎ 505/ 758-2233), dates from the mid-1800s and was once home to Taos's first doctor. It was also the first place in town to install indoor plumbing. Today, though the inn maintains much of its historic detail, it is equipped with all modern conveniences. It appears on both the State and National Registers of Historic Places.
- **Best for Families:** Without doubt, Taos's best full-service hotel for families is the **Best Western Kachina Lodge de Taos,** 413 Paseo del Pueblo Norte (☎ 505/ 758-2275). The room prices are manageable, the location is excellent, there's an outdoor pool where kids can cool off during the summer, snack machines are helpful when Mom and Dad just don't feel like taking the car out again, and there are laundry facilities on the premises.
- **Best Moderately Priced Hotel:** For rooms in this category (that are not bed-and-breakfasts), you're best off with the **Sun God,** 919 Paseo del Pueblo Sur (5513 NDCBU), Taos, NM 87571 (☎ 800/821-2437 or 505/758-3162). You'll get clean rooms with a Southwestern feel on nice, grassy grounds.
- **Best Bed-and-Breakfast:** The best bed-and-breakfast in Taos is the **Adobe and Pines Inn,** Box 837, Ranchos de Taos, NM 87557 (☎ 505/751-0947). The rooms are creatively decorated, hosts Chuck and Charil Fulkerson are gracious, and the setting is lush and peaceful.
- **Best Modern Adobe Architecture:** While you'll find many bed-and-breakfasts in New Mexico in historic or old buildings that were constructed of true adobe bricks, you'd be hardpressed to find new, modern buildings using the same materials. **Little Tree Bed and Breakfast,** P.O. Box 960, El Prado, NM 87529 (☎ 505/ 776-8467) is a beautiful adobe inn. Even inside, you can see pieces of straw and mica in the walls.
- **Best Location for Skiers:** The **Inn at Snakedance,** P.O. Box 89, Taos Ski Valley, NM 87525 (☎ 505/776-2277), is a modern hotel with ski-in/ski-out privileges.

2 The Taos Area

EXPENSIVE
HOTELS/MOTELS

Fechin Inn

227 Paseo del Pueblo Norte, Taos, NM 87571. ☎ 800/811-2933 or 505/751-1000. Fax 505/ 751-7338. Web: http://www.fechin-inn.com. 85 rooms, 14 suites. AC MINIBAR TV TEL. $109– $189 double; $199–$319 suite. Rates vary according to season and include continental breakfast. AE, DC, DISC, MC, V. Free parking.

When this luxury hotel opened in the summer of 1996, word quickly spread that it hadn't lived up to expectations, and it's no wonder. People who hoped it would

Central Taos Accommodations

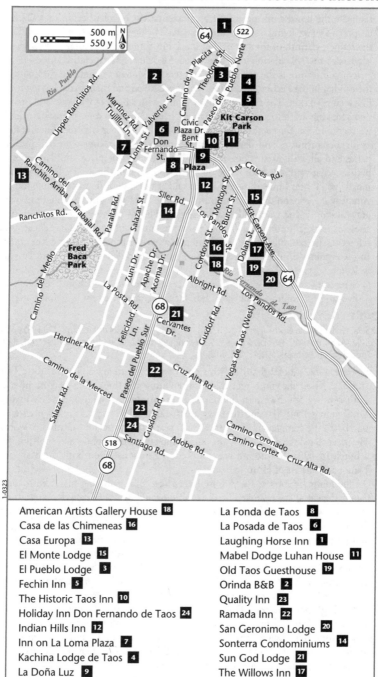

American Artists Gallery House **18**
Casa de las Chimeneas **16**
Casa Europa **13**
El Monte Lodge **15**
El Pueblo Lodge **3**
Fechin Inn **5**
The Historic Taos Inn **10**
Holiday Inn Don Fernando de Taos **24**
Indian Hills Inn **12**
Inn on La Loma Plaza **7**
Kachina Lodge de Taos **4**
La Doña Luz **9**

La Fonda de Taos **8**
La Posada de Taos **6**
Laughing Horse Inn **1**
Mabel Dodge Luhan House **11**
Old Taos Guesthouse **19**
Orinda B&B **2**
Quality Inn **23**
Ramada Inn **22**
San Geronimo Lodge **20**
Sonterra Condominiums **14**
Sun God Lodge **21**
The Willows Inn **17**

replicate the house built next door by Russian artist Nicolai Fechin in 1927 (see chapter 14) were living in the wrong century. The new hotel certainly doesn't have the refined warmth of the Inn of the Anasazi in Santa Fe. It's closer in style to the Eldorado, and considering it's a large hotel, I would call it a qualified success.

Its main feature is the carved wood by Jeremy Morelli of Santa Fe. In line with the carving in the Fechin House, much of it is beveled and then waxed, giving it the fine texture of skin. The two-story lobby is airy, with a bank of windows and French doors looking south to a portal where ristras hang and diners can eat breakfast in the warmer months. The rooms have a Southwestern motif decor with nice touches such as hickory furniture and flagstone-topped tables, many rooms with balconies or patios and all with guest robes. All the suites have kiva fireplaces and balconies or patios. What's notable here is the quiet. Surrounded on three sides by rural land, mostly you'll hear birds, though Kit Carson Park next door does send some sounds your way if you're on the south side of the hotel. The bathrooms are delightful, with gray saltillo tile and warm, adjustable lighting. There's a library where you can play chess or backgammon. Breakfast is elaborate continental with such delicacies as schnecken (pecan rolls) and almond brioche, as well as fruits, juices, and fine coffee. In the evenings drinks are served often to the sound of live entertainment. Pets are welcome.

Services: Concierge, limited room service, dry cleaning and laundry service.

Facilities: Medium-sized health club, Jacuzzi, conference rooms, Laundromat.

✪ Historic Taos Inn

125 Paseo del Pueblo Norte, Taos, NM 87571. ☎ **800/TAOS-INN** or 505/758-2233. Fax 505/ 758-5776. 36 rms. AC TV TEL. $85–$225, depending on the type of room and season. Rates include discounted breakfast. AE, DC, MC, V.

It's rare to see a hotel that has withstood the years with grace. The Historic Taos Inn has. Never do you forget that you're within the thick walls of a number of 19th-century Southwestern homes, and yet surrounded by 20th-century luxury. Dr. Thomas Paul Martin, the town's first (and for many years only) physician, purchased the complex in 1895. In 1936, a year after the doctor's death, his widow, Helen, enclosed the plaza—now the inn's darling two-story lobby—and turned it into a hotel. In 1981–82 the inn was restored; it's now listed on both the State and National Registers of Historic Places.

The lobby doubles as the **Adobe Bar,** a real local gathering place, with adobe *bancos* (benches) and a sunken fireplace, all surrounding a wishing well that was once the old town well. A number of rooms open onto a balcony that overlooks this area. If you like community and don't mind the sound of jazz and flamenco drifting up toward your room, these rooms are for you. However, if you appreciate solitude and silence, request one of the courtyard rooms downstairs. All the rooms are unique and comfortable, decorated with Spanish colonial art, Taos-style furniture, and interesting touches such as hand-woven Oaxacan bedspreads and little *nichos* often decorated with Mexican pottery, many with fireplaces; all have hair dryers.

Dining/Entertainment: Doc Martin's, with good nouveau Southwestern cuisine and a hint of Asia, is one of Taos's leading dining establishments (see chapter 13). The **Adobe Bar,** popular among Taos artists and other locals, offers live entertainment on certain weeknights as well as a full bar menu for light lunch or dinner, snacks, margaritas, and an espresso-dessert menu.

Services: Room service, baby-sitting can be arranged, free coffee or refreshments in lobby.

Facilities: VCRs on request, seasonal outdoor swimming pool, year-round Jacuzzi in greenhouse.

Quail Ridge Inn

Ski Valley Rd. (P.O. Box 707), Taos, NM 87571. ☎ **800/624-4448** or 505/776-2211. Fax 505/776-2949. 110 rms and suites. TV TEL. $75–$170 single or double; $150–$395 suite. Rates depend on season; honeymoon, bike, river raft, and golf packages available; 3% gratuity added to all rates. Extra person $10. Children under 18 stay free in parents' room. AE, CB, DISC, MC, V. Free parking.

You'll get a taste of the country at this family resort and conference center built in 1978, among the sage north of town. People enjoy its proximity both to the ski area (14 miles) and town (4 miles), as well as the indoor and outdoor tennis courts, squash and racquetball facilities, and year-round heated swimming pool, hot tubs, and saunas. Guests also enjoy the meandering walkways, surrounded by grass, that weave through this Pueblo-style complex.

Rooms are clean and nicely decorated in Southwestern style and come in a variety of types from "hotel rooms," with a queen and sleeper sofa, to fully stocked kitchenettes with stove, refrigerator, microwave oven, and dishwasher. *Beware, however:* The walls are thin, and a neighbor's television noise can blast through your quiet evening.

Dining/Entertainment: The new **Taos Kangaroo** features "a blend of east and west, a touch of down under," and is open daily for breakfast and dinner (during the off-season, the hours vary).

Facilities: Year-round heated swimming pool, health club, Jacuzzis and saunas, six outdoor and two indoor tennis courts, one squash and two racquetball courts, summer volleyball pit, Laundromat.

BED-&-BREAKFASTS

✪ Adobe and Pines Inn

NM 68 (Box 837), Ranchos de Taos, NM 87557. ☎ **800/723-8267** or 505/751-0947. Fax 505/758-8423. 4 rms, 3 casitas. TV. $95–$150 double. Rates include breakfast. MC, V.

With this inn, owners Chuck and Charil Fulkerson wanted to create a magical escape. They succeeded. Much of it is located in a 150-year-old adobe directly off NM 68, less than half a mile south of St. Francis Plaza.

The inn is set around a courtyard marked by an 80-foot-long grand portal. It's surrounded by pine and fruit trees on what's believed to be sacred land. Each room has a private entrance, fireplace (three even have a fireplace in the bathroom), and each is uniquely decorated. The theme here is the use of colors, which are richly displayed on the walls and in the furnishings. There's Puerta Azul, a cozy blue room; Puerta Verde, done in deep greens; and Puerta Turquese, a separate turquoise-painted guest cottage with a full kitchen. The two newest rooms, completed in 1996, have bold maroon and copper-yellow themes. The walls have hand prints and petroglyph motifs, and each room is furnished with rich and comfortable couches. Both have large jet tubs supplied with bubble bath and candles.

Morning brings a delicious full gourmet breakfast in front of the fire (in winter) in the glassed-in breakfast room. You can expect such delights as bansoufflé pancakes or apple caramel bread pudding; or their specialty, migas, a mixture of eggs, tortillas chips, chile, and cheese, served with salsa, turkey sausage, and a citrus fruit compote. Chuck and Charil are most gracious hosts; they will help you plan activities and make dinner reservations.

Adobe and Stars Bed and Breakfast Inn

At the corner of State Hwy. 150 and Valdez Rim Road (P.O. Box 2285), Taos, NM 87571. ☎ **800/211-7076** or 505/776-2776. 8 rms. TEL. $95–$175 double. Rates include full breakfast. AE, DISC, MC, V. Free parking.

On first appearance, this brand-new inn looks stark, sitting on the mesa between Taos town and Taos Ski Valley. However, once inside, it's apparent that the innkeeper has an eye for detail. The breakfast area and common room are sunny, with large windows facing the mountains. A few rooms are upstairs, such as La Luna, my favorite, with views in every direction and a heart-shaped Jacuzzi tub for two. All rooms have kiva fireplaces and private decks or patios. Most of the downstairs rooms open onto a por-tal. All are decorated with handcrafted Southwestern-style furniture, and many have Jacuzzi tubs. A full breakfast is served; it may vary from New Mexican dishes (such as breakfast burritos served with green chile stew) to baked goods (apple and strawberry turnovers). In the afternoons, New Mexico wines are served with the inn's special salsa and blue-corn chips, as are sweets such as chocolate cake and piñon lemon bars.

✪ Alma del Monte

372 Hondo Seco Rd., P.O. Box 1434, Taos, NM 87571. ☎ **800/273-7203**, 505/776-2721, or 505/776-8888. 5 rooms. $125–$150 single or double; prices vary with seasons and holidays. Rates include full breakfast. MC, V.

For a real Taos experience I highly recommend this bed-and-breakfast recently built on sage-covered lands bordered by fast-rising mountains and on the way to the ski area. The house is a new horseshoe-shaped, pueblo-style adobe; each room opens onto the courtyard outfitted with a fountain and hammocks hanging in the warm months. Proprietor Suzanne Head has designed and decorated the place impeccably; the living/dining room provides excellent sunrise and sunset views, and the house has saltillo tile floors and traditional antiques, as well as elegant Ralph Lauren bedding. Each room has a Jacuzzi tub, and many have picture window views and skylights; others have private gardens. Above all, the rooms are quiet, since no roads border the property. Breakfasts are equally unique: Specialties such as a deep-dish baked apple pancake or French toast stuffed with fruit and cream cheese are served on china with real silverware.

✪ Casa de las Chimeneas

405 Cordoba Lane at Los Pandos Road (Box 5303), Taos, NM 87571. ☎ **505/758-4777.** Fax 505/758-3976; e-mail: casa@newmex.com; web: http://taoswebb.com/hotel/chimneyhouse. 4 rms, 1 suite. TV TEL. $125–$140 double; $150 suite (for 2). Rates include breakfast. AE, MC, V.

This 80-year-old adobe home is a model of Southwestern elegance. Its four rooms, each with original works of art, hand-made quilts, and private entrances, look out on flower and herb gardens. Most of the rooms have dark, old vigas and are decorated with antiques and touches such as a game table complete with chess and back-gammon (in the Library Room) and talavera tile decorated with grapes (in the Blue Room). The newer Garden Room doesn't have quite the charm and elegance of the older ones, but it is sunnier. Minirefrigerators in each room are stocked with com-plimentary soft drinks, juices, and mineral water. Bathrobes are supplied in each room. A full gourmet breakfast as well as complimentary afternoon hors d'oeuvres are served daily. Massages are offered on the premises for an additional charge, and there is a large hot tub in a courtyard. Smoking is not permitted.

Inn on La Loma Plaza

315 Ranchitos Rd. (P.O. Box 4159), Taos, NM 87571. ☎ **800/530-3040** or 505/758-1717. Fax 505/751-0155. 7 rms. TV TEL. $95–$150 standard double; $170–$195 artist's studio; $320–$330 suite. Extra person $20. Children 12 and under $10 in parents' room. Discounts available. Rates include breakfast. AE, MC, V.

You may just pass by the most wonderful thing about this place if you don't look for it. The inn (formerly the Hacienda Inn) is located on a historic neighborhood plaza, complete with dirt streets and a tiny central park. It doesn't look like much, but it's a chance to glimpse a neighborhood stronghold—adobe homes built around a square, with thick outer walls to fend off marauders. The inn is made from a 200-year-old home, complete with aged vigas and maple floors, decorated tastefully with comfortable furniture and Middle Eastern rugs. Each room is unique, most with sponge-painted walls, willow shutters, and talavera tile in the baths to provide an eclectic ambience. All have fireplaces, TVs, and phones. Some have special touches, such as the Happy Trails Room, with knotty pine paneling, a brass bed and old chaps, and decorative hanging spurs. Some rooms have kitchenettes and balconies or patios. This 2-story building, a 10-minute walk from the Plaza, has a large front lawn lined with bulbs and perennials blooming in the warmer months, as well as a brick patio and Jacuzzi. Breakfast burritos and other delicacies are served in a sunroom filled with plants.

MODERATE
HOTELS

✪ Best Western Kachina Lodge

413 Paseo del Pueblo Norte (P.O. Box NN), Taos, NM 87571. ☎ **800/522-4462** or 505/758-2275. Fax 505/758-9207; e-mail: sales@kachinalodge.com; web: http://www.kachinalodge.com. 118 rms. A/C TV TEL. $80–$150 double. Extra person $8. Children under 12 stay free in parents' room. AE, DC, DISC, MC, V.

This lodge on the north end of town is an ideal spot for families and travelers. Built in the early 1960s, it has a lot of charm despite the fact that it's really a motor hotel. Rooms are placed around a grassy courtyard studded with huge blue spruce trees. In the center is a stage where a family from Taos Pueblo builds a bonfire and dances nightly in the summer and explains the significance of the dances—a real treat for anyone baffled by the Pueblo rituals.

Remodeling is ongoing in the rose-colored rooms—some have couches and most have little Taos-style *trasteros* (armoires) that hold the televisions. The rooms are solidly built and quiet, and there's plenty of outdoor space for the kids to run, as well as a Jacuzzi and a large pool area—which management assures me will get needed new carpeting by our publication date. A Laundromat, beauty salon, and shopping arcade are also on the premises, and a courtesy limo is available. No pets are allowed.

The lodge has a number of eating and entertainment options: The **Hopi Dining Room** offers a family-style menu; the **Kiva Coffee Shop** is kiva shaped (round), a unique Southwestern diner; the **Zuni Lounge** is open nightly.

Holiday Inn Don Fernando de Taos

Paseo del Pueblo Sur (P.O. Drawer V), Taos, NM 87571. ☎ **800/759-2736** or 505/758-4444. Fax 505/758-0055. 124 rms, 26 suites. A/C TV TEL. $65–$149 double. Rates depend on season; Christmas season rates higher. Extra person $10. Children 19 and under stay free in parents' room. AE, DISC, MC, V. Free parking.

If you're not a fan of surprises, this is the place to stay. Built in 1989, this hotel has the consistency you'd expect from a Holiday Inn, though it also has a few Southwestern touches. It has 2 stories, built in clusters of pueblo-style rooms, surrounded by grassy grounds. The lobby has a tacky decor accented by drums, but the rooms are more tasteful, done in warm Southwestern colors, with good beds, new carpeting, and a ski storage space in each. If you're sensitive to smells, beware of their overuse of air freshener. The hotel caters to motor coach and ski groups, conventions, as well as travelers. Each suite offers a sitting room, fireplace, and minirefrigerator.

Don Fernando's Restaurant serves a variety of regional breakfasts, lunches, and dinners. **Fernando's Hideaway** lounge, built around a large adobe fireplace, presents live entertainment and features a large hors d'oeuvre buffet.

The hotel offers room service, dry cleaning and laundry service, newspaper delivery, in-room massage, baby-sitting, secretarial services, express checkout, courtesy van, 24-hour desk, indoor swimming pool, hot tub, and tennis courts, as well as access to a nearby health club.

La Fonda de Taos

South Plaza (P.O. Box 1447), Taos, NM 87571. ☎ **800/833-2211** or 505/758-2211. Fax 505/758-8508. 24 rms, 3 suites. A/C TEL. $80 double; $120 suite. AE, MC, V.

If this hotel can make it through a difficult transition time, it could be worth the stay. Recently its flamboyant owner Saki Karavas died, and the hotel came under new ownership; the owners had big plans to remodel. Though the lobby is charming, with antique hand-carved pine furniture and bright Mexican-painted trim, all under a high ceiling with skylights, the hotel needs work. The only lodging on the Plaza, La Fonda once hosted many of Taos's most glamorous guests in the 1940s, 1950s, and 1960s. Through the years it has been known for its art collection, much of which remained with the previous owner's family. However, the strange, erotic oil paintings of D. H. Lawrence are still on view—pictures he brought to Taos in 1929 when they, along with his novel *Lady Chatterley's Lover*, were banned in England. Currently, the rooms have rather downtrodden furniture and decor; call to see if they've been upgraded, as well as if plans to reopen the bar and restaurant have taken place.

Neon Cactus Hotel

1523 Paseo del Pueblo Sur (P.O. Box 5702), Taos, NM 87571. ☎ **800/299-1258** or 505/751-1258. 4 rms. A/C. Jan, Apr–June, and Sept–Dec 18, $65–$85 double; $105 quad. Rest of the year, $85–$125 double; $125–$145 quad. Rates include continental breakfast. AE, CB, DISC, MC, V.

South of town, the Neon Cactus is a delightful change of pace from some of the other hotels and bed-and-breakfasts in New Mexico. The building, built in 1979, was remodeled in 1992 and opened as a hotel. You won't find any hint of Southwestern decor; each guest room here pays homage to a film star. For instance, there is the Marilyn Monroe Room, decorated with authentic period furnishings and photographs that document the "many moods of Marilyn." The bathroom features a heart-shaped oversized sunken tub. There's also a Rita Hayworth Room, Casablanca Room, and James Dean Room. Each is stocked with biographies of the respective star, and there's a film library downstairs in the art deco sitting room. This isn't the place for those who like newness. It has a bit of a vintage clothing store feel, and parts of it could use updating; some of the baths could use sprucing up, as could the carpet on the private decks where you do get amazing views of Taos Mountain. A Jacuzzi is available, as are discount rates at the adjacent **Taos Spa.** A newspaper is delivered, and there's free coffee and refreshments in the lobby.

Ramada Inn

615 Paseo del Pueblo Sur, Taos, NM 87571. ☎ **800/RAMADA** or 505/758-2900. Fax 505/758-1662. 124 rms. A/C TV TEL. $84–$105 double. Rollaway $15 extra. Children under 18 stay free in parents' room. AE, CB, DC, DISC, MC, V. Free parking.

This recently remodeled adobe-style hotel, a 10-minute walk from the Plaza, gives you what you'd expect from a chain hotel and that's why its very popular for travelers as well as tour groups. Built in 1995, it came under new ownership in 1996 and underwent a remodeling that gives it a warm, Southwestern feel. The rooms have new mattresses, box springs, TVs, carpet, drapes, and floral bedspreads. Though a little dark, the rooms are well constructed and quiet.

The **Fireside Cantina Restaurant and Lounge,** located just off the lobby, serves American and New Mexican food in a warm atmosphere. On Friday and Saturday evenings there's live entertainment.

The hotel offers limited room service, dry cleaning and laundry service (Monday through Friday), an indoor swimming pool, Jacuzzi, sundeck, and gift shop.

Sagebrush Inn

Paseo del Pueblo Sur (P.O. Box 557), Taos, NM 87571. ☎ **800/428-3626** or 505/758-2254. Fax 505/758-5077. 68 rms, 32 suites. A/C TV TEL. $70–$95 standard double; $90–$110 deluxe double; $95–$115 minisuite; $105–$140 executive suite. Extra person $10. Children under 12 stay free in parents' room. Rates include breakfast. AE, CB, DC, DISC, MC, V. Free parking.

Three miles south of Taos, surrounded by acres of sage, this inn is a strong example of adobe architecture, which has been added onto decade after decade, creating an interesting mix of accommodations. The edifice was built in 1929 as a stage-coach stop. The original structure had 3 floors and 12 rooms hand-sculpted from adobe; the roof was held in place with hand-hewn vigas. This part still remains, and the rooms are small but cozy—perfect for someone like Georgia O'Keeffe, who lived and worked here for 10 months in the late 1930s. But for those more accustomed to refined style, it might feel dated.

The treasure of this place is the large grass courtyard dotted with elm trees, where visitors sit and read in the warm months. Beware that some of the rooms added in the '50s through '70s have a tackiness not overcome by the vigas and tile work. More recent additions (to the west) are more skillful; these suites away from the hotel proper are spacious and full of amenities, but have noisy plumbing.

The lobby-cum-cantina has an Old West feel that livens at night with country-western dancing. Traditionally this has been a family hotel, but with a new convention center and an addition of a Comfort Suite hotel on the property, they're working to appeal to convention guests as well. With this addition there are now two outdoor pools, three Jacuzzis, tennis courts, and a business center.

Los Vaqueros Room is open for dinner daily. A complimentary full breakfast is served to guests daily in the **Sagebrush Dining Room.** The lobby bar is one of Taos's most active nightspots for live music and dancing (see chapter 14).

The hotel provides valet laundry and a courtesy van.

Sonterra Condominiums

206 Siler Rd. (P.O. Box 5244), Taos, NM 87571. ☎ **505/758-7989.** 9 suites. TV TEL. $59–$119 per day single or double, depending on the season. Extra person $10. MC, V.

This pleasant and secluded complex is located 4 blocks from the Plaza. The building was constructed in 1984 as an apartment building and converted to condominiums in 1990. Surrounded by a high adobe wall and a cedar fence, these one-story, Pueblo-style condos face a central garden courtyard with a graceful Mexican fountain. Every unit has a well-stocked full kitchen, with a refrigerator, four-burner stove, microwave oven, and coffeemaker. Each has a private outdoor patio. Three have fireplaces, for which a complimentary fire log is provided. Regional art, from paintings to Taos-style drums employed as night tables, makes each room a bit different. Some of the carpeting and bathroom tile, however, could use updating. To avoid these rooms, request one with redbrick floors. Pets are not allowed. Facilities include a hot tub and Laundromat. Daily maid service is available for an extra charge.

BED-&-BREAKFASTS

American Artists Gallery House

132 Frontier Lane (P.O. Box 584), Taos, NM 87571. ☎ **800/532-2041** or 505/758-4446. Fax 505/785-0497. 10 rms. $75–$135 double. Rates include full breakfast and afternoon snack and refreshments.

Though the exterior of this bed-and-breakfast about a 5-minute drive from the Plaza isn't much, innkeepers LeAn and Charles Clamurro make up for it on the interior, both in terms of hospitality and accommodations. Situated in a modest house, an additional small building, and a historic barn, the inn offers a variety of types of rooms. Some of the rooms, such as Gallery Rose, have rounded archways, pine floors, kiva fireplaces, and the charm only an older building can lend. The newer rooms, called Gallery A, B, and C (in the historic barn), are in much more of a new Southwestern style, with saltillo tile floors and very high ceilings. Each has a corner kiva fireplace and a Jacuzzi tub. These newer rooms don't provide the privacy I like, because the tub area is not separate from the bedroom, but instead sits up on a raised level. However, the decor is tasteful and each has a minirefrigerator and wet bar. Breakfasts are a delight here. Charles cooks up specialties such as blue-corn pancakes with blueberry sauce or a variety of egg dishes.

Casa Europa

840 Upper Ranchitos Rd. (HC 68, Box 3F), Taos, NM 87571. ☎ **800/758-9798** or 505/758-9798. 6 rms. $70–$135 double. Extra person $20. Rates include full breakfast and evening hors d'oeuvres or European pastries. MC, V.

This cream-colored Territorial-style adobe (just 1³/₄ miles west of the Plaza) under giant cottonwoods is surrounded by open pastures dotted with grazing horses and offers lovely views of the mountains in the distance. Some rooms here date from the 1700s; however, a 1983 renovation made this a contemporary 2-story luxury inn. Elegant rooms, all but one with fireplaces, each with a sitting area and full bath (two have two-person hot tubs and one has a whirlpool bath), vary in their furnishings. The regional artwork in the rooms can be purchased. There's a sitting area with a television upstairs and a common sitting room for reading and/or conversation downstairs, where coffee and pastries are offered each day between 3 and 4pm; during ski season hors d'oeuvres are served from 5 to 6pm. A full gourmet breakfast (specialties include cheese blintzes with a warm strawberry sauce and vegetarian eggs Benedict) is served each morning in the formal dining room. There's also an outdoor hot tub as well as a Swedish dry sauna. In-room massages are available at an extra charge. Smoking is not permitted.

Cottonwood Inn

2 State Rd. 230 (HCR 74, Box 24609), El Prado–Taos, NM 87529. ☎ **800/324-7120** or 505/776-5826. 7 rooms. $85–$150 double. Rates include full breakfast. MC, V.

I recommend this inn because of its cozy comfort. Built in 1947 by a flamboyant artist, it has high ceilings with vigas and almost every room has a kiva fireplace. Renovated in 1996, the inn has new owners. Kit and Bill Owen have made the place luxurious, using thick carpeting in many of the rooms, saltillo tile in the bathrooms, as well as adding Jacuzzi tubs and steam baths to some rooms. The rooms have down pillows and comforters on the beds and are decorated in a subtle, Southwestern style. The inn's location halfway between Taos and the ski area lends a pastoral quality to your stay, with a herd of sheep wandering in a meadow to the west. My favorite rooms are the ones that open into the main part of the house, but if you prefer a private entrance, you have that option too. Rooms for nonsmokers and people with disabilities are available.

Hacienda del Sol

109 Mabel Dodge Lane (P.O. Box 177), Taos, NM 87571. ☎ **505/758-0287.** Fax 505/758-5895. 9 rms. $70–$130 double. Rates include full breakfast. MC, V.

What's most unique about this property is the completely unobstructed view of Taos Mountain. The 1.2-acre property borders the Taos Pueblo; the land is pristine, and the inn has a rich history. It was once owned by art patroness Mabel Dodge Luhan, and it was here that author Frank Waters wrote *The People of the Valley*. Innkeeper Marcine Landon considers herself somewhat of a gypsy in her use of bold splashes of color throughout the place, from the gardens, where in summer tulips, pansies, and flax grow, to the rooms themselves, where bold woven bedspreads and original art lend a Mexican feel. The main house is 190 years old, so it has the wonderful curves of adobe as well as thick vigas and deep windowsills. Some guest rooms are in this section. Others range from 2 to 8 years in age. These are finely constructed and I almost recommend them over the others since they're a little more private and the bathrooms more refined. All rooms have robes and books on New Mexico. Some have minirefrigerators and cassette players. Breakfast includes specialties such as blue-corn pancakes with blueberry sauce and eggs del sol, a crustless quiche with corn and salsa, as well as wild plum piñon nut bread baked at Taos Pueblo. The Jacuzzi has a mountain view and is available for private guest use in half-hour segments. This is an excellent choice for those traveling with children—it's one of only a handful of bed-and-breakfasts that will accept children of any age.

La Doña Luz

114 Kit Carson Rd., Taos, NM 87571. ☎ **505/758-4874.** 16 rms. TV. $59–$125 double. Rates include continental breakfast. AE, DISC, MC, V.

There's a real artisan quality to this inn (formerly El Rincón Inn) just off the Plaza. The innkeepers, Nina Meyers and her son Paul "Paco" Castillo, have made it that way: Nina painted murals on doors and walls and Paco carved wood and set tile. The 200-year-old structure was once home to 19th-century cultural leader La Doña Luz Lucero de Martinez, who was most known for her hospitality. These innkeepers carry on her legacy. The inn comprises two dwellings separated by a flower-filled courtyard, where breakfast is served in warm weather. Fine art and hand-carved furnishings representing the three cultures of Taos (Indian, Spanish, and Anglo) are scattered through both houses, some of which are heirloom quality, reminiscent of the museum in the inn's adjacent store, **El Rincón** (see chapter 14).

Renovation is ongoing throughout the place. For now I recommend rooms in the main 3-story house rather than the adjacent property connected to the store. The main-house rooms are up-to-date, some with hot tubs and one with a full kitchen with amenities (stove, microwave, dishwasher, as well as blender and crockpot and a full-sized washer and dryer).

La Posada de Taos

309 Juanita Lane (P.O. Box 1118), Taos, NM 87571. ☎ **800/645-4803** or 505/758-8164. 5 rms, 1 cottage. $85–$120 double. Extra person $15. Rates include breakfast. No credit cards.

There's room for 12 guests at La Posada de Taos, a secluded adobe inn, located just 2¹/₂ blocks from the Plaza. Now owned by Bill and Nancy Swan, who have recently completed an extensive renovation, La Posada was actually the first bed-and-breakfast in the town of Taos. There are five rooms and one self-contained honeymoon cottage, all of which have been charmingly decorated in a Southwest/New England country style. International art and country pine antique furnishings predominate, and each room offers a private bath. La Casa de la Luna Miel, the cottage named for honeymooners, has its own fireplace and a queen-size bed tucked back in an alcove

t. The Beutler Room has a king-size bed, fireplace, and Jacuzzi tub. oms also have fireplaces. The sixth room has a queen-size bed, a small ...μιay, and a window that looks out to Taos Mountain. The Swans have recently added a private courtyard as well as patios with gardens for five of the six rooms. A full breakfast is served in the dining room each morning, and during the day beverages and ice (from the much-appreciated ice machine) are available.

✪ Little Tree Bed & Breakfast

P.O. Drawer II, Taos, NM 87571. ☎ **505/776-8467.** 4 rms. $80–$105 double. Rates include breakfast. DISC, MC, V.

Little Tree is one of my favorite Taos bed-and-breakfasts, partly because it's located in a beautiful, secluded setting, and partly because it's constructed with real adobe that's been left in its raw state, lending the place an authentic hacienda feel. Located 2 miles down a country road about midway between Taos and the ski area, it's surrounded by sage and piñon.

The rooms are charming and very cozy. They all have adobe floors, which are warm in the winter because radiant heat (which is healthier) has been installed here. All rooms feature queen-size beds, private baths, and access to the portal and courtyard garden, at the center of which is the "little tree" for which the inn is named. The Piñon (my favorite) and Juniper rooms are equipped with fireplaces and private entrances. The Piñon and Aspen rooms offer sunset views. The Spruce Room, Western in feeling, is decorated with beautiful quilts. All but one have a TV and VCR.

In the main building, the living room has a traditional viga-and-latilla ceiling and tierra blanca adobe (adobe that's naturally white; if you look closely at it you can see little pieces of mica and straw). Two cats entertain, and the visiting hummingbirds enchant guests as they enjoy a healthy breakfast on the portal during warmer months. On arrival, guests are treated to refreshments.

Mabel Dodge Luhan House

240 Morada Lane (P.O. Box 3400), Taos, NM 87571. ☎ **800/84-MABEL** or 505/758-9456. 18 rms (13 with bath). $75–$150 double. Extra person $17.50. Rates include breakfast. MC, V.

This inn is also called "Las Palomas de Taos" because of the throngs of doves (palomas) that live in enchanting weathered birdhouses on the property. Like so many other free spirits, they were attracted by the flamboyant Mabel Dodge (1879–1962), who came to Taos in 1916. A familiar name in these parts, she and her fourth husband, a full-blooded Pueblo named Tony Luhan, enlarged this 200-year-old home to its present size of 22 rooms in the 1920s. If you like history, and don't mind the curves and undulations it brings to a building, this is a good choice. The place has a mansion feel, evoking images of the glitterati of the 1920s—writers, artists, adventurers—sitting on the terrace under the cottonwoods drinking margaritas. The entrance is marked by a Spanish colonial–style portal.

Recently under new ownership, badly needed repairs are being made. All main rooms have thick vigas, arched Pueblo-style doorways, hand-carved doors, kiva fireplaces, and dark hardwood floors. Guest rooms in the main building feature antique furnishings. Six have fireplaces; baths are either private or shared. Eight more guest rooms were recently added with the completion of a second building. All new accommodations are equipped with fireplaces. Many educational workshops are held here throughout the year; the rooms are often reserved for participants. The Mabel Dodge Luhan House is now a National Historic Landmark.

Old Taos Guesthouse

1028 Witt Rd. (Box 6552), Taos, NM 87571. ☎ **800/758-5448** or 505/758-5448. 6 rms, 2 suites. $70–$80 double; $105–$115 suite. MC, V.

Less than 2 miles from the Plaza, this 150-year-old adobe hacienda sits on 7.5 acres and provides a cozy Northern New Mexico rural experience. Once a farmer's home, later an artist's estate, it's recently been restored by owners Tim and Leslie Reeves, who have carefully maintained the country charm: Mexican tile in the bathrooms, vigas on the ceilings, and kiva-style fireplaces in most of the rooms. Each room enters from the outside, some off the broad portal that shades the front of the hacienda, some from a grassy lawn in the back, with a view toward the mountains. Some rooms are more quaint, some more utilitarian, so make a request depending on your needs. Off the lawn is a hot-tub scheduled on the half hour. Also in back is a nature and history path, recently completed, where guests can read brief quips about chamisa bushes and acequia systems (ditch systems), and see some too. The Reeves pride themselves on their healthy breakfasts. They serve a pot of hot organic oatmeal, homemade granola (Leslie's grandmother's recipe), and baked breads and muffins, all worth lingering over in the big common room where the sun shines through wide windows. If you're lucky, Tim will pull out some of his wild skiing videos. Children are welcome; smoking is too—outside.

Orinda B&B

461 Valverde St. (4451 NDCBU), Taos, NM 87571. ☎ **800/847-1837** or 505/758-8581. Fax 505/751-4895 (call first); e-mail: orinda@newmex.com. 4 rms, 1 suite. $70–$88 double; $170 suite for 4. Extra person $15. Rates include full breakfast. AE, DISC, MC, V.

With its dramatic setting and homelike feel, this bed-and-breakfast is like staying at your aunt's house; it's not refined, but it's comfortable. It's also a 15-minute country walk to the Plaza. Surrounded by an uncommon terrain in the high desert—sprawling irrigated meadowland—the original building was constructed in the 1950s and has been added onto over the years, with remodeling ongoing. Innkeepers Cary and George Pratt share their living room (including a TV, sound system, and wood-burning stove) with guests. It has comfortable furniture and looks out on a tremendous view of llamas, horses, and Taos Mountain. The decor here, as in many of the rooms, wavers between inexpensive knickknacks that border on tacky and authentic Navajo rugs that hang from a Pueblo-style ladder near the breakfast table. The rooms are well constructed, with vigas, latillas, and saltillo tile floors, furnished in Southwestern style, some with fireplaces. A full, hot breakfast is served family style in a sunny room. Smoking is not permitted, and guests may not bring pets.

Salsa del Salto

P.O. Box 1468, El Prado, NM 87529. ☎ **800/530-3097** or 505/776-2422. Fax 505/776-2422. 10 rms. $85–$160 double. Extra person $20. MC, V.

Situated between Taos town and the Taos Ski Valley, Salsa del Salto is a good choice for those seeking a secluded retreat equipped like a country club. The main house was built in 1971 and has the openness of that era's architecture. The rooms here are tastefully decorated with pastel shades in a Southwestern motif and the beds are covered with cozy down comforters. Each room offers views of the mountains or mesas of Taos, and the private bathrooms are modern and spacious. Two new rooms are well designed, decorated in a whimsical Southwestern style, and each includes a minirefrigerator, fireplace, TV, VCR, and jet tub.

The focus here is on relaxation and outdoor activities. Salsa del Salto is the only bed-and-breakfast in Taos with a pool and hot tub as well as private tennis courts. Innkeeper Mary Hockett, a native New Mexican and avid sportswoman, is eager to share information about her favorite activities with her guests. Mary's husband, Dadou Mayer, who was born and raised in Nice, France, is an accomplished chef (and the author of *Cuisine à Taos,* a French cookbook). Dadou has also been a member of the French National Ski Team, and was named "The Fastest Chef in the

United States" when he won the Grand Marnier Ski Race. During the winter, you'll get the added bonus of an early morning briefing about ski conditions.

San Geronimo Lodge

1101 Witt Rd. (216M Paseo del Pueblo Norte #167), Taos, NM 87571. ☎ **505/751-3776.** Fax 505/751-1493. 18 rms. TV. $90–$120 double. Rates include full breakfast.

Built in 1925 in the style of a grand old lodge, this inn has high ceilings and rambling verandas, all situated on 2¹/₂ acres of grounds with views of Taos Mountain, yet it's a 5-minute drive from the Plaza. The owners—two sisters escaping the rigors of life on the East Coast—bought the lodge and did a major renovation in 1994. The only drawback here is that the structure is so immense it demands big furniture, which they haven't supplied, so there's an empty feel to some of the common areas. The guest rooms, however, are cozy and comfortable, three with kiva fireplaces, all with furniture hand-built by local craftspeople, and Mexican tile in the bathrooms. The full breakfast is tasty (though a little skimpy for my appetite), as are the snacks served in the afternoons. There's an outdoor swimming pool, Jacuzzi, as well as rooms for people with disabilities.

Willows Inn

Corner of Kit Carson and Dolan St. (NDCBU 6560), Taos, NM 87571. ☎ **800/525-8267,** 505/758-5445, or 505/758-2558. E-mail: willows@taos.newmex.com. 5 rooms. $95–$130 double. Rates include full breakfast. MC, V.

What sets this inn apart from the rest is that it sits on the estate of E. Marin Hennings (one of the Taos Society artists). Built in the 1920s, the inn has thick adobe walls and kiva fireplaces in each guest room. It lies just a half mile from the Plaza, but within a tall adobe wall it's very secluded. The grassy yard is decorated with fountains, and most impressively, shaded by the largest weeping willow trees I've ever seen.

The common area is cozy and comes equipped with a TV, VCR, and CD player with access to innkeepers Janet and Doug Camp's extensive music collection—over 450 classical and jazz CDs. The guest rooms are also cozy and have rich personality. Built around a courtyard, they have old latch doors, oak floors, and tin light fixtures. I recommend the Santa Fe Room, which has a nice sitting area in front of a fireplace and looks out toward the garden. Hennings's spacious studio still has paint splattered on the floor. It has a Jacuzzi tub in the bathroom. Smoking is permitted outdoors only.

INEXPENSIVE

Abominable Snowmansion Skiers' Hostel

Taos Ski Valley Rd., Arroyo Seco (P.O. Box 3271), Taos, NM 87571. ☎ **505/776-8298.** Fax 505/776-2107. 60 beds. $15.50 for bed; $60 private room, depending on size of accommodation and season. Rates include full breakfast in winter. DISC, MC, V.

Located in the small community of Arroyo Seco, about 8 miles north of Taos and 10 miles from the Taos Ski Valley, this lodging attracts many young people. It's clean and comfortable for those who like the advantages of dormitory-style accommodations. Toilets, shower rooms, and dressing rooms are segregated by sex. A two-story lodge room features a circular fireplace as well as a piano and games area. Lodging is also offered in traditional tepees and smaller bunkhouses in the campfire area. Tent camping is permitted outside. Guests can either cook their own meals or indulge in home-cooked fare.

El Monte Lodge

317 Kit Carson Rd. (P.O. Box 22), Taos, NM 87571. ☎ **800/828-8267** or 505/758-3171. Fax 505/758-1536. 13 rms. TV TEL. $75–$125 double. Kitchenette units $10 extra. AE, DISC, MC, V.

This charming 1930s motel set among century-old cottonwoods offers old-fashioned but homey rooms in a parklike setting. However, I'm not sure the rooms are quite worth the cost. Those farthest back from the road are nestled in woods where plenty of birds sing. These are quiet and have nice touches such as saltillo tile doorsteps. Rooms closer to Kit Carson Road will be noisier and not quite so sunny. Some have fireplaces, and some have fully stocked kitchenettes. Still, if you like a 1930s court motel feel, this is the place to stay and a good place to bring your kids since it's equipped with a Laundromat and a public barbecue grill. Pets are permitted for an extra $5.

El Pueblo Lodge

412 Paseo del Pueblo Norte (P.O. Box 92), Taos, NM 87571. ☎ **800/433-9612** or 505/758-8700. Fax 505/758-7321; e-mail: elpueblo@newmex.com; web: http://taoswebb.com/hotel/el pueblo/. 60 rms, 4 suites. TV TEL. $48 single; $65–$75 double; $105–$215 condo. Call for Christmas rates. Extra person $5–$10. Rates include continental breakfast. AE, DISC, MC, V.

Considering its location and setting, this hotel is a bargain, especially for families, although you'll want to reserve carefully. It's set on 3½ grassy, cottonwood-shaded acres, on the north end of town, a reasonable walk from the Plaza. Three buildings form a U-shape, each with its own high and low points. The oldest part to the south was once a 1950s court motel, and it maintains that cozy charm, with heavy vigas and tiny kitchenettes. The section could use some updating, but it's worth the price. To the north is a 2-story building, constructed in 1972, that seems frail and lacks charm, but provides lots of space. The newest section (to the west) is one of Taos's best bargains, with nicely constructed suites and double rooms decorated in blonde pine with kiva fireplaces and doors that open onto the center yard or, if you're on the second floor, onto a balcony. There are also fully equipped condominium units (with 1970s construction) good for large families. Pets are permitted for an extra $10. An outdoor swimming pool and hot tub are on the premises, and there's free use of laundry facilities.

Indian Hills Inn

233 Paseo del Pueblo Sur (P.O. Box 1229), Taos, NM 87571. ☎ **800/444-2346** or 505/758-4293. 55 rms. A/C TV TEL. $49–$99 double; group and package rates available. Rates include continental breakfast. AE, DISC, MC, V.

With its close location to the Plaza (3 blocks), this is a good choice if you're looking for a decent, functional night's stay. There are two sections to the hotel: one built in the 1950s, the other completed in 1996. Though the older section is currently receiving a major remodeling, at press time much of it remained mediocre, though clean and comfortable. For a few more dollars you can stay in the newer section, where the rooms are larger, the bathrooms fresher. The buildings are set around a broad lawn studded with big blue spruce trees. There are picnic tables, barbecue grills, and a pool. The hotel offers golf, ski, and rafting packages at reduced rates.

Laughing Horse Inn

729 Paseo del Pueblo Norte (P.O. Box 4889), Taos, NM 87571. ☎ **800/776-0161** or 505/758-8350. Fax 505/751-1123. 13 rms (3 with bath). TV. $49–$105 double; $105–$120 suite. AE, MC, V.

In 1924, a man named Spud Johnson bought the building now occupied by the Laughing Horse Inn (an unmistakable stucco structure with lilac-purple trim located just a mile north of the Plaza) and established a print shop known as "The Laughing Horse Press." Johnson was a central figure in the cultural development of Taos, and he often provided rooms to writers (for example, D. H. Lawrence) who needed a place to hang their hats.

> ### 🏵 Family-Friendly Hotels
>
> **Best Western Kachina Lodge** *(see page 165)*   An outdoor swimming pool and snack machines in the summer make a great late-afternoon diversion for hot, tired, and cranky kids.
>
> **El Pueblo Lodge** *(see page 173)*   A slide, year-round swimming pool, and hot tub set on 3½ acres of land will please the kids; a barbecue, some minikitchens with microwave ovens, laundry facilities, and the rates will please their parents.
>
> **Quail Ridge Inn** *(see page 163)*   The year-round swimming pool and tennis courts will keep active kids busy during the day when it's time for a parent's siesta.

Today guests who don't mind hippie charm can choose between sunny dorm rooms, cozy private rooms, a solar-heated penthouse, or guest houses. The private rooms feature sleeping lofts, cassette decks, TVs, VCRs, and some have fireplaces. Bathrooms are shared, and they are rustic, with lots of rough-sawed wood. The penthouse has a private solar bedroom and enclosed sleeping loft, a private bath, a woodstove, a big-screen TV, and video and audio decks (though my guess is this would be very hot in the summer). The guest houses include living rooms, fireplaces, private baths, and queen lofts; one has a full kitchen. These guest houses adjoin for large parties. The inn offers a common room around a big fireplace, a games room, a kitchen area where continental breakfasts (not included in the rates) are served, and a refrigerator (which is stocked for use on the honor system). There's an outdoor hot tub, a masseuse (by appointment), and mountain bikes for guests' free use. Frankly, for what you get here, the prices seem high to me.

Quality Inn

1043 Camino del Pueblo Sur, Taos, NM 87571. ☎ **800/845-0648** or 505/758-2200. Fax 505/758-9009. 99 rms, 2 suites. A/C TV TEL. $59–$99 single or double; $150 suite. Extra person $7. Children under 18 stay free in parents' room. Rates include continental breakfast. AE, CB, DC, DISC, EU, JCB, MC, V.

Another south-of-town budget hotel, this place will meet your needs while providing a relaxing stay. It's an L-shaped building built in 1974 about 2 miles from the Plaza. Remodeled in 1989, with more remodeling ongoing, it has a courtyard area with lots of grass surrounding a nice-sized pool, open year-round. The rooms have decorative touches such as mirrors framed with copperwork and R. C. Gorman prints. There are 11 units with king-size beds; two suites offer kitchenettes, and half the rooms have microwaves and minirefrigerators. Request a north-facing room for a view of grass, swimming pool, and mountains, rather than the parking lot. There's a restaurant, lounge, and hot tub.

Sun God Lodge

919 Paseo del Pueblo Sur (5513 NDCBU), Taos, NM 87571. ☎ **800/821-2437** or 505/758-3162. Fax 505/758-1716. 40 rms, 8 suites. TV TEL. $45–$99 single or double; $100–$145 casita. AE, MC, DISC, V.

For a comfortable, economical stay—with a Northern New Mexico ambience—this is my choice. This hotel, a 5-minute drive from the Plaza, has three distinct parts spread across 1½ acres of landscaped grounds. The oldest was solidly built in 1958 and has some court motel charm with a low ceiling and large windows. To update the rooms, owners added talavera tile sinks, Taos-style furniture, and new carpeting. To the south is a recently remodeled section, built in 1988. The rooms are small but have little touches that make them feel cozy, such as pink accent walls, little *nichos,* and hand-carved furnishings. In back to the east are the newest build-ings built in 1994. These are 2-story with either downstairs porches or upstairs bal-conies that run the length of the building. Some rooms have kitchenettes, which include microwaves, refrigerators, and stoves (all rooms have coffeemakers), while others have kiva fireplaces. The two rooms on the northeast corner of the property are the quietest and have the best views. There is a Jacuzzi, and pets are allowed.

Taos Motel
1798 Paseo del Pueblo Sur (P.O. Box 729F), Ranchos de Taos, NM 87557. ☎ **800/323-6009** or 505/758-2524. Fax 505/758-1989. 28 rms. TV TEL. $34–$52 double. Children under 12 stay free in parents' room. DISC, MC, V.

If you need a place to sleep and don't require much more, this is the place to stay. A strip-mall style motel with Spanish colonial touches, it was built in 1984 on the south end of town (about 3¹/₂ miles from the Plaza). Remodeling is ongoing but slow to come, so request one of the recently updated rooms. Also, street noise does carry into the rooms, so ask for one of the few in the back that are quieter. Complimen-tary coffee and continental breakfast are available. Pets are welcome.

3 Taos Ski Valley

For information on the skiing and the facilities offered at Taos Ski Valley, see "Ski-ing" in chapter 14.

EXPENSIVE
HOTELS

Alpine Village Suites
P.O. Box 917, Taos, NM 87571. ☎ **800/576-2666** or 505/776-8540. Fax 505/776-8542. 23 rms. TV TEL. Summer season, $60–$85 double (includes continental breakfast); ski season, $140–$260 double. AE, DISC, MC, V.

Alpine Village is a small village within the ski valley a few steps from the lift. Owned by John and Barbara Cottam, the complex also houses a ski shop and bar/restaurant. The Cottams began with 7 rooms, still nice rentals, above their ski shop. Each has a sleeping loft for the agile who care to climb a ladder, as well as sunny windows. The newer section has nicely decorated rooms, with attractive touches such as Mexican furniture and inventive tile work done by locals. As with most accommodations at Taos Ski Valley, the rooms are not especially quiet. Fortunately, most skiers go to bed early. All rooms have small kitchenettes equipped with stoves, microwaves, and minirefrigerators. In the newer building, rooms have fireplaces and balconies. Request a south-facing room for a view of the slopes. The hot tub has a view of the slopes as well as a fireplace.

Hotel Edelweiss
P.O. Box 83, Taos Ski Valley, NM 87525. ☎ **800/I-LUV-SKI** or 505/776-2301. Fax 505/776-2533; e-mail: edelweiss@taos.newmex.com; web: www.taosnet.com/edelweiss. 10 rms, 2 condos. TV TEL. May–Oct, hotel room $65 double; condo $125 for up to 6 people. Ski sea-son, hotel room $170 double; condo $260 for up to 6 people. AE, MC, V.

The big drawing card for this A-frame hotel built in 1973 and remodeled in 1996 is its service. The owners Timothy and Ann-Marie Wooldridge, both certified chefs, have many years of hostelry experience behind them. From the minute you drive up, they're ready to take your bags, valet park your car, check you in, and show you around. Such service extends to the meals as well—a full breakfast (included in room packages) as well as a ploughman's lunch of soup, salad, bread, and cheese, served in the sunny cafe. Families are very welcome here, and the most desirable rooms can accommodate them well, most with sleeping lofts and multiple bedrooms and living rooms. Baby-sitting can be arranged. I recommend rooms on the second and third floors, because the ground floor tends to be noisy. The upper rooms are separated from each other by stairs, so they're more quiet. All rooms have full-length robes, minirefrigerators, humidifiers, and hair dryers. There's a Jacuzzi with a view of the beginner's hill.

There is also a ski locker room with boot dryers and a full-service ski shop on the premises. Most important, with its proximity to the lifts, you can ski to and from your room. The owners, as well as others in the area, are promoting summer visits to Taos. The Edelweiss, as well as offering year-round air-shuttle service daily from Albuquerque to Taos, offers day trips to southwest destinations, such as Lake Powell, or road trips to Ojo Caliente and Bandelier National Monument.

✪ Inn at Snakedance

P.O. Box 89, Taos Ski Valley, NM 87525. ☎ **800/322-9815** or 505/776-2277. Fax 505/776-1410. 60 rms. TV TEL. June 6–Oct 6, $75 double, $95 fireplace double; Oct 7–Nov 26, closed; Nov 27–Dec 20, $125 double, $145 fireplace double; Dec 21–Dec 25, Jan 1–Jan 3, and Feb 8–Mar 29, $195 standard double, $215 fireplace double; Dec 26–Dec 31, $250 standard double, $270 fireplace double; Jan 4–Feb 7, $175 standard double, $195 fireplace double; Mar 30–Apr 5, $145 double, $165 fireplace double; April 6–June 5, closed. AE, MC, V. Free parking at Taos Ski Valley parking lot.

Located in the heart of Taos Ski Valley, the most attractive feature of this hotel is that it's literally only 10 yards from the ski lift. The original structure that stood on this site (part of which has been restored for use today) was known as the Hondo Lodge. Before there was a Taos Ski Valley, Hondo Lodge served as a refuge for fishermen, hunters, and artists. Constructed from enormous pine timbers that had been cut for a copper mining operation in the 1890s, it was literally nothing more than a place for the men to bed down for the night. The Inn at Snakedance today offers comfortable guest rooms—the quietest in the valley—many of which feature wood-burning fireplaces. All of the furnishings are modern, the decor stylish, and the windows (many of which offer mountain views) open to let in the mountain air. All rooms provide hair dryers, minirefrigerators, and wet bars (not stocked). Some rooms adjoin, connecting a standard hotel room with a fireplace room—perfect for families. Smoking is prohibited in the guest rooms and most public areas. Children over 6 are welcome.

Dining/Entertainment: The **Hondo Restaurant and Bar** offers dining and entertainment daily during the ski season (schedules vary off-season) and also sponsors wine tastings and wine dinners. Grilled items, salads, and snacks are available on an outdoor deck. The slopeside bar provides great views.

Services: Shuttle service to nearby hotels, shops, and restaurants.

Facilities: Video rentals, small health club (with hot tub, sauna, exercise equipment, and massage facilities), massage therapist on site, conference rooms, sundeck, in-house ski storage and boot dryers, convenience store (with food, sundries, video rental, and alcoholic beverages).

CONDOMINIUMS

Hacienda de Valdez

Ski Valley Rd. (Box 357), Arroyo Seco, NM 87514. ☎ **800/837-2218** or 505/776-2218. Fax 505/776-2218. 13 units. A/C TV TEL. Ski season, $150–$300; May–Oct, $85–$145. Rates are based on 4 to 6 people per unit. AE, DISC, MC, V.

Located about 8 miles from the Ski Valley, in a little river valley dotted with orchards, these pueblo-style luxury condo units are cozy and efficient. Each has a fully equipped kitchen, including microwave, stove, oven, and dishwasher. They all have fireplaces (firewood is provided), queen-size beds, and color TVs. Some are bilevel; others are trilevel. There are outdoor hot tubs and daily maid service.

Kandahar Condominiums

P.O. Box 72, Taos Ski Valley, NM 87525. ☎ **800/7556-2226** or 505/776-2226. Fax 505/776-2481. 27 units. A/C TV TEL. Ski season, $150–$350; May–Oct, $75–$125. Rates are based on 4 to 6 people per unit. AE, MC, V.

These condos have almost the highest location on the slopes—and with it, ski-in/ski-out access; you actually ski down to the lift from here. Built in the 1960s, the condos have been maintained well and are sturdy and functional. Two stories and private bedrooms allow for more privacy than most condos offer. Each unit is privately owned, so decor varies, although a committee makes suggestions to owners for upgrading. Facilities include a very small health club, a Jacuzzi, Laundromat, steam room, professional masseur, and small conference/party facility. Situated just above the children's center, it offers good access for families with young children.

Sierra del Sol Condominiums

P.O. Box 84, Taos Ski Valley, NM 87525. ☎ **800/523-3954** or 505/776-2981. Fax 505/776-2347; e-mail: sol@newmex.com; web: http://taoswebb.com/hotel/sierradelsol. 32 units. TV TEL. $65–$185. AE, DISC, MC, V. Free parking.

I have wonderful memories of these condominiums, which are just a 2-minute walk from the lift; family friends used to invite me to stay with them when I was about 10. I was happy to see that the units, built in the 1960s with additions through the years, have been well maintained. Though they're privately owned, and therefore decorated at the whim of the owners, management does inspect them every year and make suggestions. They're smartly built and come in a few sizes, which they term studio, one-bedroom, and two-bedroom. The one- and two-bedroom units have big living rooms with fireplaces and porches that look out on the ski runs. The bedroom is spacious and has a sleeping loft. There's a full kitchen with a dishwasher, stove, oven, and refrigerator. Most rooms also have microwaves, VCRs, and humidifiers, but if yours doesn't, management has some on hand to loan. There are indoor hot tubs and saunas. Two-bedroom units sleep up to six. There's also a business center, conference rooms, and a Laundromat. It's open in summer.

Twining Condominiums

P.O. Box 696, Taos Ski Valley, NM 87525. ☎ **800/828-2472** or 505/776-8873. Fax 505/776-8873. 19 units. TV TEL. $95–$355 depending on the season and number of people. AE, DISC, MC, V.

Unlike many condos, these are owned by one person, so the advantage here is consistency. These units, about a 3-minute walk from the lift, have a basic motel-room feel, with benign carpeting and decorations. There are studios—much like hotel rooms—and larger units that climb up two stories and a loft and have fireplaces. Both have fully stocked kitchens with refrigerator, stove, oven, microwave, and dishwasher. The larger units also have fireplaces and two baths. There's a Jacuzzi and sauna for use in the winter. The Twining condos are open in the summer as well.

MODERATE
LODGES & CONDOMINIUMS

Austing Haus Hotel

Taos Ski Valley Rd. (P.O. Box 8), Taos Ski Valley, NM 87525. ☎ **800/748-2932** or 505/776-2649. Fax 505/776-8751. 53 rms. TV TEL. $49–$170 double. Rates include continental breakfast. AE, CB, DC, DISC, MC, V.

About 1¹/₂ miles from the ski resort, the Austing Haus is a beautiful example of a timber-frame building. It was hand-built by owner Paul Austing, who will gladly give you details of the process. Though an interesting structure, at times it feels a bit fragile. The guest rooms have a Victorian feel; they're comfortable if a bit cutesy. Each room has its own ski locker, and there's a nice hot tub. Tasty continental cuisine is served in a sunny dining room. *Beware:* Water runs very hot from the taps—don't burn yourself.

Taos Mountain Lodge

Taos Ski Valley Rd. (P.O. Box 698), Taos Ski Valley, NM 87525. ☎ **800/530-8098** or 505/776-2229. Fax 505/776-8791. 10 suites. A/C TV TEL. Ski season, $168–$275 suite; May–Oct, $79.50–$122 suite. AE, MC, V.

About 1 mile west of the valley on the road from Taos, these loft suites (which can accommodate up to six) provide airy, comfortable lodging for a good price. Under new ownership, this lodge, built in 1990, is undergoing some renovation. I wouldn't expect a lot of privacy in these condominiums, but they're good for a romping ski vacation. Each unit has a small bedroom downstairs and a loft-bedroom upstairs, as well as a fold-out or futon couch in the living room. All rooms have microwaves and coffeemakers. Regular rooms have kitchenettes, with minirefrigerators and stoves, and deluxe rooms have more full kitchens, with full refrigerators, stoves, and ovens. There's daily maid service. The new owner hopes to add a cafe to the property by press time.

Thunderbird Lodge

P.O. Box 87, Taos Ski Valley, NM 87525. ☎ **800/776-2279** or 505/776-2280. Fax 505/776-2238. 32 rms. $99–$142 double. Seven-day Ski Week Package $1,080–$1,350 per adult, double occupancy (7 days room, 21 meals, 6 lift tickets), depending on season and type of accommodation (if lift rates go up, these rates may change a little). Seven-day Lodge Packages are also available. AE, MC, V. Free valet parking.

Owners Elisabeth and Tom Brownell's goal at this Bavarian-style lodge is to bring people together, and they accomplish it, sometimes a little too well. The lodge sits on the sunny side of the ski area. The lobby has a stone fireplace, raw pine pillars, and tables accented with copper lamps. There's a sunny room ideal for breakfast and lunch, with a bank of windows looking out toward the notorious Al's run and the rest of the ski village. Adjoining is a large bar/lounge with booths, a grand piano, and fireplace, where there's a variety of live entertainment through the winter. The rooms are small, some tiny, and noise travels up and down the halls, giving these 3 stories a dormitory atmosphere. I suggest when making reservations that you request their widest room; otherwise you may feel as though you're stuck in a train car.

Across the road, the lodge also has a chalet with larger rooms and a brilliant sunporch. Food is the big draw here. Included in your stay, you get three gourmet meals—some of the best food available in the region. For breakfast we had blueberry pancakes and bacon, with our choice from a table of continental breakfast accompaniments, and for dinner we had four courses highlighted by rack of lamb Provençale. You can eat at your own table or join the larger communal one and get to know the guests, some of whom have been returning for as many as 28 years. You'll find saunas, Jacuzzi, and a small conference/living area on the ground floor.

4 RV Parks & Campgrounds

Carson National Forest

208 Cruz Alta Rd., Taos, NM 87571. ☎ **505/758-6200.**

There are nine national forest campsites within 20 miles of Taos, all open from April or May until September or October, depending on snow conditions. For information on other public sites, contact the **Bureau of Land Management,** 224 Cruz Alta Rd., Taos, NM 87571 (☎ **505/758-8851**).

Enchanted Moon Campground
#7 Valle Escondido Rd. (on US 64 E.), Valle Escondido, NM 87571. ☎ **505/758-3338.** 69 sites. Full RV hookup, $17 per day. Closed Nov–Apr.

At an elevation of 8,400 feet, this campground is surrounded by pine-covered mountains and sits up against Carson National Forest.

Questa Lodge
Two blocks from NM 522 (P.O. Box 155), Questa, NM 87556. ☎ **505/586-0300.** 24 sites. Full RV hookup, $15 per day, $175 per month. Four cabins for rent in summer. Closed Nov–Apr.

On the banks of the Red River, this RV camp is within the small village of Questa.

Taos RV Park
Paseo del Pueblo Sur (P.O. Box 729), Ranchos de Taos, NM 87557. ☎ **800/323-6009** or 505/758-1667. Fax 505/758-1989. 29 spaces. $12 without RV hookup, $18 with RV hookup.

Very clean and nice bathrooms and showers. Two tepee rentals, as well. Senior discounts are available.

Taos Valley RV Park and Campground
120 Estes Rd. off NM 68 (7204 NDCBU), Taos, NM 87571. ☎ **800/999-7571** or 505/758-4469. Fax 505/758-4469. 92 spaces. $14.50–$15.50 without RV hookup, $18.50–$22.50 with RV hookup. MC, V.

Just 1 1/2 miles south of the Plaza, this campground is surrounded by sage, with views of the surrounding mountains. Open year-round.

13

Where to Dine in Taos

Taos is one of my favorite places to eat. Informality reigns; at a number of restaurants you can dine on world-class food while wearing jeans or even ski pants. Nowhere is a jacket and tie mandatory. This informality doesn't extend to reservations, however; especially during the peak season, it is important to make reservations well in advance and keep them or else cancel.

In the listings below, **Expensive** refers to restaurants where most main courses are $15 or higher; **Moderate** includes those where main courses generally range from $10 to $15; and **Inexpensive** indicates that most main courses are $10 or less.

1 Expensive

Doc Martin's

In the Historic Taos Inn, 125 Paseo del Pueblo Norte. ☎ **505/758-1977.** Reservations recommended. Breakfast $3.95–$7.50; lunch $4.50–$8.50; dinner $15.50–$28; early diners' menu $12.95. AE, DC, MC, V. Daily 8am–2:30pm and 5:30–9:30pm. CONTEMPORARY SOUTHWESTERN.

Doc Martin's restaurant (not to be confused with those urban-warrior boots, *Doc Martens*) comprises Dr. Thomas Paul Martin's former home, office, and delivery room. In 1912, painters Bert Philips (Doc's brother-in-law) and Ernest Blumenschein hatched the concept of the Taos Society of Artists in the Martin dining room. Art still predominates here, in both paintings that adorn the walls and the cuisine offered. The food is widely acclaimed, and the wine list has received numerous "Awards of Excellence" from *Wine Spectator* magazine.

The atmosphere is rich, with bins of yellow squash, eggplants, and red peppers set about near the kiva fireplace. Recently redecorated, it follows a Southwestern decor. Breakfast might include local favorites: huevos rancheros (fried eggs on a blue-corn tortilla smothered with chile and Jack cheese) or "The Kit Carson" (eggs Benedict with a Southwestern flair). Lunch might include a Pacific ahi tuna sandwich, Caesar salad, or for the heartier appetite, Northern New Mexican casserole (layers of blue-corn tortillas, pumpkin-seed mole, *calabacitas,* and Cheddar). For a dinner appetizer, I'd recommend one of the specials, such as black-bean cakes (on red chile sauce, with guacamole and goat cheese cream) or chile rellenos. This might be followed by seared salmon with brioche-lime dressing or the Southwest lacquered duck (poached, roasted, and grilled duck breast

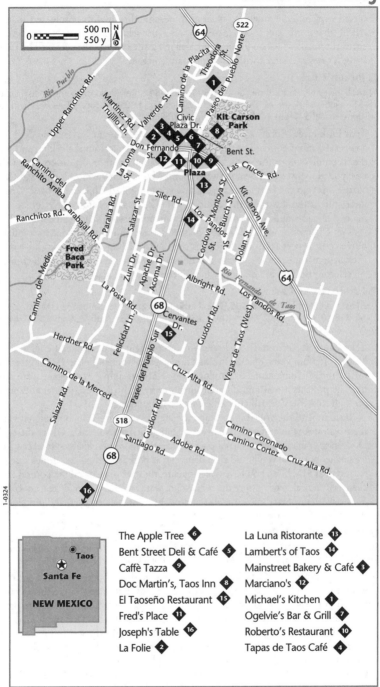

The Apple Tree 6
Bent Street Deli & Café 5
Caffè Tazza 9
Doc Martin's, Taos Inn 8
El Taoseño Restaurant 15
Fred's Place 11
Joseph's Table 16
La Folie 2

La Luna Ristorante 13
Lambert's of Taos 14
Mainstreet Bakery & Café 3
Marciano's 12
Michael's Kitchen 1
Ogelvie's Bar & Grill 7
Roberto's Restaurant 10
Tapas de Taos Café 4

NEW MEXICO

Taos
Santa Fe

served over julienne duck-leg meat and red-chile broth with posole and mango relish). If you still have room, there's always a nice selection of desserts—try Leroy's citrus cheesecake, a lemon-, lime-, and orange-flavored cheesecake perfected by baker Leroy Torres and served with an orange-tarragon sauce.

✪ Joseph's Table

4167 Highway 68, Ranchos de Taos. ☎ **505/751-4512.** Reservations recommended. Lunch $5–$12; dinner $11–$19. AE, DC, DISC, MC, V. Mon–Fri 11:30am–2:30pm; Tues–Thur and Sun 5–10pm, Fri–Sat 5–11pm. NEW AMERICAN/MEDITERRANEAN.

Taos funk meets European flair at this one-room restaurant in Ranchos de Taos, about a 10-minute drive from the Plaza. Bird cages hang from the ceiling and medieval candles adorn the walls, reminding you that you're in Taos—so expect the unexpected. Between deliberately water-damaged walls, chef/owners Joseph and Gina Wrede serve up such dishes as steak au poivre and grilled Chilean sea bass. What's interesting is the way these dishes are served. The steak sits atop a layer of smooth mashed potatoes and is crowned with an exotic mushroom salad, while the sea bass rests on potatoes, as well as a layer of mashed squash, and is surrounded by a tomato puree. For dessert try such delicacies as bread pudding or the dark-and-white-chocolate marquis, which Joseph describes as "creamy like the inside of a truffle." An eclectic selection of beers and wines by the bottle and glass is available.

✪ Lambert's of Taos

309 Paseo del Pueblo Sur. ☎ **505/758-1009.** Reservations recommended. Main courses $5.50–$18.50. AE, DC, MC, V. Sun–Thurs 5:30–9pm, Fri and Sat 5:30–9:30pm. CONTEMPORARY AMERICAN.

Zeke Lambert, a former San Francisco restaurateur who was head chef at Doc Martin's for four years, opened this fine dining establishment in late 1989 in the historic Randall Home near Los Pandos Road. Now, in simple but elegant surroundings, he presents a new and different menu every night. His eye for detail extends through to the service, which is efficient and unimposing.

I recommend the grilled lobster and asparagus salad with mango vinaigrette and the roast duck with apricot chipotle sauce and grilled polenta cake. Or try the fresh Dungeness crab cakes with Thai dipping sauce. For those who like rich food but prefer smaller portions, you'll find such delicacies as pork tenderloin and grilled salmon for less than $15, but if you have a big appetite, the chef will expand the portion (and charge $6 more). For dessert the white-chocolate ice cream and Zeke's chocolate mousse with raspberry sauce are outstanding. Espresso coffees, beer, and dinner and dessert wines are available.

✪ Stakeout Grill & Bar

Stakeout Dr., just off NM 68. ☎ **505/758-2042.** Reservations recommended. Main courses $11.95–$25.95. AE, CB, DC, DISC, MC, V. Daily 5–10pm. CONTINENTAL.

I love this restaurant. Maybe it's because you have to drive about a mile up a dirt road toward the base of the Sangre de Cristo Mountains, but once you get there (you'll think I'm insane for sending you up Stakeout Drive at night and fear you're never going to get there), you'll have one of the greatest views of Taos (and the sunset). Or maybe it's because after you're inside, you're enveloped in the warmth of its rustic decor (which is a great contrast to the almost-white exterior). There are paneled walls, creaking hardwood floors, and a crackling fireplace in the winter. The food, which focuses on steak and seafood, is marvelous. You can start with baked Brie served with sliced almonds and apples or escargots baked with walnuts, herbs, white wine, and garlic. Move on to a wonderful filet mignon, served with béarnaise sauce and cooked to your liking, or my favorite, duck Cumberland (half a duck roasted with apples and

Now that you know your way around, let's move on to something simple.

1 8 0 0
C A L L
A T T®

For card and collect calls.

1 800 CALL ATT is the only number you need to know when you're away from home. Dial it from any phone, anywhere* and your calls will always go through to AT&T.
*Available in U.S. and Canada. © 1997 AT&T

AT&T

prunes and served with an orange-currant sauce); this dish is never fatty and always leaves you feeling satisfied. Among the seafood offerings are salmon, Alaskan king crab legs (steamed and served with drawn butter), scallops, and shrimp. Finish your meal with a fresh pastry and a cappuccino. Try to time your reservation so you can see spectacular sunset.

Villa Fontana

NM 522, 5 miles north of Taos. ☎ **505/758-5800.** Reservations recommended. Main courses $19–$25. AE, CB, DC, DISC, MC, V. Mon–Sat 5:30–9:30pm. ITALIAN.

Carlo and Siobhan Gislimberti have received wide acclaim for their restaurant on the north end of town. They like to talk about "peccato di gola" (lust of the palate), which they brought with them to Taos when they left their home in the Italian Dolomites, near the Austrian border. They have their own herb garden, and Carlo, a master chef, is a member of the New Mexico Mycological Society—wild mushrooms play an important role in many of his kitchen preparations. Also something of an artist, Carlos displays his own works throughout the restaurant.

  Meals are truly gourmet, but the word around town is that they're not always consistent. I myself have had nice meals here. My favorite began with cream of wild mushroom soup (highly recommended). You also may want to try the shrimp marinara served with bruschetta. For a main course I enjoyed the red snapper with green olives, white wine, and garlic. The tenderloin of venison with wild mushrooms is also delicious. Dinners are served with fresh vegetables, and potatoes, rice, or polenta. Outdoor dining in the summer offers pleasant mountain and valley views.

2 Moderate

Apple Tree

123 Bent St. ☎ **505/758-1900.** Reservations recommended. Lunch $5.25–$9.95; dinner $10.95–$18.95. CB, DC, DISC, MC, V. Mon–Sat 11:30am–3pm; light meals and snacks daily 3–5:30pm; daily 5:30–9pm. Brunch Sun 10am–3pm. INTERNATIONAL.

Eclectic music pervades the four adobe rooms of this restaurant, a block north of the Plaza. Original paintings by Taos masters overlook the candlelit service indoors. Outside, diners sit at wooden tables on a graveled courtyard beneath a spreading apple tree.

  This restaurant is popular among locals and travelers, but it doesn't appeal much to me. The recipes try too hard. I suggest ordering what looks simplest, either from the menu or from the daily specials. The Apple Tree salad (greens sprinkled with dried cranberries, walnuts, and blue cheese served with a vinaigrette) is very good, as is the posole (hominy with chile). A very popular dish, though too sweet for me, is mango chicken enchiladas (chicken simmered with onions and spices, layered between blue corn tortillas with mango chutney, sour cream, and salsa fresca and smothered with green chile). I prefer the Thai red curry (either vegetarian or with shrimp). The best thing here is the chile-jalapeño bread, served with the meal. The Apple Tree has an award-winning wine list, and the desserts are prepared fresh daily.

Casa de Valdez

1401 South Santa Fe Rd. ☎ **505/758-8777.** Reservations recommended. Main courses $5.95–$19.95. AE, DC, DISC, MC, V. Mon–Tues and Thurs–Sat 11:30am–9:30pm; Sun 3:30–9:30pm. NEW MEXICAN.

Located 2^{1}/$_{2}$ miles south of the Plaza, Casa de Valdez has recently been renovated. The interior is warm and welcoming, and the extremely friendly staff know many of their customers by name. The menu is predictable as far as New Mexican cuisine

---

🚸 Family-Friendly Restaurants

El Taoseño Restaurant *(see page 186)*   The jukebox and games room will keep the kids happy while you wait for tacos and enchiladas at low prices.

Casa de Valdez *(see page 183)*   Parents will feel comfortable dining with children here, and the menu is simple enough to please even the pickiest of palates.

Michael's Kitchen *(see page 187)*   With a broad menu, comfy booths, and a very casual, diner-type atmosphere, both kids and their parents will feel at home here.

---

goes, with blue-corn enchiladas, bean burritos, and tamales, but everything is prepared with a concern for low fat and low cholesterol. The chile rellenos are exceptionally good. You can also order spareribs, barbecued chicken, steak, and shrimp. The sopaipillas served at the Casa de Valdez are the best around—they practically melt in your mouth.

Jacquelina's Restaurant

1541 Paseo del Pueblo Sur. ☎ **505/751-0399.** Reservations recommended. Main courses $8.25–$14.95. AE, DISC, MC, V. Tues–Fri and Sun 11am–2pm; Tues–Sun 5–9pm. CREATIVE SOUTHWESTERN.

The decor here isn't much to talk about, but the food is quite good. To start, I have enjoyed the grilled crab cakes with spicy black-bean sauce, and I particularly liked the portobello mushrooms with polenta french fries. I also tried the Southwestern-style spring rolls with a red-bell-pepper dipping sauce and was pleased. For a main course I would recommend the fresh salmon with sesame crust served with a Chinese special sauce or Jacquelina's signature dish—chipotle chile and honey-marinated pork tenderloin with roasted corn red chile. Southwestern specialties include a variety of burritos and fajitas as well as chile rellenos. In short, if you're not a fanatic about ambience, Jacquelina's will more than satisfy.

La Luna Ristorante

223 Paseo del Pueblo Sur. ☎ **505/751-0023.** Reservations recommended. Main courses $8.50–$12. MC, V. Daily 5–10pm (last order is taken at 10pm; restaurant closes at midnight). ITALIAN.

The atmosphere here is bright and colorful. Black-and-white squares have been painted in trompe l'oeil fashion on the floor to resemble tiles, and the walls, hung with interesting original works, are done in rich, warm tones. Outside, in front, there are several tables shaded by deep blue market umbrellas. The food is also attractive and tasty. Start with the focaccia de patate (pizza bread drizzled with olive oil, topped with thin slices of potato, and spiced with rosemary leaves) or—if it's listed as a special—the Brie served with toast points and peach salsa. The restaurant offers several pizzas as well as a wide variety of pasta choices. The vegetarian lasagna was tempting, and I've also enjoyed the spaghetti alla carbonara (spaghetti with Italian bacon, peas, cream, and Parmesan cheese, finished with egg yolk). There are also a couple of meat dishes on the menu. The desserts change regularly, but tiramisù is usually available.

Ogelvie's Bar & Grill

1031 East Plaza. ☎ **505/758-8866.** No reservations. Lunch $5.95–$9.50; dinner $8.50–$18.50. AE, DC, MC, V. Daily 11am–closing. INTERNATIONAL.

The only real reason to go to this restaurant is to have a drink right on the Plaza. In the warm months there's a nice balcony where diners can sit and drink a margarita and indulge in chips or other appetizers such as potato skins or even a burger. Otherwise, the food here is not flavorful, and the atmosphere inside is dated.

✪ Trading Post Café

4179 Paseo del Pueblo Sur. Ranchos de Taos. ☎ **505/758-5089.** No reservations except for parties of five or more. Menu items $5–$21. DISC, MC, V. Mon–Sat 11:30am–9:30pm. ITALIAN/ INTERNATIONAL.

Without a doubt, this continues to be Taos's hottest restaurant. In fact, it's so popular that it has undergone several expansions in the last 2 years. Be prepared to wait for a table. As you enter this bright, airy restaurant, to your right you'll see high metal stools surrounding a bar (where meals can be taken). The bar itself encloses an open exhibition kitchen. If you're dining alone or just don't feel like waiting for a table, the bar is a fun place to sit. I like the Trading Post's menu; it lists a nice variety of items without distinguishing between appetizers and main courses. This small detail speaks volumes about the restaurant. Although the focus is on the food, diners can feel comfortable here, even if trying three appetizers and skipping the main course. The Caesar salad is traditional with an interesting twist—garlic chips. You've probably never had a Caesar salad this good. If you like pasta, you'll find a nice variety on the menu. The angel-hair pasta with chicken, wild mushrooms, and Gorgonzola cream is surprisingly light and flavorful. There's also a fresh fish of the day and usually nice stews and soups at very reasonable prices. For dessert there are delicious tarts.

3 Inexpensive

Bent Street Deli & Cafe

120 Bent St. ☎ **505/758-5787.** Reservations accepted. Breakfast $1.25–$6; lunch $2.50–$8; dinner $10–$16. MC, V. Mon–Sat 8am–9pm. DELI/INTERNATIONAL.

This popular cafe is a short block north of the Plaza. Outside, a flower box surrounds sidewalk cafe–style seating that is heated in winter. Inside, baskets and bottles of homemade jam provide a homey, country feel. The menu features breakfast burritos and homemade granola in the morning; and for lunch you can choose from 18 deli sandwiches, plus a "create-your-own" column. At dinner, the menu becomes a bit more sophisticated, with dishes such as beef tenderloin medaillons served over fettuccine with a chipotle, Fontina cream, or roja shrimp (black tiger shrimp, red chile, and jicama, cilantro, and corn relish). All dinner entrees are served with a salad and freshly baked bread. If you'd like to grab a picnic to go, Bent Street Deli and Cafe offers carry-out service.

Caffè Tazza

122 Kit Carson Rd. ☎ **505/758-8706.** No reservations. All menu items under $10. No credit cards. Daily 8am–6pm (until 10pm on performance nights). CAFE.

This cozy three-room cafe is a gathering spot for local community groups, artists, performers, and poets—in fact, it's home to SOMOS, the Society of the Muse of the Southwest. Plays, films, comedies, and musical performances are held here on weekends (and some weeknights in summer) in the cafe's main room. The walls are always hung with the works of local emerging artists who have not yet made it to the Taos gallery scene. Locals and tourists alike enjoy sitting in the cafe, taking in the scene, or reading one of the assorted periodicals available while sipping a Café Mexicana (mocha coffee made with Mexican chocolate), a cappuccino, or espresso. Of course, the food is also quite good, though don't expect anything fancy. Soups (the chef invents new recipes daily), sandwiches, tamales, croissants, and pastries are all popular, as are the breakfast burritos.

El Taoseño Restaurant

819 South Santa Fe Rd. ☎ **505/758-4142.** Main courses $3.10–$11.95. AE, MC, V. Mon–Fri 6am–9pm, Sat 6:30am–10pm, Sun 6:30am–2pm. NEW MEXICAN/AMERICAN.

A long-established local diner, El Taoseño features a jukebox in the corner and local art on the walls. There are daily specials such as barbecued chicken and Mexican plates; standard fare includes everything from huevos rancheros to enchiladas and tacos. Locals flock here for the breakfast burrito. A low-fat menu is also available.

✪ Fred's Place

332 Paseo del Pueblo Sur. ☎ **505/758-0514.** Main courses $5–$12. AE, DC, DISC, MC, V. Mon–Sat 5–9:30pm. NEW MEXICAN.

I was warned by a number of locals not to put Fred's Place in this guide. This is *our* place, they pleaded. But alas, the guide wouldn't be complete without mention of Fred's. God and the devil are at odds at this New Mexican food restaurant, and judging by the food, God has won out. The atmosphere is rich, though for some it may be a bit unnerving. Walls are hung with crucifixes and a vivid ceiling mural depicts a very hungry devil that appears to swoop down toward the dining tables.

Fred's offers New Mexican food, but it's not of the greasy spoon variety. It's very refined, the flavors carefully calculated. You have to try Dee Dee's squash stew (squash, corn, beans, and vegetarian green chile, topped with cheese and fresh oregano). For me, Fred's chicken enchilada is heaven; however, you can't go wrong with a burrito either. Daily specials include grilled trout and carne asada served with a watercress salad. For dessert try the warm apple crisp with ice cream.

Guadalajara Grill

1384 Paseo del Pueblo Sur. ☎ **505/751-0063.** All items under $8. No credit cards. Mon–Sat 10:30am–9pm, Sun 11am–8pm. MEXICAN.

This excellent Mexican restaurant is in an odd location. On the south end of town, it shares a building with a car wash. Don't be deceived, however; the food here is excellent. It's Mexican rather than New Mexican, a refreshing treat. I recommend the tacos, particularly pork or chicken, served in soft homemade corn tortillas, the meat deliciously grilled and flavorful. The burritos are large and smothered in chile. Platos are served with rice and beans and half orders are available for smaller appetites.

La Folie

122 Dona Luz. ☎ **505/758-8800.** Main courses $6.95–$12.75. AE, DISC, MC, V. Tues–Fri 11am–8pm, Sat 8:30am–4:30pm, Sun 9am–3pm. FRENCH.

This new restaurant opened by Mark and Lisa Felix provides an excellent respite from all the New Mexican fare in the area. You'll experience delicate French dishes presented with flair. The atmosphere is fun. Inside you're surrounded by walls faux painted with images of pillars and birds, and outside there's a charming courtyard with a fountain. Next door there's also a cafe, where you can get sandwiches and deli items. In the restaurant, try the tuna tartare (a ceviche made with tuna, garlic, ginger, and oyster sauce). Move on to the spinach, goat cheese, tomato, and roasted garlic frittata. Or try the Pacific sea bass with Provençale ratatouille and a risotto cake. For dessert the piña colada mousse is surprisingly light. Due to the restaurant's proximity to a church, no alcoholic beverages are served.

Mainstreet Bakery & Cafe

Guadalupe Plaza, Camino de la Placita. ☎ **505/758-9610.** All items $2.95–$5.25. No credit cards. Mon–Fri 7:30am–6pm, Sat–Sun 7:30am–2pm. NATURAL FOODS.

Located about 1¹/₂ blocks west of the Plaza, this is one of Taos's biggest counter-culture hangouts. The image is fostered by the health-conscious cuisine and the wide selection of newspapers, magazines, and other reading material presenting alternative lifestyles. Coffee and pastries are served all day. Breakfast omelets, pancakes, French toast, and huevos rancheros are popular. The lunch menu focuses on sandwiches, tostadas, and garden burgers (vegetarian). In keeping with the spirit of good health promoted here, alcoholic beverages are not served and smoking is not permitted.

Marciano's

Corner of La Placita at Ledoux. ☎ **505/751-0805.** Reservations accepted. Main courses $7.50–$12. AE, CB, DC, MC, V. Wed–Mon 6–9pm. CONTINENTAL.

This is a good place to go for an inexpensive, romantic dinner. The walls are peach colored, tables are adorned with little tin candleholders and cactus plants, and the decor is rounded out by vigas and chandeliers. In the summer there's courtyard dining. Chef Skot Kirshbaum serves up mostly light pastas and some organic meats and vegetable dishes. To start, I recommend the crostini (grilled homemade bread, black olive tapenade, baby mozzarella, and roasted bell peppers). For an entree, I'd stick with the pasta dishes, such as the linguine with hot Italian sausage or the smoked turkey and chipotle pasta. Save room for plenty of the excellent homemade bread and above all for the homemade ice cream. I had a Frangelico ice cream and a coffee, both excellent. Beer and wine are available.

Michael's Kitchen

304 C Paseo del Pueblo Norte. ☎ **505/758-4178.** No reservations. Breakfast $1.55–$7.95; lunch $3.25–$9.50; dinner $5–$12. AE, DISC, MC, V. Daily 7am–8pm (except major holidays). NEW MEXICAN/AMERICAN.

A couple of blocks north of the Plaza is this eatery, a throwback to earlier days. Between its hardwood floor and viga ceiling are various knickknacks on posts, walls, and windows: a deer head here, a Tiffany lamp there, and several scattered antique woodstoves. Seating is at booths and tables. Meals, too, are old-fashioned, as far as quality and quantity for price. Breakfast dishes, including a large selection of pancakes and egg preparations (with names like the "Moofy," "Omelette Extra-ordinaire," and "Pancake Sandwich"), are served all day (because they're so good), as are lunch sandwiches (including Philly cheesesteak, tuna melt, chile burger, and a veggie sandwich). One of my favorite lunch dishes is generically (and facetiously) called "Health Food," a double order of fries with red or green chile and cheese. Dinners range from veal Cordon Bleu to knockwurst and sauerkraut, plantation-fried chicken to enchiladas rancheros. Now on Saturday evening you can get prime rib. Michael's has its own excellent full-service bakery.

Roberto's Restaurant

122 Kit Carson Rd. ☎ **505/758-2434.** Reservations recommended. Main courses $9.95–$11.50. MC, V. Summer, Wed–Mon 5:30–9pm; Winter, 5:30–9pm on weekends and holidays only. NEW MEXICAN.

Hidden within a warren of art galleries opposite the Kit Carson Museum (see chapter 14) is this joint, which for more than 25 years has focused on authentic native dishes. In this 160-year-old adobe building are three small, high-ceilinged dining rooms, each with hardwood floors and a maximum of five tables; one room has a corner fireplace. Despite having earned a considerable local following, one out-of-town visitor to the restaurant called it "relentlessly ordinary." Everyday dishes include tacos, enchiladas, tamales, and chile relleños. All meals start with sopaipillas and come with homemade refried beans and chicos (dried kernels of corn). Beer and wine are the only alcoholic beverages served. Since Roberto is an avid skier, the restaurant's winter hours are unpredictable at best.

Tapas de Taos Cafe

136 Bent St. ☎ **505/758-9670.** No reservations. Tapas $3.95–$7.95; main courses $3.95–$12.95. MC, V. Mon–Fri 11:30am–3pm; Mon–Sat 3–9:30pm, Sun 3–9pm. NEW MEXICAN/TAPAS.

If you're familiar with Mexican culture, you'll recognize the theme decor at Tapas de Taos Cafe; it represents the Mexican Day of the Dead (Día de los Muertos). The rows of black skulls lining the walls of the dining room in this 300-year-old adobe are surprisingly lively as well as a bit haunting. The tapas menu is short but fairly good; I recommend the pork-and-ginger pot stickers but wasn't impressed by the Vietnamese fried calamari. For a main course, the Yucatán shrimp is nice, though a little bland. You might try the fajitas or tacos. A wide variety of coffees is available, including café macchiata, cappuccino, caffè latte, and espresso. Outdoor patio dining is available during the warmer months. There's also a kids' menu.

What to See & Do in Taos

14

With a history shaped by pre-Colombian civilization, Spanish colonialism, and the Wild West; outdoor activities that range from ballooning to world-class skiing; and a clustering of artists, writers, and musicians, Taos has something to offer almost everybody. Its pueblo is the most accessible in New Mexico, and its museums, including the new Van Vechten Lineberry Taos Art Museum, represents a world-class display of regional history and culture.

SUGGESTED ITINERARIES

If You Have Only 1 Day

Spend at least 2 hours at the Taos Pueblo. You'll also have time to see the Millicent Rogers Museum and to browse in some of the town's fine art galleries. Try to make it to Ranchos de Taos to see the San Francisco de Asis Church and to shop on the Plaza there.

If You Have 2 Days

On the second day, explore the Kit Carson Historic Museums—the Martinez Hacienda, the Kit Carson Home, and the Ernest L. Blumenschein Home. Then head out of town to enjoy the view from the Rio Grande Gorge Bridge.

If You Have 3 Days or More

On your third day, drive the "Enchanted Circle" through Red River, Eagle Nest, and Angel Fire or head to the Van Vechten Lineberry Taos Art Museum. You may want to allow a full day for shopping or perhaps drive up to the Taos Ski Valley for a chairlift ride or a short hike. Of course, if you're here in the winter with skis, that's your first priority.

A Note on Taos Museums: If you would like to visit all seven museums that comprise the Museum Association of Taos—Blumenschein Home, Fechin Institute, Hacienda Martinez, Harwood Museum, Kit Carson Home and Museum, Millicent Rogers Museum, and Van Vechten Lineberry Taos Art Museum—it might be worthwhile to purchase a combination ticket for $20.

1 The Top Attractions

Taos Pueblo

P.O. Box 1846, Taos Pueblo, NM 87571. ☎ **505/758-1028.** Admission $5 per car. If you would like to use a still camera, the charge is $5; for a video camera $10; if you would like to sketch or paint, written permission is required. Other charges apply for commercial use of imagery. Photography is not permitted on feast days. Winter, daily 8am–4:30pm; summer, daily 8am–5pm, with a few exceptions. Closed for 1 month every year in late winter or early spring (call to find out if it will be open at the time you expect to be in Taos). Also, since this is a living community, you can expect periodic closures.

It's amazing that in our frenetic world, 200 Taos Pueblo residents still live much as their ancestors did a thousand years ago. When you enter the pueblo you'll see where they dwell, in two large buildings, each with rooms piled on top of each other, forming structures that echo the shape of Taos Mountain (which sits to the northeast). Here, a portion of Taos residents live without electricity and running water. The remaining 2,000 residents of Taos Pueblo live in conventional homes on the pueblo's 95,000 acres.

The main buildings' distinctive flowing lines of shaped mud, with a straw-and-mud exterior plaster, are typical of Pueblo architecture throughout the Southwest. It's architecture that blends in with the surrounding land—which makes sense, given that it is itself made of earth. Bright blue doors are the same shade as the sky that frames the brown buildings.

The northernmost of New Mexico's 19 pueblos, Taos has been home to the Tiwa tribes for more than 900 years. Many residents here still practice ancestral rituals. The center of their world is still nature; many still bake bread in hornos, and most still drink water that flows down from the sacred Blue Lake. Meanwhile, arts and crafts and other tourism-related businesses support the economy, along with government services, ranching, and farming.

The village looks much the same today as it did when a regiment from Coronado's expedition first came upon it in 1540. Though the Tiwa were essentially a peaceful agrarian people, they are perhaps best remembered for spearheading the only successful revolt by Native Americans in U.S. history. Launched by Pope ("Po-*pay*") in 1680, the uprising drove the Spanish from Santa Fe until 1692 and from Taos until 1698.

As you explore the pueblo, you can visit their studios, munch on homemade bread, look into the new **San Geronimo Chapel,** and wander past the fascinating ruins of the old church and cemetery. You're expected to ask permission from individuals before taking their photos; some will ask for a small payment, but that's for you to negotiate. Kivas and other ceremonial underground areas are taboo.

San Geronimo is the patron saint of the Taos Pueblo, and his feast day (September 30) combines Catholic and pre-Hispanic traditions. The **Old Taos Trade Fair** on that day is a joyous occasion, with foot races, pole climbs, and crafts booths. Dances are performed the evening of September 29. Other annual events include a **turtle dance on New Year's Day, deer or buffalo dances on Three Kings Day** (January 6), and **corn dances on Santa Cruz Day** (May 3), **San Antonio Day** (June 13), **San Juan Day** (June 24), **Santiago Day** (July 23), and **Santa Ana Day** (July 24). The **Taos Pueblo Powwow,** a dance competition and parade that brings together tribes from throughout North America, is held the weekend after July 4 on reservation land off NM 522. **Christmas Eve bonfires** mark the start of the children's **corn dance,** the **Christmas Day deer dance,** and the 3-day-long **Matachines dance.**

During your visit to the pueblo you will have the opportunity to purchase traditional fried and oven-baked bread as well as a variety of arts and crafts. If you would

Taos Attractions

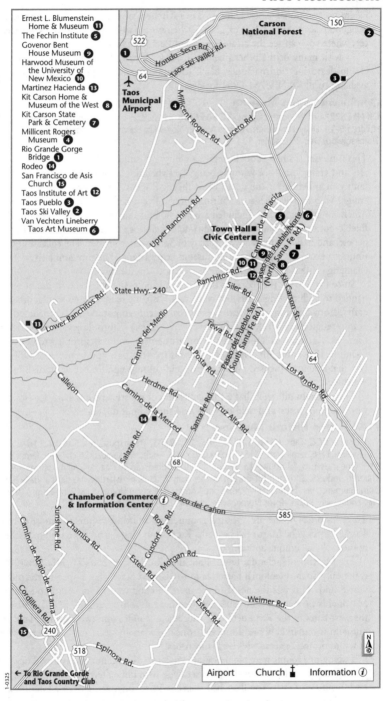

Ernest L. Blumenstein
Home & Museum ⑪

The Fechin Institute ⑤

Govenor Bent
House Museum ⑨

Harwood Museum of
the University of
New Mexico ⑩

Martinez Hacienda ⑬

Kit Carson Home &
Museum of the West ⑧

Kit Carson State
Park & Cemetery ⑦

Millicent Rogers
Museum ④

Rio Grande Gorge
Bridge ①

Rodeo ⑭

San Francisco de Asis
Church ⑮

Taos Institute of Art ⑫

Taos Pueblo ③

Taos Ski Valley ②

Van Vechten Lineberry
Taos Art Museum ⑥

Carson
National Forest

150

522

Hondo-Seco Rd.

Taos Ski Valley Rd.

64

Taos
Municipal
Airport

Millicent Rogers Rd.

Lucero Rd.

Upper Ranchitos Rd.

Camino de la Placita

Town Hall
Civic Center

Paseo del Pueblo Norte
(North Santa Fe Rd.)

Kit Carson St.

Ranchitos Rd.

Siler Rd.

State Hwy. 240

Lower Ranchitos Rd.

Camino del Medio

Tewa Rd.

La Posta Rd.

Paseo del Pueblo Sur
(South Santa Fe Rd.)

Los Pandos Rd.

64

Callejon

Herdner Rd.

Camino de la Merced

Santa Fe Rd.

Cruz Alta Rd.

Salazar Rd.

68

Chamber of Commerce
& Information Center

Paseo del Cañon

585

Sunshine Rd.

Chamisa Rd.

Roy Rd.

Gusdorf Rd.

Morgan Rd.

Estees Rd.

Camino de Abajo de la Lama

Cordillera Rd.

240

518

Espinosa Rd.

Estees Rd.

Weimer Rd.

N

Airport   Church ✝   Information ⓘ

I-0325

← To Rio Grande Gorde
and Taos Country Club

191

like to try traditional feast-day meals, **Tiwa Kitchen,** near the entrance to the pueblo, is a good place to stop. Close to Tiwa Kitchen is the **Oo-oonah Children's Art Center,** where you can see the creative works of Pueblo children.

As with many of the other pueblos in New Mexico, Taos Pueblo has opened a casino, featuring slot machines, blackjack, and poker. Free local transportation is available. Call ☎ 505/751-0991 for details.

✪ Millicent Rogers Museum

Off NM 522, 4 miles north of Taos. ☎ **505/758-2462.** Admission $6 adults, $5 students, $1 children 6–16. Daily 10am–5pm. Closed Mon in Nov–Mar, Easter, San Geronimo Day (Sept 30), Thanksgiving, Dec 25, and Jan 1.

This museum is small enough to give a glimpse of some of the finest Southwestern arts and crafts you'll see without being overwhelming. It was founded in 1953 by family members after the death of Millicent Rogers. Rogers was a wealthy Taos émigré who in 1947 began acquiring a magnificent collection of beautiful Native American arts and crafts. Included are Navajo and Pueblo jewelry, Navajo textiles, Pueblo pottery, Hopi and Zuni kachina dolls, paintings from the Rio Grande Pueblo people, and basketry from a wide variety of Southwestern tribes. The collection continues to grow through gifts and museum acquisitions. The museum also presents changing exhibitions of Southwestern art, crafts, and design.

Since the 1970s, the scope of the museum's permanent collection has been expanded to include Anglo arts and crafts and Hispanic religious and secular arts and crafts, from Spanish and Mexican colonial to contemporary times. Included are *santos* (religious images), furniture, weavings, colcha embroideries, and decorative tin work. Agricultural implements, domestic utensils, and craftspeople's tools dating from the 17th and 18th centuries are also displayed. Last year a special exhibition, "Animal Friends," focused on the artistic and cultural significance of animals in the Southwest.

The museum gift shop has a fine collection of superior regional art. Classes and workshops, lectures, and field trips are held throughout the year.

✪ Kit Carson Historic Museums

P.O. Drawer CCC, Taos, NM 87571. ☎ **505/758-0505.** Three museums, $8 adults, $6 seniors, $5 children 6–16; family rate, $15. Two museums, $6 adults, $5 seniors, $4 children; family rate, $13. One museum, $4 adults, $3 seniors, $2.50 children; family rate, $8. All museums free for children under 5. Summer: Kit Carson Home, daily 8am–6pm; Martinez Hacienda, daily 9am–5pm; Blumenschein Home, daily 9am–5pm. Winter: Kit Carson Home, daily 9am–5pm; Martinez Hacienda, daily 10am–4pm; Blumenschein Home, daily 11am–4pm.

Three historical homes are operated as museums, affording visitors a glimpse of early Taos lifestyles. The Martinez Hacienda, Kit Carson Home, and Ernest Blumenschein home each has unique appeal.

The **Martinez Hacienda,** Lower Ranchitos Road, Hwy. 240 (☎ **505/758-1000**), is the only Spanish colonial hacienda in the United States that's open to the public year-round. This was the home of the merchant and trader Don Antonio Severino Martinez, who bought it in 1804 and lived here until his death in 1827. Located on the west bank of the Rio Pueblo de Taos about 2 miles southwest of the Plaza, the museum is remarkably beautiful, with thick, raw adobe walls and no exterior windows, to protect against raids by Plains tribes.

Twenty-one rooms were built around two *placitas,* or interior courtyards. They give you a glimpse of the austerity of frontier lives, with only a few pieces of modest period furniture in each. You'll see bedrooms, servants' quarters, stables, a kitchen, and even a large fiesta room. Exhibits in one newly renovated room tell the story of the

Martinez family and life in Spanish Taos between 1598 and 1821, when Mexico gained control.

Don Antonio Martinez, who for a time was *alcalde* (mayor) of Taos, owned several caravans that he used in trade on the Chihuahua Trail to Mexico. This business was carried on by his youngest son, Don Juan Pascual, who later owned the hacienda. His eldest son was Padre Antonio José Martinez, Northern New Mexico's controversial spiritual leader from 1826 to 1867.

Kit Carson Historic Museums has developed the hacienda into a living museum with weavers, blacksmiths, and wood carvers. Demonstrations are scheduled daily, even during the **Taos Trade Fair** (the last weekend in September) when they run virtually nonstop. The Trade Fair commemorates the era when Native Americans, Spanish settlers, and mountain men met here to trade with each other. The Martinez Hacienda is now home to a new santos exhibit.

The **Kit Carson Home and Museum of the West,** East Kit Carson Road (☎ 505/758-4741), located a short block east of the Plaza intersection, is the town's only general museum of Taos history. The 12-room adobe home, with walls 2^1/$_2$ feet thick, was built in 1825 and purchased in 1843 by Carson, the famous mountain man, Indian agent, and scout, as a wedding gift for his young bride, Josefa Jaramillo. It remained their home for 25 years, until both died (exactly a month apart) in 1868.

A living room, bedroom, and kitchen are furnished as they might have been when occupied by the Carsons. The Indian Room contains artifacts crafted and utilized by the original inhabitants of Taos Valley; the Early American Room has a variety of pioneer items, including many antique firearms and trappers' implements; and the Carson Interpretive Room presents memorabilia from Carson's unusual life. In the kitchen is a Spanish plaque that reads: *Nadie sabe lo que tiene la olla mas que la cuchara que la menea* (Nobody better knows what the pot holds than the spoon that stirs it). New permanent exhibits in the Carson home include Native American prehistory and history as well as "Kit Carson: Life and Times."

The museum bookshop, with perhaps the town's most comprehensive inventory of New Mexico historical books, is adjacent to the entry.

The **Ernest L. Blumenschein Home & Museum,** 222 Ledoux St. (☎ 505/758-0505), 1^1/$_2$ blocks southwest of the Plaza, re-creates the lifestyle of one of the founders of the Taos Society of Artists (founded 1915). An adobe home with garden walls and a courtyard, parts of which date from the 1790s, it became the home and studio of Blumenschein (1874–1960) and his family in 1919. Period furnishings include European antiques and handmade Taos furniture in Spanish colonial style.

Blumenschein was born and raised in Pittsburgh. In 1898 he arrived in Taos somewhat by accident. After training in New York and Paris, he and fellow painter Bert Phillips were on assignment for *Harper's* and *McClure's* magazines of New York when a wheel of their wagon broke while they were traversing a mountain 30 miles north of Taos. Blumenschein drew the short straw and thus was obliged to bring the wheel by horseback to Taos for repair. He later recounted his initial reaction to the valley he entered: "No artist had ever recorded the New Mexico I was now seeing. No writer had ever written down the smell of this air or the feel of that morning sky. I was receiving . . . the first great unforgettable inspiration of my life. My destiny was being decided."

That spark later led to the foundation of Taos as an art colony. An extensive collection of works by early-20th-century Taos artists is on display in several rooms of the home, including some by Blumenschein's daughter, Helen. The home and museum are open daily from 9am to 5pm.

2 More Attractions

Kit Carson Park and Cemetery
Paseo del Pueblo Norte.

Major community events are held in the park in summer. The cemetery, established in 1847, contains the graves of Carson and his wife, Governor Charles Bent, the Don Antonio Martinez family, Mabel Dodge Luhan, and many other noted historical figures and artists. Their lives are described briefly on plaques.

✪ Fechin Institute
227 Paseo del Pueblo Norte (P.O. Box 832), Taos, NM 87571. ☎ **505/758-1710.** Admission $4. Oct–May, Wed–Sun 10am–2pm; May–Oct, Wed–Sun 10am–5pm.

The home of Russian artist Nicolai Fechin (Feh-shin) from 1927 until 1933, this historic building commemorates the career of a 20th-century Renaissance man. Born in Russia in 1881, Fechin came to the United States in 1923, already acclaimed as a master of painting, drawing, sculpture, architecture, and woodwork. In Taos, he renovated a huge adobe home and embellished it with hand-carved doors, windows, gates, posts, fireplaces, and other features of a Russian country home. The house and adjacent studio are now used for Fechin Institute educational activities, as well as concerts, lectures, and other programs. Fechin died in 1955.

Governor Bent House Museum
117 Bent St. ☎ **505/758-2376.** Admission $1 adults, 50¢ children. Summer, daily 9am–5pm; winter, daily 10am–4pm.

Located a short block north of the Plaza, this was the residence of Charles Bent, New Mexico Territory's first American governor. Bent, a former trader who established Fort Bent, Colorado, was murdered during the 1847 Native American and Hispanic rebellion, while his wife and children escaped by digging through an adobe wall into the house next door. The hole is still visible. Period art and artifacts are on display.

Harwood Museum of the University of New Mexico
238 Ledoux St. ☎ **505/758-9826.** Admission $3.50. Tues–Sat 10am–5pm, Sun noon–5pm.

Some of the finest works of art ever produced in or about Taos hang in this pueblo-style complex, a cultural and community center since 1923. The museum displays paintings, drawings, prints, sculpture, and photographs by Taos-area artists from 1800 to the present. Featured are paintings from the early days of the art colony by members of the Taos Society of Artists, including Oscar Berninghaus, Ernest Blumenschein, Herbert Dunton, Victor Higgins, Bert Phillips, and Walter Ufer. Also included are works by Emil Bisttram, Andrew Dasburg, Leon Gaspard, Louis Ribak, Bea Mandelman, Agnes Martin (seven new paintings in 1997), Larry Bell, and Thomas Benrimo.

On display are 19th-century *retablos,* religious paintings of saints that have traditionally been used for decoration and inspiration in the homes and churches of New Mexico. The permanent collection includes sculptures by Patrociño Barela, one of the leading Hispanic artists of 20th-century New Mexico, as well as the artists Marsden Hartley and John Marin.

The museum also schedules five or six changing exhibitions a year, many of which feature works by celebrated artists currently living in Taos. The Harwood just completed an expansive renovation that added five exhibition rooms.

D. H. Lawrence Ranch
San Cristobal. ☎ **505/776-2245.**

The shrine dedicated to this controversial early-20th-century author is a pilgrimage site for literary devotees. A short uphill walk from the ranch home, it's littered with various mementos—photos, coins, messages from fortune cookies—placed by visitors. The guest book is worth a long read.

Lawrence lived in Taos on and off between 1922 and 1925. The ranch was a gift to his wife, Frieda, from art patron Mabel Dodge Luhan. Lawrence repaid Luhan the favor by giving her the manuscript of *Sons and Lovers.* When Lawrence died in southern France in 1930 of tuberculosis, his ashes were returned here for burial. The grave of Frieda, who died in 1956, is outside the shrine.

The shrine is the only public building at the ranch, which is operated today by the University of New Mexico as an educational and recreational retreat. To reach the site, head north from Taos about 15 miles on NM 522, then another 6 miles east into the forested Sangre de Cristo Range via a well-marked dirt road.

✪ Rio Grande Gorge Bridge
US 64, 10 miles west of Taos.

This impressive bridge, west of the Taos airport, spans the Southwest's greatest river. At 650 feet above the canyon floor, it's one of America's highest bridges. If you can withstand the vertigo, it's interesting to come more than once, at different times of day, to observe how the changing light plays tricks with the colors of the cliff walls. A curious aside is that the wedding scene in the controversial movie, *Natural Born Killers,* was filmed here.

✪ San Francisco de Asis Church
P.O. Box 72, Ranchos de Taos NM, 87557. ☎ **505/758-2754.** Donations appreciated, $2 minimum. Mon–Sat 9am–noon and 1–4pm. Visitors may attend mass Sat at 6pm (rotates from this church to the 3 mission chapels) and Sun at 7 (Spanish), 9, and 11:30am.

From NM 68, about 4 miles south of Taos, this famous church appears as a modern adobe sculpture with no doors or windows. This is the image that has often been photographed (by Ansel Adams, among others) and painted (for example, by Georgia O'Keeffe). Visitors must walk through the garden on the west side of this remarkable 2-story church to enter and get a full perspective of its massive walls, authentic adobe plaster, and beauty.

The church office and gift shop are just across the driveway north of the church. A video presentation is given here every hour on the half hour. Also, displayed on the wall is an unusual painting, *The Shadow of the Cross* by Henri Ault (1896). Under ordinary light it portrays a barefoot Christ at the Sea of Galilee; in darkness, however, the portrait becomes luminescent, and the perfect shadow of a cross forms over the left shoulder of Jesus' silhouette. The artist reportedly was as shocked as everyone else to see this. The reason for the illusion remains a mystery. Several nice galleries and crafts shops surround the square.

Van Vechten Lineberry Taos Art Museum
501 Paseo del Pueblo Norte (P.O. Box 1848). ☎ **505/758-2690.** Admission $5 adults, $3 students and seniors. Tues–Fri 11am–4pm, Sat–Sun 1:30–4pm.

Taos's newest museum is the Van Vechten Lineberry Taos Art Museum. It was the brainchild of Ed Lineberry, who lives in the spectacular home adjacent to the museum; he conceived of it as a memorial to his late wife, Duane Van Vechten. An artist herself, Duane spent much time working in her studio, which now serves as the entryway to the 20,000-square-foot main gallery of the museum. The real thrill of this museum is seeing works of the Taos Society of Artists, which give a sense of what Taos was like in the late 19th and early 20th centuries. These are rich works that capture the panoramas as well as the personalities of the Native American and Hispanic villagers.

The entryway features, among other things, John Dunn's roulette wheel. Lineberry traveled throughout Europe studying techniques for preservation and storage, as well as the display space, climate control, and lighting of fine museums. As a result, the museum is state-of-the-art. There are works by Van Vechten, as well as some less accomplished local work. The museum is actively acquiring new works. Besides the main gallery space, there are smaller areas available for traveling exhibitions and a wonderful library that will be open by appointment to researchers.

ART & COOKING CLASSES

Perhaps you're visiting Taos because of its renown as an arts community, but galleries and studio visits may not satisfy your own urge to create. If you'd like to pursue an artistic adventure of your own here, check out the weeklong classes in such media as writing, sculpture, painting, jewelry making, photography, clay working, and textiles that are available at the **Taos Institute of Arts,** Box 5280 NDCBU, Taos, NM 87571 (☎ **505/758-2793;** e-mail: tia@taosnet.com; web: http://www.taosnet.com/tia/). Class sizes are limited, so if you're thinking about giving these workshops a try, call for information well in advance. The fees vary from class to class but are generally quite reasonable; however, they usually don't include the cost of materials.

3 Organized Tours

Damaso and Helen Martinez's **Pride of Taos Tours,** P.O. Box 5271, Taos, NM 87571 (☎ **800/273-8340** or 505/758-8340), offers two tours a day of the historical streets in Taos, Taos Pueblo, Martinez Hacienda, and Ranchos de Taos Church ($15 adults, $5 children 12 and under). The tour takes about $2^{1}/_{2}$ hours, and the cost includes admission to the sites.

An excellent opportunity to explore the historic downtown area is offered by **Taos Historic Walking Tours** (☎ **505/758-4020**). Call for schedule and prices.

4 Skiing

DOWNHILL

Five alpine resorts are within an hour's drive of Taos. All offer complete facilities, including equipment rentals. Although exact opening and closing dates vary according to snow conditions, the season usually begins around Thanksgiving and continues into early April.

Ski clothing can be purchased, and ski equipment rented or bought, from several Taos outlets. Among them are **Cottam's Ski & Outdoor Shops,** with four locations (call ☎ **800/322-8267** or 505/758-2822 for the one nearest you); **Taos Ski Valley Sportswear, Ski & Boot Co.,** in Taos Ski Valley (☎ **505/776-2291**); and **Looney Tunes Ski Shop,** also in the ski valley (☎ **505/776-8839**).

✪ **Taos Ski Valley** (P.O. Box 90), Taos Ski Valley, NM 87525 (☎ **505/776-2291;** web: http://taoswebb.com/nmusa/), is the preeminent ski resort in the southern Rocky Mountains. It was founded in 1955 by a Swiss-German immigrant, Ernie Blake. According to local legend, Blake searched for 2 years in a small plane for the perfect location for a ski resort comparable to what he was accustomed to in

---

Shredders (Snowboarders) Unite!

As you drive around the area you may see graffiti proclaiming "Free Taos" on the sides of buildings or roadside signs. With recent developments in Montana and Texas, you might think that these are the marks of a local separatist militia. On the contrary, they are part of a campaign by mostly young people (with many of the area's lodgers behind it as well) to open the ski area up to snowboarders. Traditional downhill skiers don't look kindly on sharing the mountain with the shredders, who they claim make the sport more dangerous. Currently, Taos is one of only a handful of ski resorts in the west that bans boarders completely from its slopes. However, many of the area's lodgers feel they are losing out on significant business from families and young adults who are into snowboarding. In the spring of 1997, the "Free Taos" message appeared in hundred-foot-high letters emblazoned in the snow across an open slope above the ski area.

---

the Alps. He found it at the abandoned mining site of Twining, high above Taos. Today, under the management of two younger generations of Blakes, the resort has become internationally renowned for its light, dry powder (320 inches annually), its superb ski school, and its personal, friendly manner.

Taos Ski Valley, however, can best be appreciated by the more experienced skier. It offers steep, high-alpine, high-adventure skiing. The mountain is more intricate than it might seem at first glance, and it holds many surprises and challenges—even for the expert. The esteemed *London Times* called the valley "without any argument the best ski resort in the world. Small, intimate, and endlessly challenging, Taos simply has no equal." And, if you're sick of dealing with yahoos on snowboards, you will be pleased to know that they're not permitted on the slopes of Taos Ski Valley (the only ski area in New Mexico that forbids them). The quality of the snow here (light and dry) is believed to be due to the dry southwestern air and abundant sunshine.

Between the 11,819-foot summit and the 9,207-foot base, there are 72 trails and bowls, and more than half of them are designated for expert and advanced skiers. Most of the remaining trails are suitable for advanced intermediates; there is little flat terrain for novices to gain experience and mileage. However, many beginning skiers find that after spending time in lessons they can enjoy the **Kachina Bowl,** which offers spectacular views as well as wide-open slopes.

The area has an uphill capacity of 15,000 skiers per hour on its five double chairs, one triple, four quads, and one surface tow. Tickets for all lifts, depending on the season, cost $27 to $40 for adults for a full day, $26 half day; $17 to $25 for children 12 or younger for a full day, $16 half day; teen ticket for 13- to 16-year-olds $27 to $30 for a full day, $21 half day; $25 seniors ages 65 to 69 for a full day; free for seniors over 70. Novice lift tickets cost $23 for adults, $17 for children. Full rental packages are $14 for adults and $7 for children. Taos Ski Valley is open daily from 9am to 4pm from Thanksgiving to the first week of April. It should be noted that Taos Ski Valley has one of the best ski schools in the country. This school specializes in teaching people how to negotiate steep and challenging runs.

With its children's ski school, Taos Ski Valley has always been an excellent location for skiing families, but with the 1994 addition of an 18,000-square-foot children's center (Kinderkäfig Center), skiing with your children in Taos is even better. Kinderkäfig offers every service imaginable, from equipment rental for children to baby-sitting services. Call ahead for more information.

Taos Ski Valley has many lodges and condominiums with nearly 1,000 beds. (See "Taos Ski Valley," in chapter 12, for details on accommodations.) All offer ski-week packages; four of them have restaurants. There are two more restaurants on the mountain in addition to the many facilities of Village Center at the base. For reservations, call the **Taos Valley Resort Association** (☎ **800/776-1111** or 505/776-2233).

Not far from Taos Ski Valley is **Red River Ski Area,** (P.O. Box 900), Red River, NM 87558 (☎ **800/348-6444** or 505/754-2223 for reservations). One of the bonuses of this ski area is the fact that lodgers at Red River can walk out their doors and be on the slopes. Two other factors make this 37-year-old, family-oriented area special: First, most of its 57 trails are geared toward the intermediate skier, though beginners and experts also have some trails; and second, good snow is guaranteed early and late in the year by snowmaking equipment that can work on 75% of the runs, more than any other in New Mexico. However, be aware that this human-made snow tends to be icy, and the mountain is full of inexperienced skiers, so you really have to watch your back. Locals in the area refer to this as "Little Texas" because it's so popular with Texans and other Southerners. A very friendly atmosphere, with a touch of red-neck attitude, prevails.

There's a 1,600-foot vertical drop here to a base elevation of 8,750 feet. Lifts include four double chairs, two triple chairs, and a surface tow, with a skier capacity of 7,920 skiers per hour. The cost of a lift ticket for all lifts is $37 for adults for a full day, $27 for a half day; $30 for teens 13 to 19 for a full day, $21 for a half day; $23 for children 12 and under and seniors (60 and over) for a full day, $16 for a half day. Full rental packages start at $14 for adults, $9 for children. Red River Ski Area is open daily from 9am to 4pm from Thanksgiving to March 30.

Also quite close to Taos is **Angel Fire Resort** (P.O. Drawer B), Angel Fire, NM 87710 (☎ **800/633-7463** or 505/377-6401). The 32 miles of ski runs here are heavily oriented toward beginning and intermediate skiers. You'll find lots of families skiing here because of special ticket rates and the gentleness of the mountain. Still, with a vertical drop of 2,180 feet to a base elevation of 8,500 feet, advanced skiers may find something to like, but if you want a challenge, you'll need to go to Taos. The area's lifts—one Poma surface tow, four doubles, and one high-speed quad— have an hourly capacity of 5,770 skiers. All-day lift tickets cost $36 for adults and $21 for children. Open from approximately Thanksgiving to March 29 (depending on the weather) daily from 8:30am to 4:30pm.

The oldest ski area in the Taos region, founded in 1952, **Sipapu Ski Area** (P.O. Box 29), Vadito, NM 87579 (☎ **505/587-2240**), is 25 miles southeast, on NM 518 in Tres Ritos canyon. It prides itself on being a small local area, especially popular with schoolchildren. It has just one triple chair and two surface lifts, with a vertical drop of 865 feet to the 8,200-foot base elevation. There are 18 trails, half classified as intermediate. Unfortunately beginners are limited to only a few runs here, as are advanced skiers. Still, it's a nice little area, tucked way back in the mountains, with excellent lodging rates. Be aware that since the elevation is fairly low, runs get very icy. All-day lift tickets are $28 for adults and $21 for children under 12. Open from Thanksgiving to April daily from 8am to 4pm.

Just south of the Colorado border is **Ski Rio** (P.O. Box 159), Costilla, NM 87524 (☎ **800/2-ASK-RIO** or 505/758-7707), a broad (and often cold and windy) ski area that can't quite get over its financial problems. In fact, you'll want to call before driving up there, as the resort may not make it into the next season. It's a pity because there's a lot on this mountain. Half of its 83 named trails are for the intermediate skier, 30% for beginners, and 20% for advanced skiers. There are also snowboard and snow skate parks, as well as 13 miles of cross-country trails. Annual snowfall here is about 260 inches, and there are three chairlifts (two triple, one double) and three tows. At the ski base you can rent skis, snowboards, snowshoes, and snow skates, as well as find lodgings, restaurants, and a sports shop. Sleigh rides, dogsled tours, and snowmobile tours are also available. The ski school offers private and group clinics (for adults and children) in cross-country and downhill skiing, snow skating, and snowboarding. Lift tickets are $34 for adults ($23–$28 during the value season), $24 for juniors 7–12, free for children 6 and under with a paying adult. Ski Rio is open daily from 9am to 4pm from November through April.

For information via the World Wide Web, try: **http://laplaza.com/tp/skirio/**.

CROSS COUNTRY

Numerous popular Nordic trails exist in Carson National Forest. If you call or write ahead, they'll send you a booklet titled "Where to Go in the Snow," which gives cross-country skiers details about the maintained trails. One of the more popular trails is **Amole Canyon,** off NM 518 near the Sipapu Ski Area, where the Taos Nordic Ski Club maintains set tracks and signs along a 3-mile loop. It's closed to snow-mobiles, a comfort to lovers of serenity. Several trails are open only to cross-country skiers.

Just east of Red River, with 31 miles of groomed trails in 600 acres of forestlands atop Bobcat Pass, is the **Enchanted Forest Cross Country Ski Area** (☎ **505/ 754-2374**). Full-day trail passes, good from 9am to 4:30pm, are $9 for adults, much less for children. Equipment rentals and lessons can be arranged at Miller's Crossing ski shop on Main Street in Red River (☎ 505/754-2374). Nordic skiers can get instruction in "skating," mountaineering, and telemarking.

Taos Mountain Outfitters, 114 South Plaza (☎ **505/758-9292**), offers telemark and cross-country sales, rentals, and guide service, as do Los Rios Whitewater Ski Shop (☎ **505/776-8854**) and **Southwest Nordic Center** (☎ **505/758-4761**).

5 More Outdoor Activities

Taos County's 2,200 square miles embrace a great diversity of scenic beauty, from New Mexico's highest mountain, 13,161-foot Wheeler Peak, to the 650-foot-deep chasm of the Rio Grande Gorge. Carson National Forest, which extends to the eastern city limits of Taos and cloaks a large part of the county, contains several major ski facilities as well as hundreds of miles of hiking trails through the Sangre de Cristo Range.

Recreation areas are mainly in the national forest, where pines and aspen provide refuge for abundant wildlife. Forty-eight areas are accessible by road, including 38 with campsites. There are also areas on the high desert mesa, carpeted by sagebrush, cactus, and frequently wildflowers. Both types of terrain are favored by hunters, fishermen, and horseback riders. Two beautiful areas within a short drive of Taos are the Valle Vidal Recreation Area, north of Red River, and the Wild Rivers Recreation Area, near Questa. For complete information, contact **Carson National Forest,** P.O. Box 558, Taos, NM 87571 (☎ **505/758-6200**), or the **Bureau of Land Management,** 224 Cruz Alta Rd. (P.O. Box 6168), Taos, NM 87571 (☎ **505/758-8851**).

BALLOONING As in many other towns throughout New Mexico, hot-air ballooning is a top attraction. Recreational trips are offered by **Paradise Hot Air Balloon Adventure** (☎ **505/751-6098**).

The **Taos Mountain Balloon Rally,** P.O. Box 3096, Taos, NM 87571 (☎ **800/ 732-8267**), is held each year the last full weekend of October. (See "Northern New Mexico Calendar of Events," in chapter 2, for details.)

BIKING Even if you're not an avid cyclist, it won't take long for you to realize that getting around Taos by bike is preferable to driving. You won't have the usual parking problems, and you won't have to sit in the line of traffic as it snakes through the center of town. If you feel like exploring the surrounding area, Carson National Forest rangers recommend several biking trails in the greater Taos area, including those in Garcia Park and Rio Chiquito for beginner to intermediate mountain bikers, and a number of Gallegos and Picuris peaks for experts. Inquire at the U.S. Forest Service office next to the Chamber of Commerce for excellent materials that map out trails; tell you how to get to the trailhead; specify length, difficulty, and elevation; and inform you about safety tips. You can also purchase the Taos Trails map (created jointly by the Carson National Forest, Native Sons Adventures, and Trail Illustrated). It's readily available at area bookstores and is designed to withstand water damage. Once you're out riding in Carson National Forest, you'll find trails marked in green (easy), blue (moderate), or gray (expert). Bicycle rentals are available from the **Gearing Up Bicycle Shop,** 129 Paseo del Pueblo Sur (☎ **505/751-0365**); daily rentals run $20 for a mountain bike with front suspension. **Hot Tracks Cyclery & Ski Touring Service,** 214 Paseo del Pueblo Sur (☎ **505/751-0949**), rents front-suspension mountain bikes and full-suspension bikes for $15/half day and $20/full day; and **Native Sons Adventures,** 715 Paseo del Pueblo Sur (☎ **800/753-7559** or 505/758-9342), rents regular (unsuspended) bikes for $15/half day and $20/full day, front-suspension bikes for $25/half day and $35/full day, and full-suspension bikes for $35/half day and $45/full day; they also rent car racks for $5. All of these prices include use of helmets and water bottles.

Annual touring events include Red River's **Enchanted Circle Century Bike Tour** (☎ **505/754-2366**) on September 13.

FISHING The fishing season in the high lakes and streams opens April 1 and continues through December, though spring and fall tend to be the best times. Naturally, the Rio Grande is a favorite fishing spot, but there is also excellent fishing in the streams around Taos. Taoseños favor the Rio Hondo, Rio Pueblo (near Tres Ritos), Rio Fernando (in Taos Canyon), Pot Creek, and Rio Chiquto. Rainbow, cutthroat, and German brown trout and kokanee (a freshwater salmon) are commonly stocked and caught. Pike and catfish have been caught in the Rio Grande as well. Jiggs, spinners, or woolly worms are recommended as lure, or worms, corn, or salmon eggs as bait, but many experienced anglers prefer fly-fishing.

Licenses are required, of course; they are sold, along with tackle, at several Taos sporting-goods shops. For backcountry guides, try **Deep Creek Wilderness Outfitters and Guides,** P.O. Box 721, El Prado, NM 87529 (☎ **505/776-8423**), or **Taylor Streit Flyfishing Service,** P.O. Box 2759 (☎ **505/751-1312**) in Taos.

FITNESS FACILITIES The **Taos Spa and Court Club,** 111 Dona Ana Dr. (☎ **505/758-1980**), is a fully equipped fitness center that rivals any you'd find in a big city. There are treadmills, step machines, climbing machines, rowing machines, exercise bikes, NordicTrack, weight-training machines, saunas, indoor and outdoor hot tubs, a steam room, and indoor and outdoor pools. Thirty-five aerobic "step" classes a week, as well as stretch aerobics, aqua aerobics, and classes specifically designed for senior citizens are also offered. In addition, there are five tennis and two racquetball courts. Therapeutic massage is available daily by appointment. Children's programs include tennis and swimming camp, and baby-sitting programs are available in the morning and evening. The spa is open Monday through Friday from 5:30am to 9pm; Saturday and Sunday from 7am to 8pm. Monthly, weekly, and daily memberships are available for individuals and families. For visitors there's a daily rate of $10.

The **Northside Health and Fitness Center,** at 1307 Paseo del Pueblo Norte, in Taos (☎ 505/751-1242), is also a full-service facility, featuring top-of-the-line Cybex equipment, free weights, and cardiovascular equipment. Aerobics and Jazzercise classes are scheduled daily, and there are indoor/outdoor pools and four tennis courts, as well as children's and senior citizens' programs.

GOLF   Since the summer of 1993 the 18-hole golf course at the **Taos Country Club,** Ranchos de Taos (☎ 800/758-7375 or 505/758-7300), has been open to the public. Located on NM 570, just 4 miles south of the Plaza, it's a first-rate championship golf course designed for all levels of play—in fact, it is ranked as the third-best course in New Mexico. It has open fairways and no hidden greens. The club also features a driving range, practice putting and chipping green, an additional 9-hole course, and instruction by PGA professionals. Greens fees in 1997 were $32 during the week, $40 on weekends (includes Friday) and holidays for 18 holes. Twilight fee is $23. Cart and club rentals are also available. It's always advisable to call ahead for tee times, but it's not unusual for people to show up unannounced and still manage to find a time to tee off.

The par-72, 18-hole course at the **Angel Fire Country Club and Golf Course** (☎ 505/377-3055) has been endorsed by the Professional Golfers Association. Surrounded by stands of ponderosa pine, spruce, and aspen, at 8,500 feet, it's one of the highest regulation golf courses in the world. It also has a driving range and putting green. Carts and clubs can be rented at the course, and the club pro provides instruction.

For 9-hole play, stop at the golf course at **Valle Escondido** residential village just off US 64. It's a par-36 course with mountain and valley views. Clubs and pull-carts are available for rental, and the clubhouse serves refreshments.

Another golf course is under construction in Red River. At press time 9 holes were open. Call **Red Eagle Golf Course** (☎ 505/754-6569).

HIKING   There are hundreds of miles of hiking trails in Taos County's mountain and high-mesa country. They're especially well traveled in the summer and fall, although nights turn chilly and mountain weather may be fickle by September.

Maps (for a nominal fee) and free materials and advice on all **Carson National Forest** trails and recreation areas can be obtained from the **Forest Service Building,** 208 Cruz Alta Rd. (☎ 505/758-6200), and from the office adjacent to the Chamber of Commerce on Paseo del Pueblo Sur. Both are open Monday through Saturday from 8am to 4:30pm. Detailed USGS topographical maps of backcountry areas can be purchased from **Taos Mountain Outfitters** on the Plaza (☎ 505/758-9292). This is also the place to rent camping gear, if you came without your own. Tent rentals are $14 and sleeping bags are $9 each per day. Backpacks can be rented for $14 a day. Ask about special deals on weekend packages.

Two wilderness areas close to Taos offer outstanding hiking possibilities. The 19,663-acre **Wheeler Peak Wilderness** is a wonderland of Alpine tundra encompassing New Mexico's highest peak (13,161 feet). The 20,000-acre **Latir Peak Wilderness,** north of Red River, is noted for its high lake country. Both are under the jurisdiction of the **Questa Ranger District,** P.O. Box 110, Questa, NM 87556 (☎ 505/586-0520).

HORSEBACK RIDING The **Taos Indian Horse Ranch,** on Pueblo land off Ski Valley Road, just before Arroyo Seco (☎ **505/758-3212**), offers a variety of guided rides. Open by appointment, the ranch provides horses for all types of riders (English, western, bareback) and ability levels. Call ahead to reserve. Rates start at $32 and go up to $65 to $95 for a 2-hour trail ride. Horse-drawn hay wagon rides are also offered in summer. From late November to March, the ranch provides afternoon and evening sleigh rides to a bonfire and marshmallow roast at $25 to $37.50 per person; ask for prices for a steak cookout. Also ask about the bed-and-breakfast special for $50 per person and the Indian minivacation, which includes riding, a raft trip, and pack trip for $495.

Horseback riding is also offered by the **Shadow Mountain Guest Ranch,** 6 miles east of Taos on US 64 (☎ **505/758-7732**); **Rio Grande Stables** (P.O. Box 2122), El Prado (☎ **505/776-5913**); and **Llano Bonito Ranch** (P.O. Box 99), Penasco, about 40 minutes from Taos (☎ **505/587-2636;** fax 505/587-2636). Rates at Llano Bonito Ranch are $22 for a 1-hour trail ride, $70 per person for a half-day ride ($95 if breakfast is included), and $100 per person for a full-day ride. In addition to trail rides, Llano Bonito Ranch offers 3-day pack trips for $595 per person. On the 3-day trip you'll spend 2 nights in the high-country wilderness, and during the day you'll ride to an altitude of 12,500 feet. Meals are included on the pack trip.

Most riding outfitters offer lunch trips and overnight trips. Call for further details.

HUNTING Hunters in Carson National Forest bag deer, turkey, grouse, band-tailed pigeons, and elk by special permit. On private land, where hunters must be accompanied by qualified guides, there are also black bear and mountain lions. Hunting seasons vary year to year, so it's important to inquire ahead with the **New Mexico Game and Fish Department in Santa Fe** (☎ **505/827-7882**).

Several Taos sporting-goods shops sell hunting licenses. Backcountry guides include **Moreno Valley Outfitters** (☎ **505/377-3512**) in Angel Fire and **Rio Costilla Park** (☎ **505/586-0542**) in Costilla.

ICE-SKATING For ice-skating, **Kit Carson Park Ice Rink** (☎ **505/758-8234**), located in Kit Carson Park, is open from Thanksgiving through February. Skate rentals are available for adults and children.

JOGGING You can jog anywhere (except on private property) in and around Taos. I would especially recommend stopping by the Carson National Forest office in the Chamber of Commerce building to find out what trails they might recommend.

LLAMA TREKKING For a taste of the unusual, **El Paseo Llama Expeditions** (☎ **800/455-2627** or 505/758-3111) utilizes U.S. Forest Service–maintained trails that wind through canyons and over mountain ridges. The llamas will carry your gear and food, allowing you to walk and explore, free of any heavy burdens. They're friendly, gentle animals that have a keen sense of sight and smell. Often, other animals, such as elk, deer, and mountain sheep, are attracted to the scent of the llamas and will venture closer to hikers if the llamas are present. Llama expeditions are scheduled from June to early October. Half-day hikes cost $49, day hikes $70, and 2- to 8-day hikes run up to $850. **Taos Llama Adventures** (☎ **800/758-LAMA** or 505/776-1044) also offers half- or full-day as well as overnight llama treks.

RIVER RAFTING Half- or full-day white-water rafting trips down the Rio Grande and Rio Chama originate in Taos and can be booked through a variety of outfitters in the area. The wild **Taos Box,** a steep-sided canyon south of the Wild Rivers Recreation Area, is especially popular. May and June, when the water is rising, is a good time to go. Experience is not required, but you will be required to wear a life jacket (provided), and you should be willing to get wet.

One convenient rafting service is **Rio Grande Rapid Transit,** P.O. Box A, Pilar, NM 87531 (☎ **800/222-RAFT** or 505/758-9700). In addition to Taos Box ($69 to $85 per person), Rapid Transit also runs the **Pilar Racecourse** ($30 to $32 per person) on a daily basis. Its headquarters is at the entrance to the BLM-administered **Orilla Verde Recreation Area,** 16 miles south of Taos, where most excursions through the Taos Box end. Several other serene but thrilling floats through the Pilar Racecourse start at this point.

Other rafting outfitters in the Taos area include **Native Sons Adventures,** 715 Paseo del Pueblo Sur (☎ **800/753-7559** or 505/758-9342), and **Far Flung Adventures** (☎ **800/359-2627** or 505/758-2628).

Safety Warning: Taos is not the place to experiment if you are not an experienced rafter. Do yourself a favor and check with the **Bureau of Land Management** (☎ **505/758-8851**) to make sure that you're fully equipped to go white-water rafting without a guide. Have them check your gear to make sure that it's sturdy enough— this is serious rafting!

SPAS   Ojo Caliente Mineral Springs, Ojo Caliente, NM 87549 (☎ **800/ 222-9162** or 505/583-2233), is on US 285, 50 miles (a 1-hour drive) southwest of Taos. This National Historic Site was considered sacred by prehistoric tribes. When Spanish explorer Cabeza de Vaca discovered and named the springs in the 16th century, he called them "the greatest treasure that I found these strange people to possess." No other hot spring in the world has Ojo Caliente's combination of iron, soda, lithium, sodium, and arsenic. The resort offers herbal wraps and massages, lodging, and meals. It's open in summer daily from 8am to 9pm; in winter (November through March) the springs are open from 8am to 8pm.

SWIMMING   The **Don Fernando Pool,** on Civic Plaza Drive at Camino de la Placita, opposite the new Convention Center, admits swimmers over age 6 without adult supervision.

TENNIS   Quail Ridge Inn (see chapter 12) has six outdoor and two indoor tennis courts. **Taos Spa and Tennis Club** (see "Fitness Facilities" above) has five courts, and the **Northside Health and Fitness Center** (see above) in El Prado has three tennis courts. In addition, there are four free public courts in Taos, two at **Kit Carson Memorial State Park,** on Paseo del Pueblo Norte, and two at **Fred Baca Memorial Park,** on Camino del Medio south of Ranchitos Road.

6 Shopping

Given the town's historical associations with the arts, it isn't surprising that many visitors come to Taos to buy fine art. Some 50-odd galleries are located within easy walking distance of the Plaza, and a couple dozen more are just a short drive from downtown. Galleries are generally open 7 days a week, especially in high season. Some artists show their work by appointment only.

The best-known artist in modern Taos is R. C. Gorman, a Navajo from Arizona who has made his home in Taos for more than two decades. Now in his 50s, Gorman is internationally acclaimed for his bright, somewhat surrealistic depictions of Navajo women. His **Navajo Gallery,** at 210 Ledoux St. (☎ **505/758-3250**), is a showcase for his widely varied work: acrylics, lithographs, silk screens, bronzes, tapestries, handcast ceramic vases, etched glass, and more.

A good place to begin exploring galleries is the **Stables Fine Art Gallery,** operated by the Taos Art Association at 133 Paseo del Pueblo Norte (☎ **505/758-2036**). A rotating group of fine arts exhibits features many of Taos's emerging and established artists. All types of work are exhibited, including painting, sculpture, printmaking, photography, and ceramics. Admission is free; it's open year-round daily from 10am to 5pm.

My favorite place to shop in Taos is in the **St. Francis Plaza** in Rancho de Taos, just a few miles south of the Plaza. This is what shopping in Northern New Mexico once was. Forget T-shirt shops and fast food. Here you'll find small shops, generally with the owner presiding. There's even a little cafe where you can stop to eat, and of course, you'll want to visit the St. Francis of Assisi church (discussed earlier in this chapter).

Here are a few shopping recommendations, listed according to their specialties:

ART

Act I Gallery
226D Paseo del Pueblo Norte. ☎ **800/666-2933** or 505/758-7831.

This gallery has a broad range of works in a variety of media; you'll find watercolors, retablos, furniture, paintings, Hispanic folk art, pottery, jewelry, and sculpture.

✪ Philip Bareiss Contemporary Exhibitions
15 Ski Valley Rd. ☎ **505/776-2284.**

The works of some 30 leading Taos artists, including sculptor Gray Mercer and watercolorist Patricia Sanford, are exhibited here. In 1995 Philip Bareiss opened "Circles and Passageways," a sculptural installation by Gray Mercer, on the 2,500-acre Romero Range located just west of Taos. This is true land art; a four-wheel-drive vehicle is recommended in order to get there.

Desurmont-Ellis Gallery
118 Camino de la Placita. ☎ **505/758-3299.**

Here you'll find abstract and impressionist oils and watercolors, sculpture, ceramics, and jewelry.

✪ El Taller de Taos Gallery
237 Ledoux St. ☎ **505/758-4887.**

Now located near the Harwood Museum, this gallery has exclusive representation of Amado Peña, as well as fine art by an excellent group of Southwestern artists and Native American art and artifacts.

Fenix Gallery
228B N. Pueblo Rd. ☎ **505/758-9120.**

The Fenix Gallery focuses on Taos artists with national and/or international collections and reputations who live and work in Taos. The work is primarily nonobjective and very contemporary. Some "historic" artists are represented as well. Recent expansion has doubled the gallery space.

Franzetti Metalworks
120-G Bent St. ☎ **505/758-7872.**

This work appeals to some more than others. The designs are surprisingly whimsical for metalwork. Much of the work is functional; you'll find laughing horse switch plates and "froggie" earthquake detectors.

Gallery A
105–107 Kit Carson Rd. ☎ **505/758-2343.**

The oldest gallery in town, Gallery A has contemporary and traditional paintings, sculpture, and graphics, including Gene Kloss oils, watercolors, and etchings.

Hirsch Fine Art
146 Kit Carson Rd. ☎ **505/758-5460.**

If you can find this gallery open, it's well worth the visit. Unfortunately, the hours posted are not always maintained. Spread through a beautiful old home are watercolors, etchings and lithographs, and drawings by early Southwestern artists, including the original Taos founders.

Lizard of Oz
156 Vista del Valle. ☎ **505/758-0708.**

This gallery has a great collection of fine Australian art, textiles, pottery, jewelry, and other items.

✪ Lumina of New Mexico
239 Morada Rd. (P.O. Box LL). ☎ **505/758-7282.**

Located in the historic Victor Higgins home, next to the Mabel Dodge estate, Lumina is one of the loveliest galleries in New Mexico. You'll find a large variety of fine art, including paintings, sculpture, and photography. This place is as much a tourist attraction as any of the museums and historic homes in town.

Look for wonderful Picasso-esque paintings of the New Mexico village life by Andrés Martinez and take a stroll through the new 3-acre outdoor sculpture garden with a pond and waterfall—where you'll find large outdoor pieces from all over the United States.

✪ New Directions Gallery
107B North Plaza. ☎ **505/758-2771.**

Here you'll find a variety of contemporary abstract works such as Larry Bell's unique mixed-media "Mirage paintings." My favorite, though, are the impressionistic works depicting Northern New Mexico villages by Tom Noble.

Quast Galleries—Taos
229 and 133 E. Kit Carson Rd. ☎ **505/758-7160** or 505/758-7779.

You won't want to miss this gallery, where you'll find representational landscapes and figurative paintings and distinguished sculpture. Rotating national and international exhibits are shown here.

R. B. Ravens
St. Francis Church Plaza (P.O. Box 850), Ranchos de Taos. ☎ **800/253-5398** or 505/758-7322.

A trader for many years, including 15 on the Ranchos Plaza, R. B. Ravens is skilled at finding incredible period artwork. Here you'll see (and have the chance to buy) a late 19th-century Plains elk teeth woman's dress, as well as moccasins, Navajo rugs, and pottery, all in the setting of an old home, with raw pine floors and hand-sculpted adobe walls.

Shriver Gallery
401 Paseo del Pueblo Norte. ☎ **505/758-4994.**

Traditional paintings, drawings, etchings, and bronze sculpture.

Taos Gallery
403 North Pueblo Rd. ☎ **505/758-2475.**

Here you'll find Southwestern impressionism, traditional Western art, contemporary fine art, and bronze sculpture.

BOOKS

Brodsky Bookshop
218 Paseo del Pueblo Norte. ☎ **505/758-9468.**

Exceptional inventory of fiction, nonfiction, Southwestern and Native American studies, children's books, topographical and travel maps, cards, tapes, and CDs.

Kit Carson Home
E. Kit Carson Rd. ☎ **505/758-4741.**

Fine collection of books about regional history.

✪ Moby Dickens Bookshop
124A Bent St. ☎ **505/758-3050.**

This is one of Taos's best bookstores. You'll find children's and adults' collections of Southwest, Native American, and out-of-print books. A renovation has added 600 square feet, much of it upstairs, where there's a comfortable place to sit and read.

Taos Book Shop
122D Kit Carson Rd. ☎ **505/758-3733.**

Founded in 1947, this is the oldest general bookstore in New Mexico. Taos Book Shop specializes in out-of-print and Southwestern titles.

CRAFTS

Clay & Fiber Gallery
126 W. Plaza Dr. ☎ **505/758-8093.**

Clay & Fiber represents over 150 artists from around the country; merchandise changes frequently, but you should expect to see a variety of ceramics, fiber arts, jewelry, and wearables.

Southwestern Arts
In the Dunn House, Bent St. ☎ **505/758-8418.**

This shop features historic and contemporary Navajo weavings, Pueblo pottery, and jewelry, as well as photography by Dick Spas.

Southwest Moccasin & Drum
803 Paseo del Pueblo Norte. ☎ **800/447-3630** or 505/758-9332.

Home of the All One Tribe Drum, this favorite local shop carries a large variety of drums in all sizes and styles, handmade by master Native American drum makers from Taos Pueblo. Southwest Moccasin & Drum also has the country's second-largest selection of moccasins, as well as an incredible inventory of indigenous world instruments and tapes, sculpture, weavings, rattles, fans, fetishes, bags, decor, and many handmade one-of-a-kind items. A percentage of the store's profits goes to support Native American causes.

✪ Taos Artisans Cooperative Gallery
107A Bent St. ☎ **505/758-1558.**

This eight-member cooperative gallery, owned and operated by local artists, sells local handmade jewelry, wearables, clay work, glass, drums, baskets, leather work, garden sculpture, and woven Spirit Women. You'll always find an artist in the shop.

Taos Blue
101A Bent St. ☎ **505/758-3561.**

This gallery has fine Native American and contemporary handcrafts; it specializes in clay and fiber work.

Twining Weavers and Contemporary Crafts
133 and 135 Paseo del Pueblo Norte. ☎ **505/758-9089** or 505/758-9000.

Here you'll find an interesting mix of handwoven wool rugs and pillows by owner Sally Bachman, as well as creations by other gallery artists in fiber, basketry, and clay.

Weaving Southwest
216 Paseo del Pueblo Norte. ☎ **505/758-0433.**

Contemporary tapestries by New Mexico artists, as well as one-of-a-kind rugs, blankets, and pillows, are the woven specialties found here.

FASHIONS

Martha of Taos
121 Paseo del Pueblo Norte. ☎ **505/758-3102.**

This boutique next to the Taos Inn is the place to go if you want to try on some elegant Southwestern style. Martha has pleated "broomstick" skirts, velvet dresses, and Navajo-style blouses.

Overland Sheepskin Company
NM 522. ☎ **505/758-8822.**

You can't miss the romantically weathered barn sitting on a meadow north of town. Inside you'll find anything you can imagine in leather: gloves, hats, slippers, coats. The coats here are exquisite, from oversized ranch styles to tailored blazers in a variety of leathers from sheepskin to buffalo hide.

FOOD

Amigos Co-op Natural Grocery
326 Paseo del Pueblo Sur. ☎ **505/758-8493.**

If you've had your fill of rich food, this is the place to find healthy treats. In the front of the store are organic fruits and vegetables. And in the back, there's a small cafe, where you can eat or take out sandwiches, healthy green chile stew, or my favorite, a huge plate of stir-fried veggies over brown rice. You'll also find baked goods such as muffins and brownies.

Casa Fresen Bakery
Ski Valley Rd., Hwy. 150. ☎ **505/776-2969.**

Located on the road to the Taos Ski Valley in Arroyo Seco, this is a wonderful place to buy fresh pastries, cakes, cheeses, pâtés, specialty meats, pastas, sauces, preserves, and oils. You can enjoy a sandwich right here or select a box lunch to take along on a picnic. It's open Wednesday through Monday from 7:30am to 5pm.

FURNITURE

Country Furnishings of Taos
534 Pueblo Norte. ☎ **505/758-4633.**

Here you'll find unique hand-painted folk-art furniture that has become popular all over the country. The pieces are as individual as the styles of the local folk artists who make them. There are also home accessories, unusual gifts, clothing, and jewelry.

Greg Flores Furniture of Taos
120 Bent St. ☎ **800/880-1090** or 505/758-8010.

This is a great little find. Greg Flores, native of Taos, fashions Southwestern furniture out of native ponderosa pine; he uses wood joinery and hand rubs each piece with an oil finish. You'll also find charming paintings by his wife Johanna Flores.

Lo Fino

201 Paseo del Pueblo Sur. ☎ **505/758-0298.**

With a name meaning "the refined," you know that this expansive showroom is worth taking time to wander through. You'll find a variety of home furnishings, from driftwood lamps and exotic masks made of dried flowers, to wagon-wheel furniture and finely painted *trasteros* (armoires), as well as handcrafted traditional and contemporary Southwestern furniture. Lo Fino specializes in custom building furniture.

Taos Company

124K John Dunn Plaza, Bent St. ☎ **800/548-1141** or 505/758-1141.

This interior design showroom specializes in unique Southwestern antique furniture and decorative accessories. Especially look for wrought-iron lamps and iron-and-wood furniture.

GIFTS & SOUVENIRS

Big Sun

66 St. Francis Church Plaza, Ranchos de Taos. ☎ **505/758-3100.**

This folk-art gallery and curio emporium is packed with beautiful objects, from rugs made by the Tarahumara Indians in Mexico to local tinwork made in Dixon and authentic Navajo throw rugs, made by beginning weavers, that cost around $70.

El Rincón

114 Kit Carson Rd. ☎ **505/758-9188.**

This shop has a real trading post feel. It's a wonderful place to find turquoise jewelry, whether you're looking for contemporary or antique. In the back of the store is a museum full of Native American and Western artifacts.

JEWELRY

Artwares Contemporary Jewelry

Taos Plaza (P.O. Box 2825). ☎ **800/527-8850** or 505/758-8850.

The gallery owners here call their contemporary jewelry "a departure from the traditional." Indeed, each piece here is a new twist on traditional Southwestern and Native American design.

Leo Weaver Gallery

62 St. Francis Plaza (P.O. Box 1596), Ranchos de Taos. ☎ **505/751-1003.**

This shop carries the work of 40 local silversmiths. There's an expansive collection of concho belts and some very fresh work, using a variety of stones from charolite to lapis.

Taos Gems & Minerals

637 Paseo del Pueblo Sur. ☎ **505/758-3910.**

This is a great place to explore; you can buy items like fetishes at prices much more reasonable than most galleries. Now in its 30th year of business, Taos Gems & Minerals is a fine lapidary showroom. You can also buy jewelry, carvings, and antique pieces at reasonable prices.

MUSICAL INSTRUMENTS

Taos Drum Company

5 miles south of Taos Plaza, off NM 68. ☎ **505/758-3796.**

Drum making is an age-old tradition that local artisans give continued life to in Taos. The drums are made of hollowed-out logs stretched with rawhide, and they come in all different shapes, sizes, and styles. Taos Drums has the largest selection of Native American log and hand drums in the world. In addition to drums, the showroom displays Southwestern and wrought-iron furniture, cowboy art, and lamps, as well as a constantly changing selection of primitive folk art, ethnic crafts, Native American music tapes, books, and other information on drumming. To find Taos Drum Company, look for the tepees and drums off NM 68. Ask about the tour that demonstrates the drum-making process.

POTTERY & TILES

Stephen Kilborn Pottery

136D Paseo del Pueblo Norte. ☎ **505/758-5760.**

Visiting this shop in town is a treat, but for a real adventure go 17 miles south of Taos toward Santa Fe to Stephen Kilborn's studio, open daily 8:30am to 5pm. There you'll see him throw, decorate, and fire pottery that's fun, fantastical, and functional.

Vargas Tile Co.

NM 68. ☎ **505/758-5986.**

Vargas Tile has a great little collection of hand-painted Mexican tiles at good prices. You'll find beautiful pots with sunflowers on them and colorful cabinet doorknobs, as well as inventive sinks.

7 Taos After Dark

For a small town, Taos has its share of top entertainment. Performers are attracted to Taos because of the resort atmosphere and the arts community, and the city enjoys annual programs in music and literary arts. State troupes, such as the New Mexico Repertory Theater and New Mexico Symphony Orchestra, make regular visits.

Many events are scheduled by the **Taos Art Association,** 145 Paseo del Pueblo Norte (P.O. Box 198), Taos, NM 87571 (☎ **505/758-2052**), at the **Taos Community Auditorium** (☎ **505/758-4677**). The TAA imports local, regional, and national performers in theater, dance, and concerts (Dave Brubeck, the late Dizzy Gillespie, the American String Quartet, and the American Festival Ballet have performed here) and offers two weekly film series, including one for children.

You can obtain information on current events in the *Taos News,* published every Thursday. The **Taos County Chamber of Commerce** (☎ **800/732-TAOS** or 505/758-3873) publishes semiannual listings of "Taos County Events," as well as an annual Taos Country Vacation Guide that also lists events and happenings around town.

THE PERFORMING ARTS

MAJOR ANNUAL PROGRAMS

Fort Burgwin Research Center

6580 NM 518, Ranchos de Taos, NM 87557. ☎ **505/758-8322.**

This historic site (of the 1,000-year-old Pot Creek Pueblo), located about 10 miles south of Taos, is a summer campus of Dallas's Southern Methodist University. From mid-May through mid-August, the SMU-In-Taos curriculum (such as studio arts, humanities, and sciences) includes courses in music and theater. There are regularly scheduled orchestral concerts, guitar and harpsichord recitals, and theater performances available to the community, without charge, throughout the summer.

Music from Angel Fire

P.O. Box 502, Angel Fire, NM 87710. ☎ **505/377-3233.**

The Major Concert & Performance Halls

Taos Civic Plaza and Convention Center, 121 Civic Plaza Dr. (☎ 505/758-4160).
Taos Community Auditorium, Kit Carson Memorial State Park (☎ 505/758-4677).

This acclaimed program of chamber music begins in mid-August with weekend concerts and continues up to Labor Day. Based in the small resort community of Angel Fire (located about 21 miles east of US 64), it also presents numerous concerts in Taos, Las Vegas, and Raton.

Taos Poetry Circus
☎ 505/758-1800. Office mailing address: 5275 NDCBU, Taos, NM 87571. Events take place at various venues around town. A good source for information is Cafe Tazza, 122 Kit Carson Rd.

Aficionados of the literary arts appreciate this annual event, held during 8 days in mid-June. Billed as "a literary gathering and poetry showdown among nationally known writers," it includes readings, seminars, performances, public workshops,and a poetry video festival. The main event is the **World Heavyweight Championship Poetry Bout,** 10 rounds of hard-hitting readings—with the last round extemporaneous.

Taos School of Music
P.O. Box 1879, Taos, NM 87571. ☎ 505/776-2388. Tickets for chamber music concerts $12 adult, $10 for children under 16.

Sponsored by the Taos Art Association, the Taos School of Music was founded in 1963. It is located at the Hotel St. Bernard in Taos Ski Valley. From mid-June to mid-August there is an intensive 8-week study and performance program for advanced students of violin, viola, cello, and piano. Students receive daily coaching by the American String Quartet and pianist Robert McDonald.

The 8-week **Chamber Music Festival,** an important adjunct of the school, offers 16 concerts and seminars for the public; performances are given by the American String Quartet, pianist Robert McDonald, guest violist Michael Tree (of the Guarneri Quartet), and the international young student artists. Performances are held at the Taos Community Auditorium and the Hotel St. Bernard.

✪ Taos Talking Picture Festival
216M North Pueblo Rd., Taos, NM 87571. ☎ 505/751-0637. Fax 505/751-7385; e-mail: ttpix@taosnet.com; web: http://www.taosnet.com/ttpix/. Individual screenings $6. Fast pass $250. In mid-April; screenings run 9am–11pm.

Filmmakers and film enthusiasts gather to view a variety of films, from serious documentaries to lighthearted comedies. You'll see locally made films as well as films involving Hollywood big-hitters. In 1997, Louis Gossett Jr. and Philip Kaufman showed their films.

THE CLUB & MUSIC SCENE

Adobe Bar
In the Historic Taos Inn, 125 Paseo del Pueblo Norte. ☎ 505/758-2233. No cover. Hours vary; be sure to call ahead.

A favorite gathering place for locals and visitors, the Adobe Bar is known for its live music series (Sunday and Wednesday) devoted to the eclectic talents of Taos musicians. The schedule offers a little of everything—classical, jazz, folk, Hispanic, and acoustic. The Adobe Bar features a wide selection of international beers, wines by the glass, light New Mexican dining, desserts, and an espresso menu.

Alley Cantina

121 Teresina Lane. ☎ **505/758-2121.** No cover.

This new bar has become the hot late-night spot in Taos. The focus is on interaction—so there's no television. Instead, patrons play shuffleboard and pool, as well as chess and backgammon in this building, which is said to be the oldest house in Taos. Pastas, sandwiches, and other dishes are served until past midnight.

Fireside Cantina

At Rancho Ramada, 615 Paseo del Pueblo Sur. ☎ **505/758-2900.** No cover.

Live entertainment during the summer and ski season Friday and Saturday nights. Call for information and schedule.

Hideaway Lounge

At the Holiday Inn, 1005 Paseo del Pueblo Sur. ☎ **505/758-4444.** No cover.

This hotel lounge, built around a large adobe fireplace, offers live entertainment and an extensive hors d'oeuvre buffet. Call for schedule.

Sagebrush Inn

Paseo del Pueblo Sur. ☎ **505/758-2254.** No cover.

This is a real hot spot for locals. The atmosphere is Old West, with a rustic wooden dance floor and plenty of smoke. Dancers generally two-step to country performers nightly, year-round, from 9pm.

Thunderbird Lodge

Taos Ski Valley. ☎ **505/776-2280.** No cover, except occasionally on holidays; then the cost varies widely.

Throughout the winter, the Thunderbird offers a variety of nightly entertainment at the foot of the ski slopes. You'll also find wine tastings and two-step dance lessons here.

8 Exploring Beyond Taos: A Driving Tour of the Enchanted Circle

The one "can't-miss" trip from Taos is an excursion around the Enchanted Circle. This 90-mile loop, a National Forest Scenic Byway, runs through the towns of Questa, Red River, Eagle Nest, and Angel Fire, incorporating portions of NM 522, NM 38, and US 64. Although one can drive the entire loop in 2 hours from Taos, most folks prefer to take a full day, and many take several days.

QUESTA   Traveling north from Taos via NM 522, it's a 24-mile drive to Questa, most of whose residents are employed at a molybdenum mine about 5 miles east of town. En route north, the highway passes near **San Cristobal,** where a side road turns off to the D. H. Lawrence Shrine and **Lama,** site of an isolated spiritual retreat.

If you turn west off NM 522 onto NM 378 about 3 miles north of Questa, you'll descend 11 miles on a gravel road into the gorge of the Rio Grande at the Bureau of Land Management–administered **Wild Rivers Recreation Area (☎ 505/758-8851).** Here, where the Red River enters the gorge, is the most accessible starting point for river-rafting trips through the infamous Taos Box. Some 48 miles of the Rio Grande, south from the Colorado border, are protected under the national Wild and Scenic River Act of 1968. Information on geology and wildlife, as well as hikers' trail maps, can be obtained at the visitor center here. Ask for directions to the impressive petroglyphs in the gorge. River-rafting trips can be booked in Taos, Santa Fe, Red River, and other communities. (See the "Outdoor Activities" sections in chapter 7 and above for booking agents in Santa Fe and Taos, respectively.)

The village of **Costilla,** near the Colorado border, is 20 miles north of Questa. This is the turnoff point for four-wheel-drive jaunts and hiking trips into **Valle Vidal,** a huge U.S. Forest Service–administered reserve with 42 miles of roads and many hiking trails.

RED RIVER   Turn east at Questa onto NM 38 for a 12-mile climb to Red River, a rough-and-ready 1890s gold-mining town that has parlayed its Wild West ambience into a pleasant resort village that's especially popular with families from Texas and Oklahoma.

This community, at 8,750 feet, is a center for skiing and snowmobiling, fishing and hiking, off-road driving and horseback riding, mountain biking, river rafting, and other outdoor pursuits. Frontier-style celebrations, honky-tonk entertainment, and even staged shootouts on Main Street are held throughout the year.

The **Red River Chamber of Commerce,** P.O. Box 870, Red River, NM 87558 (☎ **800/348-6444** or 505/754-2366), lists more than 40 accommodations, including lodges and condominiums. Some are open winters or summers only.

EAGLE NEST   About 16 miles east of Red River, on the other side of 9,850-foot Bobcat Pass, is the village of Eagle Nest, resting on the shore of Eagle Nest Lake in the Moreno Valley. Gold was mined in this area as early as 1866, starting in what is now the ghost town of Elizabethtown about 5 miles north; but Eagle Nest itself (population 200) wasn't incorporated until 1976. The 4-square-mile lake is considered one of the top trout producers in the United States and attracts ice fishermen in winter as well as summer anglers. Sailboats and windsurfers also use the lake, although swimming, waterskiing, and camping are not permitted.

If you're heading to Cimarron or Denver, proceed east on US 64 from Eagle Nest. But if you're circling back to Taos, continue southwest on US 38 and US 64 to Agua Fría and Angel Fire.

Shortly before the Agua Fría junction, you'll see the **DAV Vietnam Veterans Memorial.** It's a stunning structure with curved white walls soaring high against the backdrop of the Sangre de Cristo Range. Consisting of a chapel and underground visitor center, it was built by Dr. Victor Westphall in memory of his son, David, a marine lieutenant killed in Vietnam in 1968. The chapel has a changing gallery of photographs of Vietnam veterans who gave their lives in the Southeast Asian war, but no photo is as poignant as this inscription written by young Davis Westphall, a promising poet:

> Greed plowed cities desolate.
> Lusts ran snorting through the streets.
> Pride reared up to desecrate
> Shrines, and there were no retreats.
> So man learned to shed the tears
> With which he measures out his years.

ANGEL FIRE   The year-round, full-service resort community of **Angel Fire,** approximately 150 miles north of Albuquerque, 21 miles east of Taos, and 2 miles south of the Agua Fría junction on NM 38, dates only from the late 1960s but already has a wide variety of lodging choices—from condos and lodges to hotels and cabins. Winter Nordic and alpine skiing and summer golf are the most popular activities, but there's also ample opportunity for sailing and fishing on Eagle Nest Lake, tennis, hiking, snowmobiling, mountain biking, river rafting, and horseback riding. A variety of special events attracts visitors to the area throughout the year, including a hot-air balloon festival, Winterfest, and concerts of both classical and popular music. The

Taos Area (Including Enchanted Circle)

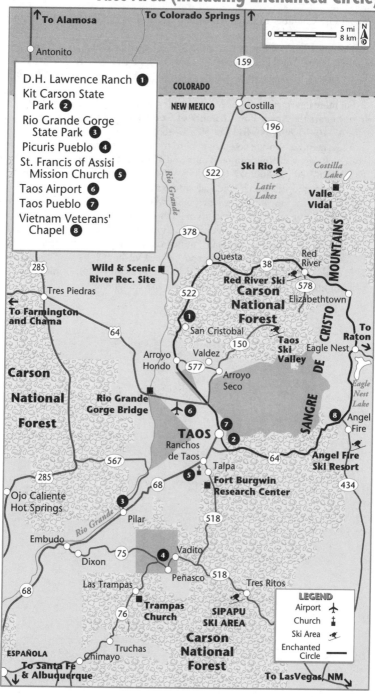

To Alamosa

To Colorado Springs

0 — 5 mi / 8 km

N

Antonito

D.H. Lawrence Ranch **1**

Kit Carson State Park **2**

Rio Grande Gorge State Park **3**

Picuris Pueblo **4**

St. Francis of Assisi Mission Church **5**

Taos Airport **6**

Taos Pueblo **7**

Vietnam Veterans' Chapel **8**

159

COLORADO

NEW MEXICO

Costilla

196

Ski Rio

Costilla Lake

Latir Lakes

Valle Vidal

Rio Grande

522

378

Questa

38

Red River

Red River Ski

578

Carson National Forest

Elizabethtown

285

Wild & Scenic River Rec. Site

522

To Raton

Tres Piedras

To Farmington and Chama

64

1

San Cristobal

150

Taos Ski Valley

Eagle Nest

Eagle Nest Lake

Carson

National

Forest

Arroyo Hondo

Valdez

577

Arroyo Seco

SANGRE DE CRISTO MOUNTAINS

8

Angel Fire

Rio Grande Gorge Bridge

6

7

TAOS

2

Ranchos de Taos

567

64

Angel Fire Ski Resort

285

5

Talpa

Fort Burgwin Research Center

434

Ojo Caliente Hot Springs

3

Pilar

68

518

Embudo

75

4

Vadito

Dixon

Peñasco

518

Tres Ritos

68

Las Trampas

76

Trampas Church

SIPAPU SKI AREA

Carson

National

Forest

ESPAÑOLA

To Santa Fe & Albuquerque

Truchas

Chimayo

To LasVegas, NM

LEGEND

Airport ✈

Church ✝

Ski Area 🎿

Enchanted Circle ──

1-0326

Angel Fire Chamber of Commerce can be reached at ☎ **800/446-8117** for more information.

The unofficial community center is the **Legends Hotel and Conference Center,** North Angel Fire Road (P.O. Drawer B), Angel Fire, NM 87710 (☎ **800/633-7463**), a 139-room hotel with rates starting at $75 in the summer, $105 in the winter.

For more information on the Moreno Valley, including full accommodations listings, contact the **Angel Fire Chamber of Commerce,** P.O. Box 547, Angel Fire, NM 87710 (☎ **800/446-8117** or 505/377-6353; fax 505/377-3034).

A fascinating adventure you may want to try here is a 1-hour, 1-day, or overnight horseback trip with **Roadrunner Tours and Elkhorn Lodge Ltd.,** Box 274 Angel Fire, NM 87710 (☎ **800/377-6416,** 505/377-6416, or 505/377-2811). From Angel Fire, Nancy and Bill Burch guide adventurers on horseback trips through private ranchland to taste the life of the lonesome cowboy. The cattle drive trip is no bland trail ride. The first day you'll travel 15 miles through ponderosa forests, across meadows of asters and sunflowers, with bald peaks in the distance. Once at camp, riders bed down in an authentic mountain cowboy cabin. The second day, you'll move as many as 300 cows through the Moreno Valley. One-hour rides are $20; day rides $95; cattle drives $184 (includes overnight stay in a cow camp). Cattle drives take place in July and August; book early, because space is limited.

Albuquerque 15

Albuquerque is the gateway to Northern New Mexico, the portal through which most domestic and international visitors pass before traveling on to Santa Fe and Taos. Though Albuquerque is a big city, it's worth stopping in for a day or two in order to get a feel of the whole history of this area.

From the rocky crest of Sandia Peak at sunset, one can see the lights of this city of almost half a million people spread out across 16 miles of high desert grassland. As the sun drops beyond the western horizon, it reflects off the Rio Grande, flowing through Albuquerque more than a mile below.

This waterway is the bloodline for the area, what allowed a city in this vast desert to spring up, and it continues to be at the center of the area's growth. Farming villages that line its banks are being stampeded by expansion. As the west side of the river sprawls, the means for transporting traffic across it have had to be built, breaking up the pastoral valley area.

The railroad, which set up a major stop here in 1880, prompted much of Albuquerque's initial growth, but that economic explosion was nothing compared with what has happened since World War II. Designated a major national center for military research and production, Albuquerque became a trading center for this state, whose populace is spread widely across the land. That's why the city may strike visitors as nothing more than one big strip mall. Look closely and you'll see ranchers, Native Americans, and Hispanic villagers stocking up on goods to take back to the New Mexico boot heel or the Texas panhandle.

Climbing out of the valley is Route 66, well worth a drive, if only to see the rust that time has left. Old court hotels still line the street, many with their funky '50s signage. One enclave on this route is the University of New Mexico district, with a number of hippie-ish cafes and shops.

Farther down the route you'll come to downtown Albuquerque. During the day, this area is all suits and heels, but at night it becomes a hip nightlife scene. People from all over the state come to Albuquerque to check out the live music and dancing clubs, most within walking distance from each other.

The section called Old Town is worth a visit. Though it's the most touristy part of town, it's also a unique Southwestern village with a beautiful and intact plaza. Also in this area are Albuquerque's new Aquarium and botanical gardens, as well as its continually upgrading zoo.

1 Orientation

ARRIVING

Since Albuquerque is the transportation hub for New Mexico, getting in and out of town is easy. For more detailed information, see "Getting There," in chapter 2.

BY PLANE The **Albuquerque International Airport** is in the south-central part of the city, between I-25 on the west and Kirtland Air Force Base on the east, just south of Gibson Boulevard. Sleek and efficient, the airport is served by eight national airlines and two local ones.

Most hotels have courtesy vans to meet their guests and take them to their respective destinations. In addition, **Shuttlejack** (☎ **505/243-3244**) and **Checker Airport Express** (☎ **505/765-1234**) run services to and from city hotels. **Sun Tran** (☎ **505/843-9200**), Albuquerque's public bus system, also makes airport stops. There is efficient taxi service to and from the airport, plus numerous car-rental agencies.

BY TRAIN Amtrak's "Southwest Chief" arrives and departs daily from and to Los Angeles and Chicago. The station is at 214 First St. SW, 2 blocks south of Central Avenue (☎ **800/USA-RAIL** or 505/842-9650). *Note:* A new train station is currently in the planning stage, so call ahead to make sure the address listed here is still current.

BY BUS **Greyhound/Trailways** (☎ **800/231-2222** for schedules, fares, and information) and **TNM&O Coaches** (☎ **505/243-4435**) arrive and depart from the Albuquerque Bus Transportation Center, 300 Second St. SW (near the train station).

BY CAR If you're driving, you'll probably arrive via either the east–west Interstate 40 or the north–south Interstate 25. Exits are well marked. For information and advice on driving in New Mexico, see "Getting There," in chapter 2.

VISITOR INFORMATION

The main office of the **Albuquerque Convention and Visitors Bureau** is at 20 First Plaza NW (☎ **800/284-2282** or 505/842-9918). It's open Monday through Friday from 8am to 5pm. There are information centers at the airport, on the lower level at the bottom of the escalator, open daily from 9:30am to 8pm; and in Old Town at 303 Romero St. NW (Suite 107), open daily from 9am to 5pm. Tape-recorded information about current local events is available from the bureau after 5pm weekdays and all day Saturday and Sunday. Call ☎ 800/284-2282. If you have access to the World Wide Web, the address for the Albuquerque Convention and Visitors Bureau is **http://www.abqcvb.org**.

CITY LAYOUT

The city's sprawl takes a while to get used to. A visitor's first impression is of a grid of arteries lined with shopping malls and fast-food eateries, with residences tucked behind on side streets.

If you look at a map of Albuquerque, the first thing you'll notice is that it lies at the crossroads of Interstate 25 north–south and Interstate 40 east–west. Refocus your attention on the southwest quadrant: Here you'll find both downtown Albuquerque and Old Town, site of many tourist attractions. Lomas Boulevard and Central Avenue, the old "Route 66" (US 66), flank downtown on the north and south. They come together 2 miles west of downtown near the Old Town Plaza, the historical and

Greater Albuquerque

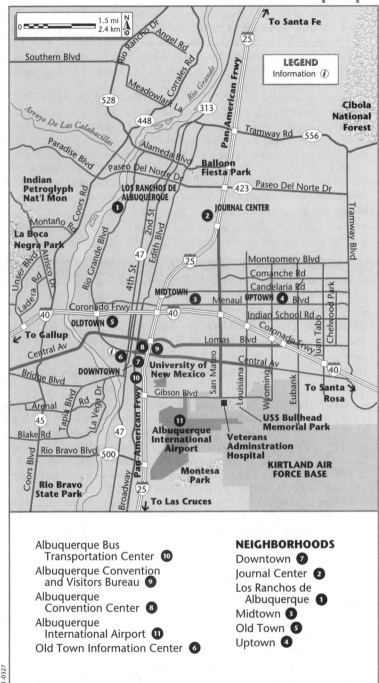

1-0327

217

spiritual heart of the city. Lomas and Central continue east across I-25, staying about half a mile apart as they pass by the University of New Mexico and the New Mexico State Fairgrounds. The airport is directly south of the UNM campus, about 3 miles via Yale Boulevard. Kirtland Air Force Base—site of Sandia National Laboratories and the National Atomic Museum—is an equal distance south of the fairgrounds on Louisiana Boulevard.

Roughly paralleling I-40 to the north is Menaul Boulevard, the focus of midtown and uptown shopping as well as the hotel districts. As Albuquerque expands northward, the Journal Center business park area, about 4^1/$_2$ miles north of the freeway interchange, is getting more attention. East of Eubank Boulevard are the Sandia Foothills, where the alluvial plain slants a bit more steeply toward the mountain.

When looking for an address, it is helpful to know that Central Avenue divides the city into north and south, and the railroad tracks—which run just east of First Street downtown—comprise the dividing line between east and west. Street names are followed by a directional: NE, NW, SE, or SW.

MAPS   The most comprehensive Albuquerque street map is the one published by First Security Bank and distributed by the Convention and Visitors Bureau.

2 Getting Around

Albuquerque is easy to get around, thanks to its wide thoroughfares and grid layout, combined with its efficient transportation systems.

BY PUBLIC TRANSPORTATION   **Sun Tran of Albuquerque** (☎ **505/ 843-9200**) cloaks the arterials with its city bus network. Call for information on routes and fares.

BY TAXI   **Yellow-Checker Cab** (☎ **505/765-1234**) serves the city and surrounding area 24 hours a day.

BY CAR   The Yellow Pages list more than 30 car-rental agencies in Albuquerque. Among them are the following well-known national firms: **Alamo,** 2601 Yale Blvd. SE (☎ 505/842-4057); **Avis,** at the airport (☎ 505/842-4080); **Budget,** at the airport (☎ 505/768-5900); **Dollar,** at the airport (☎ 505/842-4304); **Hertz,** at the airport (☎ 505/842-4235); **Rent-A-Wreck,** 501 Yale Blvd. SE (☎ 505/242-9556); and **Thrifty,** 2039 Yale Blvd. SE (☎ 505/842-8733). Those not located at the airport itself are close by and can provide rapid airport pickup and delivery service.

Parking is generally not difficult in Albuquerque. Meters operate weekdays from 8am to 6pm and are not monitored at other times. Only the large downtown hotels charge for parking. Traffic is a problem only at certain hours. Avoid I-25 and I-40 at the center of town around 5pm.

FAST FACTS: Albuquerque

Airport   See "Orientation," above.

American Express   The American Express office is at 6600 Indian School Road (☎ **800/219-1023** or 505/883-3677; fax 505/884-0008). To report lost credit cards, call ☎ 800/528-4800.

Area Code   The telephone area code for all of New Mexico is 505.

Car Rentals   See "Getting There," in chapter 2, or "Getting Around," above.

Climate   See "When to Go," in chapter 2.

Currency Exchange   Foreign currency can be exchanged at any of the branches of Sun West Bank (its main branch is at 303 Roma St. NE; ☎ 505/765-2211); or at any of the branches of First Security Bank (its main office is at Twenty-First Plaza; ☎ 505/765-4000).

Dentists   Call the Albuquerque District Dental Society at ☎ **505/260-7333** for emergency service.

Doctors   Call the University of New Mexico Medical Center Physician Referral Service at ☎ **505/843-0124** for a recommendation.

Embassies/Consulates   See "Fast Facts: For the Foreign Traveler," in chapter 3.

Emergencies   For police, fire, or ambulance, dial ☎ **911.**

Hospitals   The major facilities are Presbyterian Hospital, 1100 Central Ave. SE (☎ **505/841-1234,** 505/841-1111 for emergency services); and University of New Mexico Hospital, 2211 Lomas Blvd. NE (☎ **505/843-2411** for emergency services).

Liquor Laws   See "Fast Facts: Santa Fe," in chapter 4.

Newspapers & Magazines   The two daily newspapers are the *Albuquerque Tribune,* published mornings, and the *Albuquerque Journal,* published evenings.

Police   For emergencies, call ☎ **911.** For other business, contact the Albuquerque City Police (☎ **505/768-1986**) or the New Mexico State Police (☎ **505/841-9256**).

Post Offices   The Main Post Office, 1135 Broadway NE (☎ **505/245-9561**), is open daily from 7:30am to 6pm. There are 19 branch offices, with another 13 in surrounding communities.

Radio/TV   Albuquerque has some 30 local radio stations catering to all musical tastes. Albuquerque television stations include KOB, Channel 4 (NBC affiliate); KOAT, Channel 7 (ABC affiliate); KGGM, Channel 13 (CBS affiliate); KNME, Channel 5 (PBS affiliate); and KGSW, Channel 14 (Fox and independent). There are, of course, numerous local cable channels as well.

Taxes   In Albuquerque, the hotel tax is 10.5625%; it will be added to your bill.

Taxis   See "Getting Around," above.

Time Zone   Albuquerque is on Mountain Time, 1 hour ahead of the West Coast and 2 hours behind the East Coast.

Transit Information   Sun Tran of Albuquerque is the public bus system. Call ☎ **505/843-9200** for schedules and information.

Useful Telephone Numbers   For time and temperature, call ☎ **505/247-1611;** for road information, call ☎ **800/432-4269;** and for emergency road service (AAA), call ☎ **505/291-6600.**

3 Where to Stay

Albuquerque's hotel glut is good news to travelers looking for quality rooms at a reasonable cost. Except during peak periods—specifically, the New Mexico Arts and Crafts Fair (late June), the New Mexico State Fair (September), and the Kodak Albuquerque International Balloon Fiesta (early October)—most of the city's hotels have vacant rooms, so guests can frequently request and get a lower room rate than the one posted.

In the following listing, hotels are categorized by price range: **Expensive** means that a double room costs $110 or more per night; **Moderate** includes doubles for $75 to $110; and **Inexpensive** refers to doubles for $75 and under.

A tax of 10.81% is added to every hotel bill. All hotels listed offer rooms for non-smokers and travelers with disabilities; all the bed-and-breakfasts do as well.

EXPENSIVE

Crowne Plaza Pyramid

5151 San Francisco Rd. NE, Albuquerque, NM 87109. ☎ **800/544-0623** or 505/821-3333. Fax 505/822-8115. 311 rms, 54 suites. A/C TV TEL. $112–$142 single; $132–$162 double; $138 and up, suite. Ask about special weekend and package rates. AE, CB, DC, MC, V. Free parking.

About a 15-minute drive from Old Town and downtown is this Aztec pyramid-shaped structure reached via the Paseo del Norte exit (exit 232) from I-25. Previously the Holiday Inn Pyramid, this structure, built in 1986, has recently come under the Crowne Plaza name and with it received a $4.5 million renovation. The 10 guest floors are grouped around a "hollow" skylit core. Vines drape from planter boxes on the balconies, and water falls five stories to a pool between the two glass elevators.

The rooms are spacious, though not extraordinary, all with picture windows and ample views. The renovation has added coffeemakers, hair dryers, makeup mirrors, irons, and ironing boards. The morning newspaper is delivered to your door. Rooms on the 10th-floor executive level offer more space and a few more amenities.

With lots of convention space at the hotel, you're likely to encounter name-tagged conventioneers here, though the service seems to be good enough to handle the crowds without inconvenience to you.

Dining/Entertainment: The Terrace Restaurant offers American cuisine with a Southwestern flair. There are also two lounges on the premises.

Services: Concierge, room service, valet laundry, newspaper delivery, express checkout.

Facilities: Spectravision movie channels, indoor/outdoor pool, medium-sized health club, Jacuzzi, sauna, jogging track, business center, conference rooms, sundeck.

✪ Hyatt Regency Albuquerque

330 Tijeras Ave. NW, Albuquerque, NM 87102. ☎ **800/233-1234** or 505/842-1234. Fax 505/842-1184. 395 rms, 14 suites. A/C TV TEL. Weekdays, $150–$175 double; weekends, $89 double; $310–$725 suite. AE, CB, DC, DISC, MC, V. Self-parking $8, valet $10.

If you like luxury and want to be right downtown, this is the place to stay. This $60 million hotel, which opened in 1990, is pure shiny gloss and art deco. The lobby features a palm-shaded fountain beneath a pyramidal skylight, and throughout the hotel's public areas is an extensive art collection, including original Frederic Remington sculptures. The spacious guest rooms enhance the feeling of richness with mahogany furnishings, full-length mirrors, and big views of the mountains. This is a nice location if you want to sample Albuquerque's nightlife as well as seasonal events on the recently renovated Civic Plaza. Within the hotel are many shops; it is also right next door to the Galeria, a functional shopping area. All rooms have coffeemakers and hair dryers.

Dining/Entertainment: McGrath's serves three meals daily in a setting of forest-green upholstery and black-cherry furniture. **Bolo Saloon** is noted for its whimsical oils of "where the deer and the antelope play" (at the bar).

Services: Concierge, room service, dry cleaning, laundry service, newspaper delivery, baby-sitting, secretarial services, express checkout, valet parking.

Central Albuquerque Accommodations

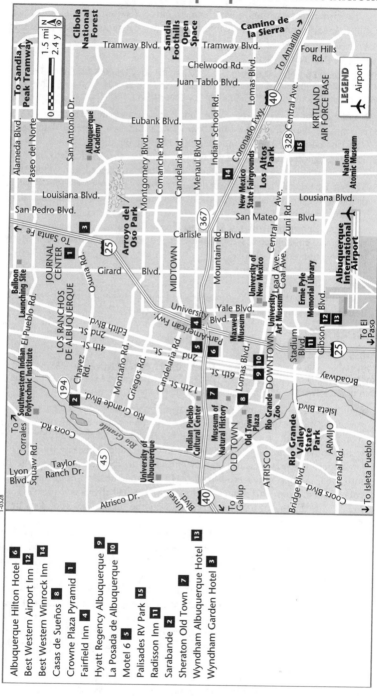

LEGEND
✈ Airport

Albuquerque Hilton Hotel **6**
Best Western Airport Inn **12**
Best Western Winrock Inn **14**
Casas de Sueños **8**
Crowne Plaza Pyramid **1**
Fairfield Inn **4**
Hyatt Regency Albuquerque **9**
La Posada de Albuquerque **10**
Motel 6 **5**
Palisades RV Park **15**
Radisson Inn **11**
Sarabande **2**
Sheraton Old Town **7**
Wyndham Albuquerque Hotel **13**
Wyndham Garden Hotel **3**

Facilities: Spectravision movie channels, outdoor swimming pool, small health club, sauna, business center, conference rooms, car-rental desk, beauty salon, boutiques.

Sheraton Old Town

800 Rio Grande Blvd. NW, Albuquerque, NM 87104. ☎ **800/325-3535** or 505/843-6300. Fax 505/842-9863. 190 rms, 20 suites. A/C TV TEL. $120–$130 double; $150 suite. Children stay free in parents' room. AE, CB, DC, DISC, MC, V. Free parking.

No Albuquerque hotel is closer to top tourist attractions than the Sheraton. It's only a 5-minute walk from the Old Town Plaza and two important museums. Constructed in 1975 and remodeled in 1993, the building has mezzanine-level windows lighting the adobe-toned lobby, an airiness that carries into the rooms. They have unique handmade furniture such as *trasteros* accented with willow shoots. Request a south-side room and you'll get a balcony. These overlook Old Town and the pool, which is heated year-round. All rooms now offer coffeemakers, hair dryers, and irons and ironing boards.

Dining/Entertainment: The **Customs House Restaurant,** specializing in seafood and regional cuisine, serves weekday lunches and nightly dinners. The **Café del Sol** is the Sheraton's coffee shop.

Services: Concierge, room service, valet laundry, secretarial and baby-sitting services.

Facilities: Old Town Place, an attached shopping center, has arts-and-crafts dealers, a bookstore, beauty salon, and manicurist.

MODERATE

Albuquerque Hilton Hotel

1901 University Blvd. NE, Albuquerque, NM 87102. ☎ **800/27-HOTEL** or 505/884-2500. Fax 505/889-9118. 264 rms, 7 suites. A/C TV TEL. $95–$135 double; $375–$475 suite. Weekend discounts are available. AE, CB, DC, DISC, MC, V. Free parking.

Equidistant from Old Town and downtown, this 12-story hotel is a good choice if you don't mind conventioneers. With its large pool and nicely lit and decorated rooms—most with views of the mountains or river—the hotel does its best to cater to travelers as well as those attending conventions. It supplies the city with one of its larger meeting spaces, so count on moments of hordes coming and going. The 10th through 12th floors were remodeled in 1997 and come with more amenities such as coffeemakers, irons, and ironing boards, as well as hair dryers and robes. Cabana rooms with 15-foot-high cathedral ceilings surround the outdoor pool. Pets are not permitted.

Dining/Entertainment: The **Ranchers Club,** built like a luxury hunting lodge transported to the high plains, is one of Albuquerque's finest restaurants. **Casa Chaco** is open daily for three meals. **The Cantina,** with its fajitas grill, has flamenco guitar Wednesday through Sunday nights.

Services: Room service (during restaurant hours), dry cleaning, laundry service, newspaper delivery, in-room massage, secretarial services, express checkout, valet parking, free coffee or refreshments in the lobby, shoe-shine stand.

Facilities: Indoor and outdoor swimming pools, medium-sized health club, Jacuzzi, sauna, business center, conference rooms, Laundromat, car-rental desk, gift shop.

La Posada de Albuquerque

125 Second St. NW (at Copper Ave.), Albuquerque, NM 87102. ☎ **800/777-5732** or 505/ 242-9090. Fax 505/242-8664. 114 rms, 3 suites. A/C TV TEL. $92–$102 double; $175–$225 suite. AE, DISC, MC, V. Valet parking $5.

Built in 1939 by Conrad Hilton as the famed hotelier's first inn in his home state of New Mexico, this hostelry on the National Register of Historic Places feels like old Spain. Though remodeled in 1996, the owners have kept the finer qualities. An elaborate Moorish brass-and-mosaic fountain stands in the center of the tiled lobby floor; old-fashioned tin chandeliers hang from the two-story ceiling. The lobby is surrounded on all sides by high archways, creating the feel of a 19th-century hacienda courtyard.

As in the lobby, all guestroom furniture is handcrafted, but here it's covered with cushions of Southwestern design. There are spacious rooms with big windows looking out across the city and toward the mountains. If you want a feel for downtown Albuquerque as well as easy access to the Civic Plaza, nightclubs, and Old Town, this hotel will suit you well.

Dining/Entertainment: Conrad's Downtown (see "Where to Dine" later in this chapter), La Posada's elegantly redesigned restaurant, features Spanish/Yucatán cuisine. The **Lobby Bar** is a favorite gathering place and has entertainment Thursday through Saturday evenings.

Services: Room service, dry cleaning, laundry service, newspaper delivery, express checkout, valet parking.

Facilities: Spectravision movie channels, video rentals, access to nearby health club, conference rooms, beauty salon.

Radisson Inn

1901 University Blvd. SE, Albuquerque, NM 87106. ☎ **800/333-3333** or 505/247-0512. Fax 505/843-7148. 148 rms. A/C TV TEL. $75–$85 double. AE, CB, DC, DISC, MC, V. Free parking.

The Spanish colonial–style Radisson, a mile from the airport, was built in 1986 and remodeled in 1994. Though south of the city, it has good freeway access, allowing you to be in the university district, downtown, or in Old Town in less than 15 minutes. The rooms are a bit narrow, but they're richly decorated. Request a courtyard view and you'll feel a bit of respite from the hubbub of the city. Each room is equipped with a hair dryer, iron and ironing board, coffeemaker, and voice-mail phone systems with data ports. Small pets are accepted.

Dining/Entertainment: Diamondback's Restaurant specializes in Southwestern and American cuisine, while **Coyote's Cantina** is a popular watering hole. Live jazz and blues bands play on the patio Wednesday night in the summer.

Services: Limited room service, valet laundry, 24-hour courtesy van.

Facilities: Rooms for nonsmokers, year-round outdoor swimming pool and whirlpool, guest use of a nearby health club.

Wyndham Albuquerque Hotel

2910 Yale Blvd. SE, Albuquerque, NM 87106. ☎ **800/227-1117** or 505/843-7000. Fax 505/843-6307. 266 rms. A/C TV TEL. $69–$139 double. AE, DC, DISC, EU, JCB, MC, V. Free parking.

No accommodations are closer to the airport than the Wyndham, just north of the main terminal. It caters to air travelers with a 24-hour desk, shuttle service, and overnight valet laundry. But it could also be a wise choice for a few days of browsing around Albuquerque, as it has good access to freeways and excellent views. Of course, you will hear some jet noise. The hotel recently came under the Wyndham name and with it a multimillion-dollar remodel.

This reasonably priced hotel now has a classy feel. The grill and lounge areas employ a lot of sandstone, wood, copper, and tile to lend an Anasazi feel. This carries into the rooms of the 14-story hotel. Each room has Nintendo on the television, a hair dryer, coffeemaker, iron, and ironing board. Room service, courtesy van, valet laundry, and complimentary shoe shine are also available. Pets are not permitted.

Facilities include an outdoor swimming pool, self-service Laundromat, coed sauna, two all-weather tennis courts, business center, and gift shop. The Rojo Grill serves a variety of American and Southwestern dishes, from tostadas to Ostrich medaillons.

INEXPENSIVE

Best Western Airport Inn

2400 Yale Blvd. SE, Albuquerque, NM 87106. ☎ **800/528-1234** or 505/242-7022. Fax 505/243-0620. 120 rms, 3 suites. A/C TV TEL. $63–$69 double. Rates include continental breakfast. AE, CB, DC, DISC, MC, V.

A landscaped garden courtyard behind the hotel is a lovely place to relax on cloudless days. The rooms, while not the most attractive in town, contain standard furnishings and offer free local phone calls. Deluxe units are equipped with refrigerators. Continental breakfast is served in the rooms, or guests can request a coupon good for $3 off their morning meal at the adjacent Village Inn. A courtesy van is available from 6am to midnight, and the hotel also offers valet laundry service. There are rooms for nonsmokers and travelers with disabilities, and guests can enjoy an outdoor swimming pool and Jacuzzi.

Best Western Winrock Inn

18 Winrock Center NE, Albuquerque, NM 87110. ☎ **800/528-1234** or 505/883-5252. Fax 505/889-3206. 173 rms. A/C TV TEL. $69–$135 double. Rates include breakfast. AE, CB, DC, DISC, MC, V. Free parking.

Located just off I-40 at the Louisiana Boulevard interchange, the Winrock is attached to one of Albuquerque's largest shopping centers: Winrock Center. This is a mall-like area featuring movies and discount-store shopping—and lots of traffic. Built in 1962 and remodeled in 1995, its two separate buildings are wrapped around a garden and private lagoon with Mandarin ducks, giant koi (carp), and an impressive waterfall. The comfortable rooms, many of which provide private patios overlooking the lagoon (request a ground-floor room, poolside; this will help minimize traffic noise from the freeway), feature a pastel Southwestern-motif decor. A full, hot breakfast is served in the Club Room every morning. Valet laundry service is available. On the premises you will find a heated outdoor pool and guest laundry.

Fairfield Inn

1760 Menaul Rd. NE, Albuquerque, NM 87102. ☎ **800/228-2800** or 505/889-4000. 188 rms. A/C TV TEL. $62.95 double. Extra person $6. Children 18 and under stay free in parents' room. AE, CB, DC, DISC, MC, V.

Owned by Marriott, this hotel has exceptionally clean rooms and a location with easy access to freeways that can quickly get you to Old Town, downtown, or the heights. Ask for an east-facing room to avoid the noise and a view of the highway. Complimentary coffee, tea, and continental breakfast are served in a sunny new breakfast room off the lobby. Local phone calls are free and valet service is available. There's an indoor swimming pool with saunas and a Jacuzzi as well as a medium-sized health club. You probably couldn't get more for your money (in a chain hotel) anywhere else. There are rooms for both nonsmokers and travelers with disabilities.

Wyndham Garden Hotel

6000 Pan American Fwy. NE (I-25 at San Mateo Blvd.), Albuquerque, NM 87109. ☎ **800/996-3426** or 505/821-9451. Fax 505/858-0239. 151 rms. A/C TV TEL. $59–$99 single or double. AE, CB, DC, DISC, MC, V.

From the outside, this five-story stucco structure doesn't look like much, but a recent remodeling has done wonders with the inside. Located off I-25, it's about 15 minutes from Old Town and downtown, but has good access to many restaurants on San Mateo, just minutes away. The hotel does a lot of corporate and some group business, but is still a good spot for travelers. It features a lobby atrium decorated in a Southwest fishing lodge decor that carries into the rooms, which are richly decorated in warm tones. Some rooms are quite large and they cost the same as the others, so request one of them. All rooms have private balconies (patios on the ground floor), city or mountain views, On Command movies, coffeemakers, safes, irons and ironing boards, and data ports. *USA Today* is delivered to guest rooms weekdays.

Dining/Entertainment: The **Garden Cafe** is open daily for a full breakfast buffet (included with room rate weekdays), lunch, and dinner. The **Atrium Lounge** is a quiet, comfortable place to enjoy cocktails or after-dinner drinks.

Services: Limited room service, complimentary airport shuttle (from 7am to 11pm), guest laundry, fax and copy service.

Facilities: Small health club, Jacuzzi, sauna, indoor/outdoor heated swimming pool.

BED-&-BREAKFASTS
EXPENSIVE

Casas de Sueños
310 Rio Grande Blvd. SW, Albuquerque, NM 87104. ☎ **800/CHAT-W/US** or 505/247-4560. 19 rms. TV TEL. $85–$250 single or double. Rates include breakfast and afternoon snacks. AE, CB, DC, DISC, MC, V.

The principal attraction of this B&B is its location. It's within walking distance from the Plaza, the aquarium, and botanical garden—even the zoo, though it's farther. You'll recognize Casas de Sueños by the bright sign and the snail-shaped front of the main building, which was designed by famed renegade architect Bart Prince. The buildings that comprise the Casas were once private homes—a compound that was part of a gathering place for artists and their admirers. Most of them face a garden courtyard filled with roses and exotic sculptures.

Each of the rooms has an individual theme; one room is designed to follow the color schemes of Monet's paintings, while another is reminiscent of an English drawing room. The rooms are interesting though not pristine, better decribed as artsy in their design and upkeep. Some are equipped with kitchens, a few have their own hot tubs, and a number have fireplaces. Outside the garden area are two newer rooms that I found spacious and efficient though lacking in charm. Every accommodation has its own entrance.

A delicious full breakfast is served in the main building every morning. Massage therapists are available. No smoking is permitted indoors. Children 12 and older are welcome, but pets are not accepted.

MODERATE

Casa del Granjero
414 C de Baca Lane NW, Albuquerque, NM 87114. ☎ **800/701-4144** or 505/897-4144. Fax 505/897-4144. 7 rms. $79–$149 double. Extra person $20. Rates include breakfast. DISC, MC, V.

From the pygmy goats to the old restored wagon out front, Casa del Granjero ("The Farmer's House") is true to its name. Located north of town—about a 15-minute drive to Old Town—it is quiet and has a rich, homey feeling. Butch and Victoria Farmer have transformed this residence—the original part of which is 120 years old—into a fine bed-and-breakfast. The Great Room has an enormous sculptured adobe fireplace, comfortable bancos for lounging, a library, and is almost cluttered with Southwestern artifacts. There's a 52-inch television in the den.

The guest rooms, with Spanish nicknames, are beautifully furnished and decorated. Most have fireplaces. Cuarto del Rey features Mexican furnishings and handmade quilts and comforters. Cuarto de Flores has French doors that open onto a portal, and Cuarto Alegre has a king-size canopy bed done up in lace and satin. The newer Guest House has comfortable rooms and access to a kitchen, but a less luxurious and Southwestern feel. In the morning, a full breakfast is served at long tables decorated with colorful Mexican cloths or on the portal. Catered lunches and dinners are also available by arrangement. There's a new organic garden and raspberry patch, as well as a new Jacuzzi, conference room, and office center equipped with fax and computer. Smoking is permitted outdoors only, and pets are not permitted. There are accommodations for travelers with disabilities. *A warning to women:* These are country folk, and the men will tend to call you "hon" and "darlin'."

○ Hacienda Antigua

6708 Tierra Dr. NW, Albuquerque, NM 87107. ☎ **800/201-2986** or 505/345-5399. 5 rms. A/C. $85–$150 double. Extra person $25. Rates include breakfast. AE, DISC, MC, V.

Located on the north side of Albuquerque, just off Osuna Road, is Hacienda Antigua, a 200-year-old adobe home that was once the first stagecoach stop out of Old Town in Albuquerque. When Ann Dunlap and Melinda Moffit bought it, they were careful to preserve the building's historic charm while transforming it into an elegant bed-and-breakfast. The beautifully landscaped courtyard, with its large cottonwood tree and abundance of greenery (including a large raspberry patch), offers a welcome respite for today's tired travelers.

The rooms are gracefully and comfortably furnished with antiques. There's the Don Pablo Suite with a king-size bed (covered with a stunning blue quilt), a sitting room with a kiva fireplace, and a bathroom with a wonderful old pedestal bathtub/shower; and La Capilla, the home's former chapel, which is furnished with a queen-size bed, a fireplace, and a beautiful carving of San Ysidro (the patron saint of farmers). All the rooms have such regional touches. They are also equipped with Caswell Massey soaps and unstocked minirefrigerators. A gourmet breakfast is served in the garden during warm weather and by the fire in winter. Guests also have use of the pool and hot tub. Just a 20-minute drive from the airport, Hacienda Antigua is a welcome change from the anonymity of the downtown Albuquerque high-rise hotels, and Ann and Melinda are terrific hosts.

Sarabande

5637 Rio Grande Blvd. NW, Albuquerque, NM 87107. ☎ **800/506-4923** or 505/345-4923. Fax 505/345-9130. 3 rms. A/C TV TEL. $85–$125 double. Rates include breakfast. MC, V. Free parking.

A bit of grandmotherly comfort describes this place situated in the North Valley, a lovely 10-minute drive from Old Town. Once you pass through the front gate and into the beautifully tended courtyard gardens with fountains, you'll forget that you're staying on the fringes of a big city. With cut-glass windows, lots of pastels, traditional antiques, and thick carpet (in all but the poolside room), you'll be well pampered here. Betty Vickers and Margaret Magnussen have filled the home with fine art as well as comfortable modern furniture. The Rose Room has a Japanese soaking tub and kiva fireplace. The Iris Room, with its stained-glass window depicting irises, has a king-size bed. Both rooms open onto a wisteria-shaded patio where breakfast can be taken in the morning. Out back are a 50-foot heated lap pool and a hot tub (which can be used through the winter). There is a library stocked with magazines, books by local authors, and books about New Mexico (including local sports and recreation). Betty and Margaret are avid hikers and will be happy to recommend hiking options for you. All-terrain bikes are available for guest use free of charge. Breakfast (fresh fruit, fresh squeezed juice, coffee, and homemade breads) may be served in the courtyard or the dining room. Don't miss the chocolate-chip cookies offered in the afternoon.

NEAR ALBUQUERQUE
MODERATE

Hacienda Vargas

El Camino Real (P.O. Box 307), Algodones/Santa Fe, NM 87001. ☎ **800/261-0006** or 505/867-9115. Fax 505/867-1902; web: http://www.swcp.com.hacvar//. 7 rms. AC. $79–$149 double. Extra person $15. Rates include breakfast. MC, V.

Unassuming in its elegance, Hacienda Vargas is located right on old Route 66. Owned and operated by the DeVargas family, the inn is situated in the small town of Algodones (about 20 miles from Albuquerque) and is a good place to stay if you're planning to visit both Santa Fe and Albuquerque but don't want to stay in one of the downtown hotels in either city. There's a real Mexican feel to the decor, with brightly woven place mats in the breakfast room and Spanish suits of armor hanging in the common area. Each guest room has a private entrance, many opening onto a courtyard. All rooms are furnished with New Mexico antiques, are individually decorated, and have handmade kiva fireplaces. Many have Jacuzzi tubs. Each of the four suites has a Jacuzzi tub, fireplace, and private patio. Hosts Jule and Paul DeVargas are extremely gracious and helpful—they'll make you feel right at home. A full breakfast is served every morning in the dining room. The only drawback here is the train tracks near the back of the house, and during my stay the last train went by around midnight. At all other times the inn is quiet and restful.

La Hacienda Grande

21 Baros Lane, Bernalillo, NM 87004. ☎ **800/353-1887** or 505/867-1887. Fax 505/771-1436. 6 rms. A/C. $99–$129 double. Extra person $15. Rates include breakfast. AE, DC, DISC, MC, V. Free parking.

A 25-minute drive from Old Town and downtown, this inn offers a quiet country experience. Opened in 1993, it's run by a brother-and-sister team, Daniel Buop and Shoshana Zimmerman. The completely restored adobe home, with 2-foot-thick walls, is more than 250 years old and sits on 4 acres of land that was once part of the original 100-square-mile Spanish land grant. The courtyard, surrounded by high adobe walls, has a vortex, which had special significance to the local native tribespeople who often came here to pray and hold ceremonies.

The guest rooms, all featuring custom-crafted furniture, are comfortable and inviting, with a hand-rounded softness that only old adobe walls can provide. Much of the furniture is made of bent willow, which lends a rustic air. All rooms have small sitting areas and Southwestern-style armoires, and five have wood-burning kiva fireplaces (one is a lovely freestanding clay fireplace). Brick or tile floors are covered with throw rugs. Televisions and VCRs are available in the rooms upon request. Early each morning, thermoses of coffee are left outside each guest room; later on, breakfast is served in the dining room. Favorite breakfast entrees include amaretto French toast or cheese strata. Smoking is prohibited except on the patio.

Sandhill Crane Bed & Breakfast

389 Camino Hermosa, Corrales, NM 87048 ☎ **800/375-2445** or 505/898-2445. Fax 505/898-2445. 3 rms. A/C TV TEL. $75–$145 double. Rates include breakfast. AE, MC, V. Free parking.

This lovely bed-and-breakfast, run by Carol Hogan and Phil Thorpe, is about 20 minutes from Albuquerque in the sleepy little town of Corrales. It's a great place to stay if you want to explore the city but don't want to be right downtown.

Wisteria-draped walls surround the renovated adobe hacienda, and each room is uniquely decorated in an elegant, traditional Southwestern style. For families or friends traveling together, the ominously named Outlaw Wing (two rooms with connecting bath, small kitchen, and private entrance) is a great choice. All rooms have cable TV and phone jacks for those who want a TV or telephone. Be aware that this is a home, and with it comes close quarters coziness, not necessarily the choice for those who like their own outdoor entrance. Carol has decorated the cozy guest rooms with her charming collection of bird decoys and birdhouses, while Phil is responsible for breakfasts that include fruit drinks, bagels, muffins, or homemade bread, as well as a special hot entree such as his frittata (an Italian omelet) served with focaccia. In warmer weather, breakfast is served on the patio, where you're likely to see a roadrunner pass by. Currently plans are in the works to add three new rooms. Rooms for nonsmokers and travelers with disabilities are available.

RV PARKS

Albuquerque Central KOA

12400 Skyline Rd. NE, Albuquerque, NM 87123. ☎ **800/562-7781** or 505/296-2729. DISC, MC, V.

This RV park sits in the foothills east of Albuquerque. It features a bathhouse, Laundromat, outdoor swimming pool (open summers only), and convenience store. Cabins are available.

Albuquerque North KOA

555 Hill Rd., Bernalillo, NM 87004. ☎ **505/867-5227**. DISC, MC, V.

At the foot of the mountains 14 miles from Albuquerque, this campground has a Laundromat, outdoor swimming pool (open May to October), playground, convenience store, cafe, and free outdoor movies. There's a free pancake breakfast daily. Reservations are recommended.

Palisades RV Park

9201 Central Ave. NW, Albuquerque, NM 87121. ☎ **505/831-5000**. Fax: 505/833-0804. 110 sites. $20 per day; $100 per week; $200 plus electricity per month. MC, V.

The owner is on site at this RV park on the west mesa, about a 10-minute drive from Old Town. There's a bathhouse, Laundromat, reception room, small convenience store, and propane is available.

4 Where to Dine

In these listings, the following categories define price ranges: **Expensive,** most dinner main courses are priced over $15; **Moderate,** most dinner main courses $10 to $15; **Inexpensive,** $10 and under.

IN OR NEAR OLD TOWN
EXPENSIVE

Antiquity

112 Romero NW (in Old Town). ☎ **505/247-3545**. Reservations recommended. Main courses $16.95–$24.95. AE, DC, DISC, MC, V. Daily 5–9pm. FRENCH/CONTINENTAL.

Antiquity is something of a surprise in this Old Town neighborhood, which is filled with snack shops and Mexican restaurants that have been around for decades, but it's a nice surprise. Antiquity, in business about 11 years, now is decorated in Southwestern style with Mexican touches, making this an intimate spot for a romantic dinner. Classical music adds a finishing touch. Appetizers include standard, well-prepared

Central Albuquerque Dining

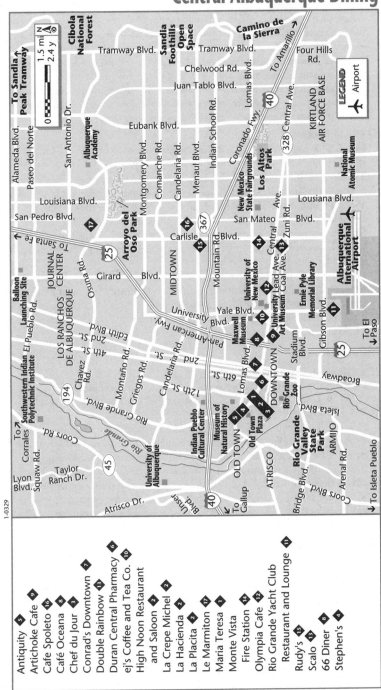

Antiquity 5
Artichoke Cafe 9
Cafe Spoleto 16
Café Oceana 6
Chef du Jour 3
Conrad's Downtown 7
Double Rainbow 13
Duran Central Pharmacy 6
ej's Coffee and Tea Co. 10
High Noon Restaurant and Saloon 2
La Crepe Michel 2
La Hacienda 2
La Placita 2
Le Marmiton 17
Maria Teresa 1
Monte Vista Fire Station 14
Olympia Cafe 12
Rio Grande Yacht Club Restaurant and Lounge 11
Rudy's 15
Scalo 13
66 Diner 8
Stephen's 4

229

dishes, such as escargots and French onion soup, as well as specialties such as Southwestern pasta (basil, tomato, and piñon nuts over linguine tossed with olive oil and garlic), or spiced shrimp (sautéed in butter and beer with red chile cream sauce). For an entree I recommend scallops with jalapeño sauce or chicken Madagascar (breast of chicken sautéed in a cream sauce with green peppercorns and brandy. Or you may want to try the steak au poivre, a specialty here. For dessert, the crème de caramel is excellent, but as long as you're splurging, go for the polyczenta (ground walnuts with cream and rum wrapped in a crepe with hot chocolate sauce).

○ Maria Teresa

618 Rio Grande Blvd. NW. ☎ **505/242-3900.** Reservations recommended. Lunch $7–$13; dinner $15–$23. AE, MC, V. Daily 11am–2:30pm and 5–9pm. NEW MEXICAN/CONTINENTAL.

The city's most beautiful and classically elegant restaurant, Maria Teresa is located in the 1840s Salvador Armijo House, a National Historic property furnished with Victorian antiques and paintings. Built with 32-inch-thick adobe walls, the house exemplifies 19th-century New Mexican architecture, when building materials were few and defense a prime consideration. The house had 12 rooms, 7 of which (along with a large patio) are now reserved for diners. Another room is home to the 1840 bar and lounge.

The menu features salads, pastas, sandwiches, seafood, and New Mexican specialties for lunch; and a wide choice of traditionally continental gourmet dinners, from seared filet mignon with béarnaise sauce or seared New York cut venison with sundried tomatoes and olives to poached salmon with Dijon basil hollandaise. There's a full bar.

MODERATE

High Noon Restaurant and Saloon

425 San Felipe St. NW. ☎ **505/765-1455.** Reservations recommended. Lunch $4.50–$8.95; dinner $9.50–$24.95. AE, CB, DC, DISC, MC, V. Mon–Sat 11am–3pm and 5–10pm, Sun noon–9pm. STEAKS/SEAFOOD/NEW MEXICAN.

One of Albuquerque's oldest buildings, this restaurant (now in business for over 20 years) boasts a 19th-century saloon atmosphere with stuccoed walls, high and low ceiling beams, and historical photos on the walls. One photo depicts the original 1785 structure, which now comprises the building's foyer and santo room. The dinner menu offers a choice of beef dishes, such as the house-specialty pepper steak sautéed with brandy; fish dishes, including red trout amandine; game dishes, including buffalo, venison, and caribou; and regional favorites, such as burritos, enchiladas, and fajitas. There's a full bar.

○ La Crêpe Michel

400 San Felipe C2. ☎ **505/242-1251.** Reservations accepted. Main courses $3.30–$14.50. MC, V. Tues–Sun 11:30am–2pm; Tues–Sat 6–9pm. FRENCH.

This small find is tucked away in a small walkway not far from the Plaza. Run by chef Claudie Zamet-Wilcox from France, it has a cozy European feel, though it's very informal, with checked table coverings and simple furnishings. You can't miss with any of the crepes. I found the crepe aux fruits de mer (blend of sea scallops, bay scallops, and shrimp in a velouté sauce with mushrooms) especially nice, as is the crepe à la volaille (chunks of chicken in a cream sauce with mushrooms and Madeira wine). For a heartier meal, try one of the specials listed on a board on the wall. My mahi basquaise came in a light vegetable sauce, and my companion's filet de boeuf had a delicious béarnaise. Both were served with vegetables cooked just enough to leave them crisp and tasty. For dessert, don't leave without having a crepe aux fraises (strawberry crepe). Because of its proximity to a church, no alcoholic beverages are served.

La Hacienda Restaurant

302 San Felipe St. NW (at North Plaza). ☎ **505/243-3131.** Reservations recommended for large parties. Lunch $4.50–$9.95; dinner $7.25–$14.95. AE, CB, DC, MC, V. Daily 11am–9:30pm. NEW MEXICAN/AMERICAN.

This restaurant, like its neighbor La Placita, offers more atmosphere than flavor. Appealing mostly to tourist traffic, the food is a muted version of real Northern New Mexican cuisine. If you want the real thing, I suggest Duran Central Pharmacy, a few blocks from the Plaza.

Still, this place is full of history. A mural girding La Hacienda's outer wall depicts the establishment of Albuquerque and the construction of this Villa de Albuquerque at the turn of the 18th century. Diners enter the cozy restaurant through a large gift shop. The interior, with an intimate, laid-back atmosphere, is adorned with hanging plants and chile ristras. The menu has such house specialties as beef or chicken fajitas and carne adovada (sautéed dried pork smothered in red chile). I recommend the tostadas compuestas (a corn tortilla cup topped with beef, beans, melted cheese, and red or green chile), and the tortilla-chile soup. Steaks, shrimp, and other American meals are also offered. Desserts include the traditional flan and fried ice cream.

INEXPENSIVE

Chef du Jour

119 San Pasquale Ave. SW. ☎ **505/247-8998.** Reservations recommended. Menu items $2.50–$7.50. No credit cards. Mon–Fri 11am–2pm. ECLECTIC.

The decor certainly isn't much to talk about at Chef du Jour, but that's not why people flock to this popular lunch spot located just a block from Old Town. Once you get a taste of what's on the menu (which changes every week), you'll be coming back for more. Take special note of the condiments, all of which—from the ketchup to the salsa—are homemade. Recent menu offerings included spicy garlic soup, a great garden burger, green-corn tamales (served with Southwest mango salsa), and marinated grilled chicken breast (on whole-wheat fruit-and-nut bread with Jamaican banana ketchup). There are also sandwiches and a salad du jour. Chef du Jour has both indoor and some outdoor tables; if you call in advance, the restaurant will fax you a copy of their current menu.

Duran Central Pharmacy

1815 Central NW. ☎ **505-247-4141.** Menu items $3.90–$6.90. No credit cards. Mon–Fri 9am–6:30pm, Sat 9am–2pm. NEW MEXICAN.

Sounds like an odd place to eat, I know. Although you could go to one of the touristy New Mexican restaurants in the middle of Old Town and have lots of atmosphere and mediocre food, you can also come here, where locals eat, and feast. It's a few blocks up Central, east of Old Town. On your way through the pharmacy, you may want to stock up on specialty soaps; there's a pretty good variety here.

The restaurant itself is plain, with a red tile floor and small tables, as well as a counter. For years I used to come here for a bowl of green chile and a homemade tortilla, which is still an excellent choice. Now I go for the full meals, such as the blue-corn enchilada plate or the huevos rancheros (eggs over corn tortillas, smothered with chile). The menu is short, but you can count on authentic Northern New Mexican food. No smoking is permitted.

La Placita

208 San Felipe St. NW (Old Town Plaza). ☎ **505/247-2204.** Reservations recommended for large parties. Lunch $4.25–$7.25; dinner $6.25–$14.50. AE, DISC, MC, V. Daily 11am–9pm (until 8:30pm in winter). NEW MEXICAN/AMERICAN.

Native American artisans spread their wares on the sidewalk outside the old Casa de Armijo, built by a wealthy Hispanic family in the early 18th century. The 283-year-old adobe hacienda, which faces the Old Town Plaza, features hand-carved wooden doorways, deep-sunken windows, and an ancient patio. Fine regional art and furnishings decorate the five dining rooms and an upstairs gallery. You won't find many locals frequenting this restaurant; it's a tourist spot, and though the atmosphere is lively, the food is mediocre. The house favorite is a full Mexican dinner, which includes an enchilada colorado de queso, chile relleños, taco de carne, frijoles con queso, arroz español, ensalada, and two sopaipillas. There is also a variety of beef, chicken, and fish selections.

DOWNTOWN
EXPENSIVE

Conrad's Downtown

125 Second St. NW (at Copper Ave.) (in La Posada de Albuquerque). ☎ **505/242-9090.** Reservations recommended. Lunch $5.95–$7.25; dinner $12.50–$18.75. AE, CB, DC, DISC, MC, V. Daily 6:30am–2pm and 5:30–10pm. Tapas bar daily 11am–10pm. SPANISH/MEXICAN.

Conrad's is located on the first floor of the historic La Posada hotel, just off the lobby. A large, classic bar is the focal point around which diners sample Spanish and Mexican specialties such as tapas and paella. Though the selections are interesting, I found the ingredients here less fresh than they might have been. Still, the paella is a unique blend of fish, fowl, and sausage flavors and the cordoorniz negrado con chipotle y mantequilla con tequila (blackened quail with a mixture of lime and tequila, chipotle, and cilantro in a creamy butter sauce) can be tasty. A few beef and pork dishes are offered as well. The atmosphere is casual. Conrad's offers complimentary valet parking.

✪ Stephens

1311 Tijeras Ave. NW (at 14th St. and Central Ave.). ☎ **505/842-1773.** Reservations recommended. Main courses $18.95–$24.95. AE, DC, DISC, MC, V. Sun–Thurs 5:30–9:30pm, Fri–Sat 5:30–10:30pm. CONTEMPORARY AMERICAN.

This is the place many Albuquerqueans go to celebrate. My father usually takes me here on my birthday. The decor by noted designer Richard Worthen is very traditional, with lots of fine wood and brass, yet the atmosphere is comfortable, not stuffy. French windows provide a view of lush gardens surrounding a fountain. For an appetizer I've enjoyed the baked brie with almonds, apples, and honey in a puff pastry as well as the smoked salmon. Entrees include a number of meat dishes as well as some chicken, polenta, and pasta. I'm partial to the steak Diane tenderloin. There's a spa menu, with entrees lower in fat and calories, which is also tasty. I enjoy the ruby red trout, grilled, with piñones and julienne vegetables. The award-winning wine list features more than 300 choices. There are daily dessert specials.

MODERATE

✪ Artichoke Cafe

424 Central Ave. SE. ☎ **505/243-0200.** Reservations recommended. Main courses $6.95–$24.95. AE, DISC, MC, V. Mon–Fri 11am–2:30pm; Mon–Sat 5:30–10pm. CONTINENTAL.

Locals recommend this restaurant for fine food in a no-frills atmosphere. The decor here is clean and tasteful, with modern art prints on azure walls, white linens on tables shaded by standing plants, and classical music playing in the background. A recent renovation to the patio has made it a good choice for warmer months. Start your meal with an artichoke appetizer or with roasted garlic served with New Mexico goat cheese. The lunch menu features many salads made with organic greens and treats such as Italian sausage served over white cannellini beans. Sandwiches and entrees such as Thai-marinated tenderloin are also available at lunch. For dinner, there are pastas, polentas, and richer entrees. My favorite is the polenta with fresh spinach, roasted vegetables, and Parmesan and Romano cheeses. Or the pecan-crusted rack of lamb "served on a shadow of orange, olive, and basil sauce" is also excellent. Meals are served with a salad, so you can count on plenty of food here.

THE NORTHEAST HEIGHTS
EXPENSIVE

High Finance Restaurant & Tavern

40 Tramway Rd. NE (atop Sandia Peak). ☎ **505/243-9742.** Reservations recommended. Lunch $4.95–$7.95; dinner $14.95–$34.95. Tramway, $10 with dinner reservations, $13.50 without. AE, CB, DC, DISC, MC, V. Daily 11am–3pm and 4:30–9:30pm. CONTINENTAL AND NEW AMERICAN.

Perched atop Sandia Peak, 2 miles above Albuquerque and the Rio Grande Valley, High Finance offers diners a breathtaking panorama of New Mexico's largest city. The atmosphere inside is elegant yet casual.

The lunch menu features sandwiches, pastas, and New Mexico specialties. A recommendation from that menu is linguine Sandia (linguine tossed with bacon, green chile, corn, sage, and fresh Parmesan). Dinner brings such entrees as chicken fettuccine with sun-dried tomatoes, fresh basil, and cream; chile-dusted seared beef tips; and marinated pork chops served with apple green-chile chutney, garlic potato puree, and grilled squash. I always enjoy High Finance's slow-roasted prime rib with ancho chile jus, roasted potatoes, and horseradish cream. There are nightly specials and a fish of the day. Many tram riders just drop in for the view and a drink at the casual full-service bar. Smoking is permitted only in the bar.

✪ Le Marmiton

5415 Academy Blvd. NE. ☎ **505/821-6279.** Reservations recommended. Main courses $13.95–$18.95. AE, DC, DISC, MC, V. Mon–Thurs 5–9pm, Fri–Sat 5:30–9:30pm, Sun 5:30–9pm. FRENCH.

The name means "The Apprentice," but there's nothing novice about the food or presentation. The 15 tables seat 45 people in a romantic French provincial atmosphere, accented with lace curtains and a lovely collection of antique plates.

Recommended main courses include fantaisie aux fruits de mer, a mixture of shrimp, scallops, and crab sautéed with shallots, basil, white wine, and mushrooms and finished with a cream sauce, served on a puff pastry shell; and cailles, two whole quail, lightly seasoned, sautéed, and finished with a shallot-sherry cream sauce. For vegetarians, the menu now includes a vegetarian appetizer and entree. If you arrive early for dinner (between 5 and 6pm) Monday through Thursday, you can take advantage of the fixed-price light dinners, which include soup or salad, a choice of main course, and coffee or tea for $7.50. There's a long wine list, and the crème aux framboises makes a great dessert.

MODERATE

Cafe Spoleto

2813 San Mateo Blvd. NE. ☎ **505/880-0897.** Reservations recommended, especially on weekends. Main courses $12–$18. AE, DISC, MC, V. Tues–Sat 6–9:30pm, Sun 5:30–9pm. MEDITERRANEAN.

If you've grown a bit weary of New Mexican cuisine (if that's possible) and are looking for a nice, unpretentious Mediterranean restaurant with a casual atmosphere, try Cafe Spoleto. Salads include a grilled raddichio wrapped in pancetta with melon, which is quite nice. My pasta choice was the penne with house-made Italian sausage, roasted garlic, tomato, capers, olives, and goat cheese. The natural chicken under a brick with a squash gratin and salsa verde is also very tasty, and the grilled mahi mahi is definitely worth trying. The menu changes every 2 weeks, so you'll get to sample plenty of seasonal dishes.

County Line

9600 Tramway Blvd. NE. ☎ **505/856-7477.** Main courses $8.95–$14.95. AE, CB, DC, DISC, MC, V. Mon–Thurs 5–9pm, Fri–Sat 5–10pm, Sun 4–9pm. BARBECUE.

Although this extremely popular spot doesn't take reservations, if you call before you leave your hotel they'll put your name on the waiting list; by the time you get there you'll probably be next in line. If not, you can always wait at the ever-crowded bar. The restaurant is loud and always busy, but it has a spectacular view of the city lights, great food, and new decor that has a fun roadhouse cafe feel.

When you finally get a table, you'll be given a Big Chief Writing Tablet menu offering great Southwestern barbecue at very reasonable prices. You might opt for barbecued chicken or a steak grilled to perfection, along with a baked potato (with your choice of toppings), beans, and coleslaw. They've also added grilled fish to the menu. If you're not very hungry you should probably consider going somewhere else.

INEXPENSIVE

Rudy's Country Store and Bar-B-Q

2321 Carlisle Blvd. NE (at I-40; with a second location at 10136 Coors Rd. NW). ☎ **505/884-4000.** Main courses $3.75–$10.95. AE, DISC, MC, V. Daily 10am–10pm. BARBECUE.

Don't be put off by the picnic tables in this modest barbecue spot. There are three counters where you can order your meal. Start at the first counter by ordering your side dishes (for example, traditional pinto beans, potato salad, and creamed corn) by the pint. The second counter is where you order what you came for—delectable barbecue brisket, spare ribs, short ribs, chicken, and turkey, among other things. Finally, head to the third counter for beer, soda, and iced tea. The sauce, which is quite thin and spicy with lots of vinegar, is up to you (bottles are on each table). The dessert selections aren't worth mentioning, and chances are you won't have any room left anyway.

UNIVERSITY & NOB HILL
MODERATE

Cafe Oceana

1414 Central Ave. SE. ☎ **505/247-2233.** Reservations recommended. Main courses $11.95–$17.95. AE, DC, DISC, MC, V. Mon–Thurs 11am–11pm, Fri 11am–11:30pm; Sat 5–11:30pm. Oyster hour, Mon–Fri 3–6:30pm, Sat 5–7pm. SEAFOOD.

The Cafe Oceana is and has long been Albuquerque's favorite oyster bar and fresh seafood cafe. In a New Orleans–style dining room with high ceilings and hardwood floors, you can enjoy fresh oysters, fresh fish daily, scallops, crab rellenos (for the New Mexican touch), and the house special—beer-batter-fried shrimp Oceana. The daily oyster hour features two-for-one oysters and boiled shrimp. If you're really in the mood for New Orleans cuisine, you can also order red beans and rice here.

Monte Vista Fire Station

3201 Central Ave. NE (Nob Hill). ☎ **505/255-2424.** Reservations recommended. Lunch $5.95–$8.95; dinner $12.95–$19.95. AE, CB, DC, DISC, MC, V. Mon–Fri 11am–2:30pm;

Sun–Thurs 5–10:30pm, Fri–Sat 5–11pm. Bar, Mon–Fri 11am–2am, Sat noon–2am, Sun noon–midnight. NEW AMERICAN.

This restaurant is onto a very novel idea. Each month it brings in a chef from a different city. You may savor New Orleans–style Cajun food in April and New American food with a dash of San Francisco–Asian flavor in July. The setting is unique too. It actually was a fire station, built in Pueblo Revival style in 1936. Now the decor is art deco, a little cold feeling, perhaps, but fun. There are menu standards such as the cornmeal-dusted calamari on wild baby greens with chipotle chile aioli, a starter. From visiting chef Barbara Miachika from San Francisco, I tried a risotto of sun-dried tomatoes, porcini, and fresh mushrooms, but I've also enjoyed a number of good pastas and chicken or meat dishes. There's fish as well. If you want something smaller, there are also sandwiches, including the fire burger, served with caramelized red onions, green chile, and mozzarella cheese on homebaked bread. For dessert you have to try the Monte Vista chocolate "bombe" with gold leaf and crème anglaise. There's a popular singles bar on the second-floor landing.

✪ Scalo

3500 Central Ave. SE (Nob Hill). ☎ **505/255-8782.** Reservations recommended. Lunch $5.75–$9; dinner $7.95–$17.95. AE, CB, DC, DISC, MC, V. Mon–Sat 11:30am–2:30pm and 5–11pm; Sun 5–9pm. Bar, Mon–Sat 11:30am–1am. NORTHERN ITALIAN.

Scalo has a simple bistro-style elegance, with white linen-clothed tables indoors, plus outdoor tables in a covered, temperature-controlled patio. The kitchen, which makes its own pasta and breads, specializes in contemporary adaptations of classical Northern Italian cuisine.

Seasonal menus focus on New Mexico–grown produce. Featured appetizers include calamaretti fritti (fried baby squid served with a spicy marinara and lemon aioli) and caprini con pumante (local goat cheese with fresh focaccia, capers, tapenade, and a roasted garlic spread). There's a selection of pastas for lunch and dinner, as well as meat, chicken, and fish dishes. The filetto con salsa balsamica (grilled filet of beef with rosemary, green peppercorns, garlic, and a balsamic demiglace sauce) is one of my favorites, and the Battuta di Vitello Mandorlata (veal scaloppine prepared with toasted almonds, sun-dried cranberries, and Pinot Grigio) is also quite good. Dessert selections change daily.

INEXPENSIVE

Double Rainbow

3416 Central Ave. SE. ☎ **505/255-6633.** No reservations. All menu items under $8. AE, DC, DISC, MC, V. Daily 6:30pm–midnight. CAFE/BAKERY.

Albuquerque's literati hang out at this cafe, as do university professors and hippies. They come for the many coffee drinks such as chocolate cappuccino or iced java (made with cream and chocolate), as well as for salads and sandwiches. My favorite is the tandoori chicken sandwich, a marinated breast, grilled and served on a bun with cucumbers and tomatoes. There are heartier meals too, such as lasagna and quiche. For dessert, try one of the elaborate baked goods such as a tricolor mousse with white, milk, and dark chocolate, or sample from a variety of ice-cream flavors. One of the best things about this place is its enormous selection of magazines. In fact, there are more than 700 titles, ranging from comic books to gay and lesbian travel magazines and unusual titles like *Tricycle* (a Buddhist review). The selection of newspapers spans the globe. Double Rainbow has another branch at 4501 Juan Tabo NE (☎ 505/275-8311), which features a large outdoor patio, a slightly more extensive menu, and live music on Thursday nights as well as Saturday and Sunday mornings.

Olympia Cafe

2210 Central Ave. SE. ☎ **505/266-5222.** Menu items $1.50–$8.95. AE, DC, DISC, MC, V. Mon–Fri 11am–10pm, Sat noon–10pm. GREEK.

Ask any Northern New Mexico resident where they go for Greek food, and the hands-down favorite is Olympia. Though it's very informal, right across from the university, diners eat there at all times of day. It has a lively atmosphere, with the sound bursts of enthusiastic Greek emanating from the kitchen. With a full carry-out menu, it's also a great place to grab a meal on the run. In the summer months I like to get the Greek salad, served with fresh pita bread, and white bean soup. A standard is the falafel sandwich with tahini. The restaurant is well known for its gyros (slices of beef and lamb broiled on a vertical spit wrapped in pita), and I hear the moussaka is excellent. For dessert try the baklava.

66 Diner

1405 Central Ave. NE. ☎ **505/247-1421.** Menu items $3–$7.95. AE, DISC, MC, V. Mon–Thurs 11am–11pm, Fri 11am–midnight, Sat 8am–midnight, Sun 8am–10pm. AMERICAN.

Like a trip back in time to the days when Martin Milner and George Maharis got "their kicks on Route 66," this thoroughly 1950s-style diner comes complete with Seeburg jukebox and full-service soda fountain. The white caps make great green-chile cheeseburgers, along with meat-loaf sandwiches, grilled liver and onions, and chicken-fried steaks. Ham-and-egg and pancake breakfasts are served every morning. Beer and wine are available.

SOUTHEAST: NEAR THE AIRPORT

Rio Grande Yacht Club Restaurant & Lounge

2500 Yale Blvd. SE. ☎ **505/243-6111.** Reservations recommended at dinner. Main courses $4.25–$7.95 at lunch, $9.95–$32.95 at dinner. AE, CB, DC, DISC, MC, V. Mon–Fri 11am–2pm, daily 5:30–10:30pm. SEAFOOD.

Red, white, and blue sails are draped beneath the skylight of a large room dominated by a tropical garden. The walls, hung with yachting prints and photos, are of wood strips, like those of a ship's deck. The lunch menu features burgers, sandwiches, salads, and a few New Mexican specialties. At dinner, however, fresh fish is the main attraction. Catfish, whitefish, bluefish, sole, salmon, grouper, mahi mahi, and other denizens of the deep are prepared broiled, poached, blackened, teriyaki, Vera Cruz, au gratin, mornay, amandine, and more. If you'd rather have something else, the chef here also prepares certified Angus beef, shrimp, Alaskan king crab, several chicken dishes, and even barbecued baby back pork ribs. If you find it difficult to choose one of these, you might want to try a steak and seafood combination.

OUT OF TOWN

Prairie Star

1000 Jemez Canyon Dam Rd., Bernalillo. ☎ **505/867-3327.** Reservations recommended. Main courses $15–$24. AE, DISC, MC, V. Daily 5–10pm (lounge opens at 4pm); brunch Sun 11am–2:30pm. CONTEMPORARY REGIONAL.

A sprawling adobe home, with a marvelous view across the high plains and a golf course, is the site of this intimate dining experience about 30 minutes north of Albuquerque. The 6,000-square-foot house, on a rural site leased from Santa Ana Pueblo, was built in the 1940s in Mission architectural style. Exposed vigas and full latilla ceilings, as well as hand-carved fireplaces and bancos, complement the thick adobe walls in the dining room. The art displayed on the walls is for sale. There is a lounge at the top of the circular stairway.

Diners can start with smoked quail or baked cheese in puff pastry (a blend of mascarpone and Gruyère cheese scented with toasted hazelnuts and fresh tarragon). Main courses include shrimp margarita (shrimp sautéed and flambéed with tequila, fresh tomatoes, and roasted poblano chiles, and finished with lime juice and butter), veal sweetbreads (served in puff pastry with a tarragon and wild-mushroom cream sauce), lamb loin (stuffed with roast garlic, pine nuts, basil, and goat cheese), and pan-fried Truchas trout with piñon nuts. There are daily specials.

✪ Range Cafe

925 Camino del Pueblo, Bernalillo. ☎ **505/867-1700.** No reservations. Breakfast/lunch $2.95–$10.95; dinner $6.50–$15.95. AE, DISC, MC, V. Mon–Fri 7:30am–3pm, Sat–Sun 8am–3pm; daily 5pm until closing. NEW MEXICAN/NEW AMERICAN.

This cafe on the main drag of Bernalillo, about 15 minutes north of Albuquerque, is a perfect place to stop on your way out of town. However, the food's so good you may just want to make a special trip here. Housed in what was once an old drugstore, the restaurant has tin molding on the ceiling and is decorated with western touches, such as cowboy boots and period photos. There's a soda fountain in the center of the large space, and all the tables and chairs are hand-painted with whimsical stars and clouds.

The food ranges from New Mexican enchiladas and burritos to chicken-fried steak to sandwiches and elegantly prepared meals. The proprietors here have come from such notable restaurants as Scalo in Albuquerque and Cafe Escalera in Santa Fe, and the new chef was quickly nabbed from the Double A in Santa Fe after it closed, so you can count on exquisite food. For breakfast try the pancakes or the breakfast burrito. For lunch or dinner, I recommend Tom's meat loaf, served with roasted garlic mashed potatoes, mushroom gravy, and sautéed vegetables. For dinner, you might try Range scallops, with a grilled tomato and cilantro cream sauce with pine nuts over red chile linguine. For dessert they serve Taos cow ice cream in cones or malts, shakes, or sundaes, as well as baked goods and specialty coffees. No smoking is permitted.

5 What to See & Do

Albuquerque's original town site, known today as Old Town, is the central point of interest for visitors. Here, grouped around the Plaza, are the venerable Church of San Felipe de Neri and numerous restaurants, art galleries, and crafts shops. Several important museums are situated close by. Within a few blocks are the recently completed Albuquerque Aquarium and the Rio Grande Botanic Garden (near Central Avenue and Tingley Drive NW). The project includes a 25,000-square-foot aquarium and a 50-acre botanical garden, both well worth the visit.

But don't get stuck in Old Town. Elsewhere you will find the Sandia Peak Tramway, Kirtland Air Force Base and the National Atomic Museum, the University of New Mexico with its museums, and a number of natural attractions. Within day-trip range are several pueblos and a trio of significant monuments (see "Exploring Nearby Pueblos & Monuments," below).

THE TOP ATTRACTIONS

Old Town

Northeast of Central Ave. and Rio Grande Blvd. NW.

A maze of cobbled courtyard walkways leads to hidden patios and gardens, where many of Old Town's 150 galleries and shops are located. Adobe buildings, many refurbished in the Pueblo Revival style in the 1950s, are grouped around the tree-shaded **Plaza,** created in 1780. Pueblo and Navajo artisans often display their pottery, blankets, and silver jewelry on the sidewalks lining the Plaza.

The buildings of Old Town once served as mercantile shops, grocery stores, and government offices, but the importance of Old Town as Albuquerque's commercial center declined after 1880, when the railroad came through 1 1/4 miles east of the Plaza and businesses relocated to be closer to the trains. Old Town clung to its historical and sentimental roots, but the quarter fell into disrepair until the 1930s and 1940s, when it was rediscovered by artisans and other shop owners, and the tourism industry burgeoned.

When Albuquerque was established in 1706, the first building erected by the settlers was the **Church of San Felipe de Neri,** which faces the Plaza on its north side. It's a cozy church with wonderful stained-glass windows and vivid retablos. This house of worship has been in almost continuous use for about 290 years.

It's sad to see the changes the past 10 years or so have wrought on Old Town shopping. When I was growing up in the area, this was the place to go to buy gifts. Now, many of the interesting shops (such as the basket shop right on the Plaza, which used to be packed with thousands of dusty baskets) have become trinket stores. However, you can still find good buys from the Native Americans selling jewelry on the Plaza. Look especially for silver bracelets and strung turquoise. If you want to take something fun home and spend very little, buy a dyed corn necklace. Your best bet when wandering around Old Town is just to peek into shops, but there are a few places you'll definitely want to spend time in. See the shopping section for a list of recommendations.

✪ Indian Pueblo Cultural Center

2401 12th St. NW. ☎ **800/766-4405** or 505/843-7270. Admission $3 adults, $2 seniors, $1 students, free for children 4 and under. Daily 9am–5:30pm; restaurant daily 7:30am–4pm. Closed Jan 1, Thanksgiving, Dec 25. From Lomas Blvd., turn north on 12th St. The Cultural Center is on the left, just beyond the I-40 underpass. From midtown, head west on Menaul Blvd. and turn left onto 12th St.; the center will be on the right.

Owned and operated as a nonprofit organization by the 19 pueblos of New Mexico, this is a fine place to begin exploring local Native American culture. Situated about a mile northeast of Old Town, this museum—modeled after Pueblo Bonito, a spectacular 9th-century ruin in Chaco Culture National Historic Park—consists of several parts.

In the basement, a permanent exhibit depicts the evolution (from prehistory to the present) of the various pueblos, including displays of the distinctive handcrafts of each community. Note, especially, how pottery differs in concept and design from pueblo to pueblo. The displays include a series of remarkable photographs of Pueblo tribe members taken between 1880 and 1910, a gift of the Smithsonian Institution. The Pueblo House Children's Museum, also on the premises, is a hands-on museum that gives children the opportunity to learn about the evolution of Pueblo culture.

Upstairs is an enormous gift shop—a fine place to check the price of the Pueblo people's colorful ceramics, weavings, and paintings before you barter with private artisans. The Indian Arts and Crafts Association's code of ethics guarantees that the work here is stylistically authentic and that everything was made by Native Americans only. A gallery displays a variety of ancient and modern works from different pueblos, and the exhibits change monthly.

Local Native American dancers perform and artisans demonstrate their craft expertise in an outdoor arena surrounded by original murals. **Craft fairs** are held on July 4th weekend, the first weekend in October, and Thanksgiving weekend.

Central Albuquerque Attractions

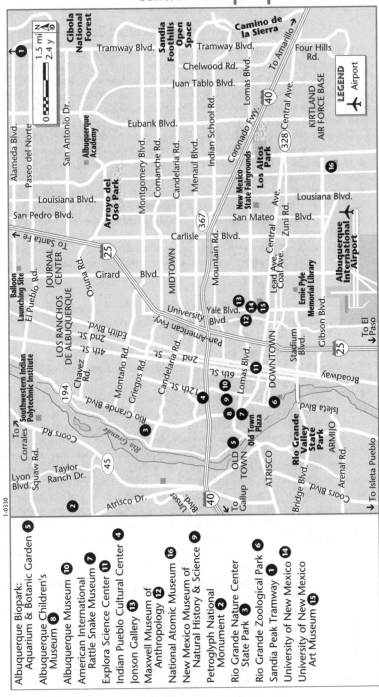

LEGEND
✈ Airport

Albuquerque Biopark: Aquarium & Botanic Garden **5**
Albuquerque Children's Museum **8**
Albuquerque Museum **10**
American International Rattle Snake Museum **7**
Explora Science Center **11**
Indian Pueblo Cultural Center **4**
Jonson Gallery **13**
Maxwell Museum of Anthropology **12**
National Atomic Museum **16**
New Mexico Museum of Natural History & Science **9**
Petroglyph National Monument **2**
Rio Grande Nature Center State Park **3**
Rio Grande Zoological Park **6**
Sandia Peak Tramway **1**
University of New Mexico **14**
University of New Mexico Art Museum **15**

1-0330

Traditional Native American Bread Baking

While visiting the pueblos in New Mexico, you'll probably notice outdoor ovens (they look a bit like giant ant hills), known as hornos, which Native Americans have used to bake bread for hundreds of years. For Native Americans, making bread is more than simply preparing a staple of daily life; it is a tradition that links them directly to their ancestors. The long process of mixing and baking also brings mothers and daughters together for what today we might call "quality time."

Usually in the evening the bread dough (ingredients: white flour, lard, salt, yeast, and water) is made and kneaded, and the loaves are shaped. They are then allowed to rise overnight. In the morning, the oven is stocked with wood and a fire lighted. After the fire burns down to ashes and embers, the oven is cleared, and the ashes are shoveled away. Unlike modern ovens, hornos don't come equipped with thermometers, so the baker must rely on her senses to determine when the oven is the proper temperature. At that point the loaves are placed into the oven with a long-handled wooden paddle. They bake for about an hour.

If you would like to try a traditional loaf, you can buy one at the Indian Pueblo Cultural Center in Albuquerque (and elsewhere throughout the state).

A **restaurant,** open for breakfast and lunch daily from 7:30am to 4pm, emphasizes the cornmeal-based foods of the Pueblo people. An ample meal might consist of posole, treated dried corn with beef, chile, and oven-fried bread.

✪ Albuquerque Museum

2000 Mountain Rd. NW. ☎ **505/243-7255.** Free admission, but donations are appreciated. Tues–Sun 9am–5pm. Closed major holidays.

The largest U.S. collection of Spanish colonial artifacts is housed here. Included are arms and armor used during the Hispanic conquest, medieval religious artifacts and weavings, maps from the 16th to 18th centuries, and coins and domestic goods traded during that era. A multimedia audiovisual presentation, *Four Centuries: A History of Albuquerque,* depicts the history of the mid–Rio Grande region from the Spanish conquest to the present. "History Hopscotch" is a hands-on history exhibit designed specifically for children. There's also a gallery of early and modern New Mexico art, with permanent and changing exhibits; a major photo archive; and a sculpture garden. Exhibits scheduled for 1998 include: *The Life and Art of Patrociño Barela, Photographs of Ghost Ranch* by Janet Russek and David Scheinbaum, and *Colonial and Modern Images from Guatemala and New Mexico.* From time to time children's exhibits and classes are offered. A gift shop sells a variety of souvenirs and other wares.

Rio Grande Nature Center State Park

2901 Candelaria Rd. NW. ☎ **505/344-7240.** Admission $1 adults, 50¢ children 6 and older, free for children under 6. Daily 10am–5pm. Closed Jan 1, Thanksgiving, Dec 25.

Located on the Rio Grande Flyway, an important migratory route for many birds, this wildlife refuge extends for nearly a mile along the east bank of the Rio Grande. Numerous nature trails wind through the cottonwood bosque, where a large variety of native and migratory species can be seen at any time of year. The center publishes a checklist to help visitors identify them, as well as several self-guiding trail brochures. Recently they've added guided "bird walks" every Saturday, year-round. Call for times.

Housed in a unique building constructed half above ground and half below, the visitor center houses classrooms, laboratory space, a library, a gift shop, and exhibits describing the history, geology, and ecology of the Rio Grande Valley. Informative hikes are scheduled every weekend.

See the "Especially for Kids" section for more museums and attractions in the Old Town area.

✪ Sandia Peak Tramway

10 Tramway Loop NE. ☎ **505/856-7325.** Admission $14 adults, $10 seniors and children 5–12, free for children under 5. AE, DISC, MC, V. Memorial Day–Labor Day, daily 9am–10pm; spring and fall, Thurs–Tues 9am–8pm, Wed 5–8pm; ski season, Mon–Tues and Thurs–Fri 9am–8pm, Wed noon–8pm, Sat–Sun 8:30am–8pm. To reach the base of the tram, take I-25 north to Tramway Rd. (exit 234), then proceed east about 5 miles on Tramway Rd. (NM 556); or take Tramway Blvd., exit 167 (NM 556), north of I-40 approximately 8 $^1/_2$ miles. Turn east the last half mile on Tramway Rd.

The Sandia Peak tram is a "jigback"; in other words, as one car approaches the top, the other nears the bottom. The two pass halfway through the trip, in the midst of a 1 $^1/_2$-mile "clear span" of unsupported cable between the second tower and the upper terminal.

Several hiking trails are available on Sandia Peak; one of them—La Luz Trail— is partly flat and quite easy. The views in all directions are extraordinary. *Note:* The trails on Sandia may not be suitable for children.

There is a popular and expensive restaurant at High Finance, Sandia's summit, (see "Where to Dine," earlier in this chapter). Special tram rates apply with dinner reservations.

OTHER ATTRACTIONS

✪ National Atomic Museum

Wyoming Blvd. and K St. (P.O. Box 5800), Kirtland Air Force Base. ☎ **505/284-3243.** Free admission. Visitors must obtain passes (and a map) at the Wyoming or Gibson Gate of the base. Children under 12 not admitted without parent or adult guardian. Daily 9am–5pm. Closed Jan 1, Easter, Thanksgiving, Dec 25.

This museum is the next-best introduction to the nuclear age after the Bradbury Science Museum in Los Alamos. It traces the history of nuclear weapons development beginning with the top-secret Manhattan Project of the 1940s, including a copy of the letter Albert Einstein wrote to President Franklin D. Roosevelt suggesting the possible development of an atomic bomb. A 51-minute film, *Ten Seconds That Shook the World,* is shown every hour.

There are full-scale models of the "Fat Man" and "Little Boy" bombs, and displays and films on the peaceful application of nuclear technology and other alternative energy sources. Fusion is explained in a manner that laypeople can understand; other exhibits deal with the problem of nuclear waste. Outdoor exhibits include a B-52 "Stratofortress," an F-1015D "Thunderchief," and a 280mm atomic cannon. An hour-long tour takes visitors through the development of the first nuclear weapons to today's technology. You'll also see a solar-powered TV and be able to test your budgeting skills as you use the Energy Environment Simulator to manipulate energy allocations (in your quest to make the planet's resources last longer). The museum is directly across the street from the International Nuclear Weapons School—do you think they offer summer courses?—adjacent to Sandia National Laboratory.

Petroglyph National Monument

4735 Unser Blvd. NW (west of Coors Rd.). ☎ **505/839-4429** or 505/899-0205. Admission $1 per vehicle weekdays, $2 weekends. MC, V. Summer, daily 9am–6pm; winter, daily 8am–5pm. Closed Jan 1, Dec 25.

Albuquerque's western city limits are flanked by five extinct volcanoes. Adjacent lava flows became a hunting and gathering area for prehistoric Native Americans, who lived here and left a chronicle of their beliefs etched and chipped in the dark basalt boulders. Some 15,000 of these petroglyphs have been found in concentrated groups at this archaeological preserve. Plaques interpret the rock drawings—animal, human, and ceremonial forms—for visitors, who have a choice of four hiking trails, ranging from easy to moderately difficult, winding through the lava. The 45-minute Mesa Point trail is the most strenuous but also the most rewarding.

Camping is not permitted in the park; it's strictly for day use, with picnic areas, drinking water, and rest rooms provided.

University of New Mexico
Yale Blvd. NE (north of Central Ave.). ☎ **505/277-0111.**

The state's largest institution of higher learning stretches across an attractive 70-acre campus about 2 miles east of downtown Albuquerque, north of Central Avenue and east of University Boulevard. The five campus museums, none of which charges admission, are constructed (like other UNM buildings) in a modified Pueblo style. Popejoy Hall, in the south-central part of the campus, hosts many performing-arts presentations, including those of the New Mexico Symphony Orchestra; other public events are held in nearby Keller Hall and Woodward Hall.

The **Maxwell Museum of Anthropology,** situated on the west side of the campus on Redondo Drive at Ash Street NE (☎ **505/277-4404**), is an internationally acclaimed repository of Southwestern anthropological finds. It's open Tuesday to Friday from 9am to 4pm, Saturday from 10am to 4pm, and Sunday from noon to 4pm; closed holidays.

The **University of New Mexico Art Museum** (☎ **505/277-4001**) is located in the UNM Center for the Arts, just north of Central Avenue and Cornell Street. The museum features changing exhibitions of 19th- and 20th-century art. Its permanent collection includes Old Masters paintings and sculpture, significant New Mexico artists, Spanish colonial artwork, the Tamarind Lithography Archives, and one of the largest university-owned photography collections in the country. It's open Tuesday to Friday from 9am to 4pm; Tuesday evening from 5 to 8pm; and Sunday from 1to 4pm; closed holidays. A gift shop offers a variety of gifts and posters. Admission is free.

A branch of the University of New Mexico's Art Museum, **University Art Museum Downtown** (☎ **505/242-8244**), is located at 516 Central Ave. SW. Its two floors of exhibition space feature changing exhibits of 19th- and 20th-century art and a variety of featured artists. It's open Tuesday to Saturday from 11am to 4pm (closed holidays). There is a gift shop offering cards, posters, jewelry, and art gift items. Admission is free.

The intimate **Jonson Gallery** at 1909 Las Lomas Blvd. NE (☎ **505/277-4967**), on the north side of the central campus, displays more than 2,000 works by the late Raymond Jonson, a leading modernist painter in early 20th-century New Mexico, as well as works by contemporary artists. The gallery is open Tuesday to Friday from 9am to 4pm and Tuesday evening from 5 to 8pm.

In Northrop Hall (☎ **505/277-4204**), about halfway between the Maxwell Museum and Popejoy Hall in the southern part of the campus, the adjacent **Geology Museum and Meteorite Museum** (☎ **505/277-1644**) cover the gamut of recorded time from dinosaur bones to moon rocks. The 550 meteorite specimens here comprise the sixth-largest collection in the United States. The Geology Museum is open Monday to Friday from 8am to 5pm; the Meteorite Museum, Monday to Friday from 9am to noon and 1 to 4pm.

Finally, the **Museum of Southwestern Biology,** in the basement of the Biology Department at Castetter Hall (☎ 505/277-5340), has few displays but rather extensive research holdings of global flora and fauna, especially representative of Southwestern North America, Central America, South America, and parts of Asia. Call for times and other information.

COOKING SCHOOLS

If you've fallen in love with New Mexican and Southwestern cooking during your stay (or even before you arrived), you might like to sign up for cooking classes with Jane Butel, a leading Southwest cooking authority and author of 14 cookbooks. At **Jane Butel's Cooking School,** 125 Second St. NW (La Posada de Albuquerque), (☎ **800/472-8229** or 505/243-2622; fax 505/243-8297), you'll learn the history and techniques of Southwestern cuisine and have ample opportunity for hands-on preparation. If you choose the weeklong session, you'll start by learning about chiles. The second and third days you'll try your hand at native breads and dishes, the fourth focuses on more innovative dishes, and the fifth and last day covers appetizers, beverages, and desserts. Weekend sessions are also available. Call or fax for current schedules and fees.

6 Especially for Kids

✪ Albuquerque Biopark: Aquarium and Botanic Garden

2601 Central Ave. SW. ☎ **505/764-6200.** Admission $2.50 for ages 3–5 and over 65; $4.50 for ages 16–64; free for children 2 and under. Mon–Fri 9am–5pm, Sat–Sun during June, July, and August 9am–6pm. Ticket sales stop at 4:30pm to allow time to view the facilities. Closed Jan 1, Thanksgiving, Dec 25.

For those of us born and raised in the desert, this attraction quenches years of soul thirst. The self-guided aquarium tour begins with a beautifully produced 9-minute film that describes the course of the Rio Grande River from its origin to the Gulf Coast. Then you'll move on to the touch pool, where at certain times of day children can gently touch hermit crabs and star fish. You'll pass by a replica of a salt marsh, where a gentle tidal wave moves in and out, and you'll explore the eel tank, through which you get to walk since it's an arched aquarium over your path. There's a colorful coral reef exhibit, as well as the culminating show, in a 285,000-gallon shark tank, where many species of fish and 15 to 20 sand tiger, brown, and nurse sharks swim around looking ominous.

Within a state-of-the art 10,000-square-foot conservatory, you'll find the botanical garden split into two sections. The smaller one houses the desert collection and features plants from the lower Chihuahuan and Sonoran deserts, including unique species from Baja California. The larger pavilion exhibits the Mediterranean collection and includes many exotic species native to the Mediterranean climates of southern California, South Africa, Australia, and the Mediterranean Basin. Allow at least 2 hours to see both parks. There is a restaurant on the premises.

Albuquerque Children's Museum

800 Rio Grande Blvd. NW, Suite 10. ☎ **505/842-1537.** Admission $3 ages 2–12, $1 ages 13 and up, free for children under 2. Mon–Sat 10am–5pm, Sun noon–5pm.

At the Albuquerque Children's Museum there's something for everyone: bubbles, whisper disks, a puppet theater, a giant loom, a dress-up area, zoetropes, a capture-your-shadow wall, art activities, science demonstrations, and a giant pin-hole camera. The museum also sponsors wonderful educational workshops. "The Me I Don't Always See" was a health exhibit designed to teach children about the mysteries of the human body, and a Great Artists Series featured live performances about artists' lives and work followed by a related art activity.

American International Rattlesnake Museum

202 San Felipe St. NW. ☎ **505/242-6569.** Admission $2 adults, $1 children. AE, DISC, MC, V. Daily 10am–6:30pm.

This unique museum, located just off Old Town Plaza, has living specimens of common, uncommon, and very rare rattlesnakes of North, Central, and South America in naturally landscaped habitats. Oddities such as albino and patternless rattlesnakes are included, with a popular display for youngsters—baby rattlesnakes. More than 30 species can be seen, followed by a 7-minute film on this contributor to the ecological balance of our hemisphere. Throughout the museum are rattlesnake artifacts from early American history, Native American culture, medicine, the arts, and advertising.

You'll also find a gift shop that specializes in Native American jewelry, T-shirts, and other memorabilia related to the natural world and the Southwest, all with an emphasis on rattlesnakes.

Explora Science Center

40 1st Plaza/Galleria #68 (at Second St. and Tijeras Ave. NW). ☎ **505/842-6188.** Admission $2 adults, $1 children 5–17 and seniors 62 and older, free for children 4 and under. Wed–Sat 10am–5pm, Sun noon–5pm.

Children and adults alike will enjoy a trip to the Explora Science Center, a hands-on science and technology museum. Kids will learn about air pressure by flying a model plane and floating a ball on a stream of air. The light and electricity exhibits allow museum-goers to create their own laser shows and freeze their shadows. Furthermore, there are motion, health and body, fluid, and sound exhibits—all hands on.

✪ New Mexico Museum of Natural History & Science

1801 Mountain Rd. NW. ☎ **505/841-2800.** Admission $5.25 adults, $4.20 seniors, $2.10 children 3–11. Museum and Dynamax, $8.40 adults, $6.30 seniors, $3.15 children 3–11. Children under 12 must be accompanied by an adult. DISC, MC, V. Daily 9am–5pm. Closed Mon in Jan and Sept, and Dec 25.

Two life-size bronze dinosaurs stand outside the entrance to this modern museum, opposite the Albuquerque Museum. Inside, innovative video displays, polarizing lenses, and black lighting enable visitors to stroll through geologic time. You can walk a rocky path through the Hall of Giants, as dinosaurs fight and winged reptiles swoop overhead; step into a seemingly live volcano, complete with simulated magma flow; or share an Ice Age cave, festooned with stalagmites, with saber-toothed tigers and woolly mammoths. Hands-on exhibits in the Naturalist Center permit use of a video microscope, viewing of an active beehive, and participation in a variety of other activities. For an additional charge, the giant-screen Dynamax Theater puts you into the on-screen action.

There's a gift shop on the ground floor and a Subway sandwich shop on the mezzanine. A new space for changing exhibits is being completed.

Rio Grande Zoological Park

903 10th St. SW. ☎ **505/764-6200.** Admission $4.25 adults, $2.25 children and seniors, free for children 2 and under. Children under 12 must be accompanied by an adult. Mon–Fri 9am–5pm, Sat–Sun during June, July, and August 9am–6pm. Closed Jan 1, Thanksgiving, Dec. 25.

Open-moat exhibits with animals in naturalized habitats are a treat for zoo-goers. Major exhibits include the giraffes, sea lions (with underwater viewing), the cat walk, the bird show, and ape country with its gorilla and orangutans. More than 1,200 animals of 300 species live on 60 acres of riverside bosque among ancient cottonwoods. The zoo has an especially fine collection of elephants, mountain lions, koalas, reptiles, and native Southwestern species. A children's petting zoo is open during the summer. There are numerous snack bars on the zoo grounds, and La Ventana Gift Shop carries film and souvenirs.

7 Outdoor Activities

BALLOONING   Visitors not content to watch the colorful craft rise into the clear-blue skies have a choice of several hot-air balloon operators; rates start at about $135 per person:

> **Braden's Balloons Aloft,** 3212 Stanford Dr. NE (☎ **505/281-2714**).
> **Rainbow Ryders,** 10305 Nita Pl. NE (☎ **505/293-0000**).
> **World Balloon Corporation,** 4800 Eubank Blvd. NE (☎ **505/293-6800**).

The annual **Kodak Albuquerque International Balloon Fiesta** is held the first through second weekends of October (see "Frommer's Favorite Northern New Mexico Experiences," in chapter 1, and "Northern New Mexico Calendar of Events," in chapter 2, for details).

BIKING   Albuquerque is a major bicycling hub in the summer, both for road racers and mountain bikers. Bikes can be rented from **Rio Mountain Sport,** 1210 Rio Grande NW (☎ **505/766-9970**), and they come with helmets, maps, and locks. A great place to bike is Sandia Peak in Cíbola National Forest. You can't take your bike on the tram, but chairlift no. 1 is available for up- or downhill transportation with a bike. If you'd rather not rent a bike from the above-mentioned sports store, bike rentals are available at the top and bottom of the chairlift. The lift ride one-way with a bike is $6; all day with a bike will cost you $10. Helmets are mandatory. Bike maps are available; the clearly marked trails range from easy to very difficult. Mountain Bike Challenge Events are held on Sandia Peak in May, July, and August.

Down in the valley, there's a bosque trail that runs along the Rio Grande and is easily accessible to Old Town and the Biopark. To the east, the Foothills Trail runs along the base of the mountains. Across the Rio Grande, on the west mesa, Petroglyph National Park is a nice place to ride. If you're looking for more technical mountain biking, head up through Tijeras Canyon to Cedro Peak. For information about other mountainbike areas, contact the Albuquerque Convention and Visitors Bureau.

FISHING   There are no real fishing opportunities in Albuquerque as such, but there is a nearby fishing area known as **Shady Lakes.** Nestled among cottonwood trees, it's located near I-25 on Albuquerque's north side. The most common catches are rainbow trout, black bass, bluegill, and channel catfish. To reach Shady Lakes, take I-25 north to the Tramway exit. Follow Tramway Road west for a mile and then go right on NM 313 for a half mile. Call ☎ **505/898-2568** for information. **Sandia Lakes Recreational Area** (☎ **505/897-3971**), also located on NM 313, is another popular fishing spot. There is a bait and tackle shop there.

GOLF   There are quite a few public courses in the Albuquerque area. The **Championship Golf Course at the University of New Mexico,** 3601 University Blvd. SE (☎ **505/277-4546**), is one of the best in the Southwest and was rated one of the country's top 25 public links by *Golf Digest.* **Paradise Hills Golf Course,** 10035

Country Club Lane NW (☎ **505/898-7001** for tee times and information), is a popular 18-hole golf course that has recently been completely renovated.

Other Albuquerque courses to check with for tee times are **Ladera,** located at 3401 Ladera Dr. NW (☎ **505/836-4449**); **Los Altos** at 9717 Copper Ave. NE (☎ **505/298-1897**); **Puerto del Sol,** 1800 Girard Blvd. SE (☎ **505/265-5636**); and **Arroyo del Oso,** 7001 Osuna Rd. NE (☎ **505/888-8115**).

If you're willing to drive a short distance just outside Albuquerque, you can play at the **Santa Ana Golf Club at Santa Ana Pueblo,** 288 Prairie Star Rd. (P.O. Box 1736), Bernalillo, NM 87004 (☎ **505/867-9464**), which was rated by the *New York Times* as one of the best public golf courses in the country. Rentals are available (call for information), and greens fees range from $25 to $35.

In addition, **Isleta Pueblo,** 4000 Hwy. 47, has recently completed building an 18-hole golf course (☎ **505/869-0950**).

HIKING   The 1.6-million-acre **Cíbola National Forest** offers ample opportunities. In the Sandia Ranger District alone there are 16 recreation sites, though only two allow overnight camping. For details, contact **Sandia Ranger Station,** NM 337 south toward Tijeras (☎ **505/281-3304**).

Elena Gallegos/Albert G. Simms Park, near the base of the Sandia Peak Tramway at 1700 Tramway Blvd. NE (☎ **505/291-6224** or 505/768-3550), is a 640-acre mountain picnic area with hiking-trail access to the Sandia Mountain Wilderness.

HORSEBACK RIDING   There are a couple of places in Albuquerque that offer guided or unguided horseback rides. At **Sandia Trails Horse Rentals,** 10601 N. 4th St. (☎ **505/898-6970**), you'll have the opportunity to ride on Sandia Indian Reservation land along the Rio Grande. The horses are friendly and are accustomed to children. In addition, **Turkey Track Stables, Inc.,** 1306 US 66 E. Tijeras (☎ **505/281-1772**), located about 15 miles east of Albuquerque, offers rides on trails in the Manzano foothills. Riding lessons are available.

RIVER RAFTING   This sport is generally practiced farther north, in the area surrounding Santa Fe and Taos.

In mid-May each year, the **Great Race** takes place on a 14-mile stretch of the Rio Grande through Albuquerque. Eleven categories of craft, including rafts, kayaks, and canoes, and homemade craft, race down the river. Call ☎ **505/768-3490** for details.

SKIING   The **Sandia Peak Ski Area** is a good place for family skiing. There are plenty of beginner and intermediate runs. However, if you're looking for more challenge or more variety, you'd better head north to Santa Fe or Taos. The ski area has twin base-to-summit chairlifts to its upper slopes at 10,360 feet and a 1,700-foot vertical drop. There are 30 runs (35% beginner, 55% intermediate, 10% advanced) above the day lodge and ski-rental shop. Four chairs and two pomas accommodate 3,400 skiers an hour. All-day lift tickets are $32 for adults, $22 for children; rental packages are $15 for adults, $12 for kids. The season runs from mid-December to mid-March. Contact **10 Tramway Loop NE** (☎ **505/242-9133**) for more information, or call the hotline for ski conditions (☎ **505/242-9052**).

Cross-country skiers can enjoy the trails of the Sandia Wilderness from the ski area, or they can go an hour north to the remote Jemez Wilderness and its hot springs.

TENNIS   There are 29 public parks in Albuquerque with tennis courts. Because of the city's size, your best bet is to call the Albuquerque Convention and Visitors Bureau to find out which park is closest to your hotel.

8 Spectator Sports

BASEBALL   The Albuquerque Dukes, 1994 champions of the Class AAA Pacific Coast League, are a farm team of the Los Angeles Dodgers. They play 72 home games from mid-April to early September in the city-owned 10,500-seat Albuquerque Sports Stadium, 1601 Stadium Blvd. SE (at University Boulevard; ☎ **505/243-1791**).

BASKETBALL   The University of New Mexico team, nicknamed "The Lobos," plays an average of 16 home games from late November to early March. Capacity crowds cheer the team at the 17,121-seat University Arena (fondly called "The Pit") at University and Stadium boulevards. The arena was the site of the National Collegiate Athletic Association championship tournament in 1983.

FOOTBALL   The UNM Lobos football team plays a September to November season, usually with five home games, at the 30,000-seat University of New Mexico Stadium, opposite both Albuquerque Sports Stadium and University Arena at University and Stadium boulevards.

HORSE RACING   The **Downs at Albuquerque,** New Mexico State Fairgrounds (☎ **505/266-5555** for post times), is near Lomas and Louisiana boulevards NE. Racing and betting—on thoroughbreds and quarter horses—take place on weekends from October to December and during the state fair in September. The Downs has a glass-enclosed grandstand, exclusive club seating, valet parking, and complimentary racing programs and tip sheets. General admission is free; reserved second-floor seating is $2.

9 Shopping

Visitors seeking regional specialties will find many local artists and galleries of interest, although neither group is as concentrated as in Santa Fe and Taos. The galleries and regional fashion designers around the Plaza in Old Town comprise a kind of a shopping center for tourists, with more than 40 merchants represented. The Sandia Pueblo people run their own crafts market at their reservation off I-25 at Tramway Road, just beyond Albuquerque's northern city limits.

Albuquerque has the three largest shopping malls in New Mexico, two within two blocks of each other on Louisiana Boulevard just north of I-40—Coronado Center and Winrock Center. The other is on the west mesa at 10,000 Coors Blvd. NW (☎ **505/899-SHOP**).

Business hours vary, but shops are generally open Monday to Saturday from 10am to 6pm; many have extended hours; some have reduced hours; and a few, especially in shopping malls or during the high tourist season, are open on Sunday.

The Albuquerque sales tax is 5.5625%.

BEST BUYS

The best buys in Albuquerque are Southwestern regional items, including **arts and crafts** of all kinds—traditional Native American and Hispanic as well as contemporary works. In local Native American art, look for silver and turquoise jewelry, pottery, weavings, baskets, sand paintings, and Hopi kachina dolls. Hispanic folk art—handcrafted furniture, tinwork and retablos, and religious paintings—is worth seeking out. The best contemporary art is in paintings, sculpture, jewelry, ceramics, and fiber art, including weaving.

Other items of potential interest are fashions in Southwestern print designs; gourmet foods/ingredients, including blue-corn flour and chile ristras; and unique regional souvenirs, especially local Native American and Hispanic creations.

By far the greatest concentration of **galleries** is in Old Town; others are spread around the city, with smaller groupings in the university district and the northeast heights. Consult the brochure published by the Albuquerque Gallery Association, "A Select Guide to Albuquerque Galleries," or Wingspread Communications' annual *The Collector's Guide to Albuquerque,* widely distributed at shops. Once a month, usually from 5 to 9pm on the third Friday, the Albuquerque Art Business Association (☎ **505/842-9918** for information) sponsors an **ArtsCrawl** to dozens of galleries and studios. If you're in town, it's a great way to meet the artists.

You'll find some interesting shops in the Nob Hill area, which is just west of the University of New Mexico. This whole area has an art deco feel. Look for interesting finds such as **Peacecraft, Inc.,** 3107 Central NE (☎ **505/255-5229**), a nonprofit gallery where you'll find ethnic, folk, and tribal art from around the world; and **Ooh! Aah! Jewelry,** 3205 Central NE, Suite 101 (☎ **505/265-7170**), where you'll find silver jewelry made by local artisans.

ARTS & CRAFTS

Amapola Gallery
2045 S. Plaza St. (Old Town). ☎ **505-/242-4311.**

Fifty artists and craftspeople show their talents at this lovely cooperative gallery off a cobbled courtyard. You'll find pottery, paintings, textiles, carvings, baskets, jewelry, and other items.

La Piñata
No. 2 Patio Market. (Old Town). ☎ **505/242-2400.**

This shop features what else?—piñatas, in shapes from dinosaurs to parrots to pigs, as well as paper flowers, puppets, toys, and crushable bolero hats decorated with ribbons.

Mariposa Gallery
113 Romero St. NW (Old Town). ☎ **505/842-9097.**

Fine contemporary crafts, including fiber arts, jewelry, clay works, sculptural glass, and other media, are sold here.

Mineral and Fossil Gallery
2011 Mountain Rd. NW, Suite E-1. ☎ **800/354-6213** or 505/843-8297.

A great place to find natural art, from fossils to geodes to cave bear skeletons.

Tanner Chaney Galleries
410 Romero NW (Old Town). ☎ **800/444-2242** or 505/247-2242.

In business since 1875, this gallery has fine jewelry, pottery, and rugs.

V. Whipple's Old Mexico Shop
400-E San Felipe St. NW (Patio del Norte; Old Town). ☎ **505/243-6070.**

Here you'll find a variety of Mexican imports, including folk art, tin mirrors, talavera ceramic works, and glassware.

FOOD

The Candy Lady
524 Romero NW (Old Town). ☎ **800/214-7731** or 505/243-6239.

Making chocolate for over 17 years, The Candy Lady is especially known for 21 varieties of fudge, including jalapeño flavor.

SOUTHWESTERN APPAREL

Jeanette's Original's
205 San Felipe NW, #7 (Old Town). ☎ **505/243-5905.**

Stop by here for those all-important, matching mother-daughter fashions, as well as men's Navajo and ribbon shirts.

10 Albuquerque After Dark

Albuquerque has an active performing-arts and nightlife scene, as befits a city of half a million people. As also befits this area, the performing arts are multicultural, with Hispanic and (to a lesser extent) Native American productions sharing stage space with Anglo works, including theater, opera, symphony, and dance. Albuquerque also attracts many national touring companies. Country music predominates in nightclubs, though aficionados of rock, jazz, and other forms of music can find them here as well.

Complete information on all major cultural events can be obtained from the **Albuquerque Convention and Visitors Bureau** (☎ **800/284-2282** for recorded information after 5pm). Current listings appear in the two daily newspapers; detailed weekend arts calendars can be found in the Thursday evening *Tribune* and the Friday morning *Journal.* The monthly *On the Scene* also carries entertainment listings.

Tickets for nearly all major entertainment and sporting events can be obtained from TicketMaster, 4004 Carlisle Blvd. NE (☎ **505/884-0999** for information, or 505/842-5387 to place credit- or charge-card orders on American Express, MasterCard, or Visa). Discount tickets are often available for midweek and matinee performances. Check with specific theaters or concert halls.

THE PERFORMING ARTS
CLASSICAL MUSIC

Chamber Orchestra of Albuquerque
2730 San Pedro Dr. NE, Suite H-23. ☎ **505/881-0844.** Tickets, $12–$22, depending on seating and performance.

This 31-member professional orchestra, conducted by music director David Oberg, performs from September to June, primarily at St. John's United Methodist Church, 2626 Arizona St. NE. There is a subscription series of six classical concerts (in October, November, January, March, May, and June), an all-baroque concert in February, concerts for children in February and April, and a joint concert with the University of New Mexico Chorus. The orchestra regularly features guest artists of national and international renown.

New Mexico Symphony Orchestra
3301 Menaul Blvd. NE, Suite 4. ☎ **800/251-6676** or 505/881-9590 for tickets and information. Ticket prices vary by concert; call for details. AE, MC, V.

NMSO musicians may be the busiest performing artists in New Mexico. During its 1997–98 season, the orchestra will perform about 30 classical, pops, baroque, and family concerts from September to May. Concert venues range from Popejoy Hall on the University of New Mexico campus to the 2,500-seat Hoffmantown Baptist Church and the outdoor band shell at the Rio Grande Zoo.

DANCE

New Mexico Ballet Company
3620 Wyoming Blvd. NE (P.O. Box 21518). ☎ **505/292-4245.** Tickets, $10–$16 adults, $5–$8 students.

The Major Concert & Performance Halls

Keller Hall, University of New Mexico, Cornell Street at Redondo Drive South (☎ 505/277-4569).

KiMo Theatre, 423 Central Ave. (☎ 505/764-1700).

Popejoy Hall, University of New Mexico, Cornell Street at Redondo Drive South (☎ 505/277-4569).

South Broadway Cultural Center, 1025 Broadway Blvd. (☎ 505/848-1320).

Founded in 1972, the state's oldest ballet company performs an October-to-April season at Popejoy Hall. Typically there is a fall production such as *The Legend of Sleepy Hollow,* a December performance of *The Nutcracker* or *A Christmas Carol,* and a contemporary spring production.

THEATER

Albuquerque Civic Light Opera Association

4201 Ellison Rd. NE. ☎ **505/345-6577.** Tickets, $10–$18.50 adults, $8–$16.50 students and seniors.

Light opera here is in the form of five major Broadway musicals presented each year at Popejoy Hall and the Hiland Theater during a March-to-December season. Each production is staged for two to three consecutive weekends, including two Sunday matinees.

Albuquerque Little Theatre

224 San Pasquale Ave. SW. ☎ **505/242-4750.**

The Albuquerque Little Theatre has been offering a variety of productions ranging from comedies to dramas to musicals since 1930. Six plays are presented here annually during a September-to-May season. Located across from Old Town, Albuquerque Little Theatre offers plenty of free parking.

La Compañía de Teatro de Albuquerque

518 First St. NW. ☎ **505/242-7929.** Tickets, $9 adults Thurs and Sun, $10 Fri–Sat; $8 students, seniors, and children Thurs and Sun, $9 Fri–Sat.

One of the few major professional Hispanic companies in the United States and Puerto Rico, La Compañía stages a series of productions every year between October and June. Comedies, dramas, and musicals are offered, along with one Spanish-language play a year.

Vortex Theatre

Buena Vista (just South of Central Ave.). ☎ **505/247-8600.** Tickets, $8 adults, $7 students and seniors, $6 children 13 and under; $6 for everyone on Sun.

A 19-year-old community theater known for its innovative productions, the Vortex is Albuquerque's "Off-Broadway" theater, presenting a range of plays from classic to original. The company mounts 10 shows a year. Performances take place on Friday and Saturday at 8pm and on Sunday at 6pm. The black-box theater seats 90.

THE CLUB & MUSIC SCENE
COMEDY CLUBS/DINNER THEATER

Laffs Comedy Caffé

3100-D Juan Tabo Blvd. (at Candelaria Rd. NE). ☎ **505/296-5653.**

Top acts from each coast, including comedians who have appeared on *The Late Show with David Letterman* and HBO, are booked at Albuquerque's top comedy club. Shows Wednesday, Thursday, and Sunday begin at 8:30pm ($5 per person with a two-item minimum purchase); on Friday and Saturday, shows start at 8:30 and 10:30pm ($7 per person with a two-item minimum purchase). Wednesday is non-smoking night. The club serves a full dinner menu with all items under $10. You must be 21 or older to attend.

Mystery Cafe
In La Posada de Albuquerque, 125 Second St. NW. ☎ **505/237-1385.**

If you're in the mood for a little interactive dinner theater, the Mystery Cafe might be just the ticket. You'll help the characters in this ever-popular, delightfully funny show solve the mystery as they serve you a four-course meal. Reservations are a must. Performances are Friday and Saturday evenings at 7pm and cost approximately $35.

COUNTRY MUSIC

Midnight Rodeo
4901 McLeod Rd. NE (near San Mateo Blvd.). ☎ **505/888-0100.** No cover Sun–Thurs, $3 Fri–Sat.

The Southwest's largest nightclub of any kind, there are bars in all corners of this enormous venue; it even has its own shopping arcade, including a boutique and gift shop. A deejay spins records daily until closing; the hardwood dance floor is so big (5,000 square feet) that it resembles an indoor horse track. Free dance lessons are offered on Sunday from 5:30 to 7pm and Thursday from 7 to 8pm. A busy kitchen serves simple but hearty meals to dancers who work up appetites.

ROCK/JAZZ

In recent years an interesting club scene has opened up downtown. Almost any night of the week, but particularly Thursday through Saturday nights, the place is hopping, as people wander from one club to another. The 20-something crowd should try **The Zone and The Z-Pub** (that's one joint) at 120 Central SW (☎ **505/343-7933**).

Brewsters Pub
312 Central Ave. SW (Downtown). ☎ **505/247-2533.**

Wednesday through Saturday nights, Brewsters Pub offers live blues or light rock entertainment in a "sports bar"–type setting. There are 29 beers on tap, as well as a wide variety of bottled beer. Sports fans can enjoy the game on a big-screen TV. Barbecue is served at lunch and dinner.

The Cooperage
7220 Lomas Blvd. NE. ☎ **505/255-1657.** Cover $3–$5.

Jazz, rhythm and blues, rock, and salsa keep dancers hopping on Friday and Saturday nights inside this gigantic wooden barrel.

✪ Dingo Bar
313 Gold Ave. SW (Downtown). ☎ **505/243-0663.** Cover charge varies with performance, but can run up to $20 per person.

The Dingo Bar is one of Albuquerque's premier rock clubs. Nightly live entertainment runs from punk rock to classic rock 'n' roll and jazz. The atmosphere is earthy, with patrons ranging from hippies to suits.

MORE ENTERTAINMENT

Albuquerque's best nighttime attraction is the **Sandia Peak Tramway** (see "What to See & Do," above) and the restaurant High Finance at the summit (see "Where to

Dine," above). Here you can enjoy a view nonpareil of the Rio Grande Valley and the city lights.

The best place to catch foreign films, art films, and limited-release productions is the **Guild Cinema,** 3405 Central Ave. NE (☎ **505/255-1848**). For film classics, check out the **Southwest Film Center,** on the UNM campus (☎ **505/277-5608**), with double features Wednesday through Sunday, changing nightly.

Major Albuquerque first-run theaters include the **Coronado Six Theater,** 6401 Uptown Blvd. NE (☎ **505/881-5266**); **Del Norte Cinema Four,** 7120 Wyoming Blvd. NE (☎ **505/823-6666**); **United Artists Four Hills 10 Theater,** 13160 Central Ave. SE, at Tramway Boulevard (☎ **505/275-2114**); **Ladera Six Cinema,** 3301 Coors Blvd. NW (☎ **505/836-5606**); **General Cinemas San Mateo Cinema 8,** 631 San Mateo Blvd. NE (☎ **505/889-3051**); **United Artists 8 at High Ridge,** Tramway Boulevard (at Indian School Road; (☎ **505/275-0038**); and **Winrock 6 UA Cinema,** 201 Winrock Center NE (☎ **505/883-6022**).

The **Isleta Gaming Palace,** 11,000 Broadway SE (☎ **800/460-5686** or 505/ 869-2614), is a luxurious, air-conditioned casino (blackjack, poker, slots, bingo, and keno) with a full-service restaurant, no-smoking section, and free bus transportation on request. It's open 24 hours a day.

11 Exploring Nearby Pueblos & Monuments

Ten Native American pueblos are located within an hour's drive from central Albuquerque. One national and two state monuments preserve another five ancient pueblo ruins.

The active pueblos nearby include Acoma, Cochiti, Isleta, Jemez, Laguna, Sandia, San Felipe, Santa Ana, Santo Domingo, and Zia. Of these, Acoma is the most prominent.

When you visit pueblos, it is important to observe certain **rules of etiquette:** Remember to respect the pueblos as people's homes; don't peek into doors and windows or climb on top of the buildings. Stay out of cemeteries and ceremonial rooms (such as kivas), since these are sacred grounds. Don't speak during dances or ceremonies or applaud after their conclusion; silence is mandatory. Most pueblos require a permit to carry a camera or to sketch or paint on location. Several pueblos prohibit picture taking at any time.

✪ Acoma Pueblo

To reach Acoma from Albuquerque, drive west on I-40 approximately 52 miles to the Acoma–Sky City exit (exit 108), then about 12 miles southwest.

The spectacular "Sky City," a walled adobe village perched high atop a sheer rock mesa 365 feet above the 6,600-foot valley floor, is believed to have been inhabited at least since the 11th century—the longest continuously occupied community in the United States. Native legend claims that it has been inhabited since before the time of Christ. Both the pueblo and **San Estevan del Rey Mission** are National Historic Landmarks.

The Keresan-speaking Acoma (Ack-oo-mah) Pueblo boasts about 6,005 inhabitants, but only about 50 people reside year-round on the 70-acre mesa top. They make their living from tourists who come to see the large church containing examples of Spanish colonial art and to purchase the pueblo's thin-walled white pottery with polychrome designs.

The pueblo's address is P.O. Box 309, Acoma, NM 87034 (☎ **800/747-0181** or 505/470-4966). The admission charge is $7 for adults, $6 for seniors, $5 for children

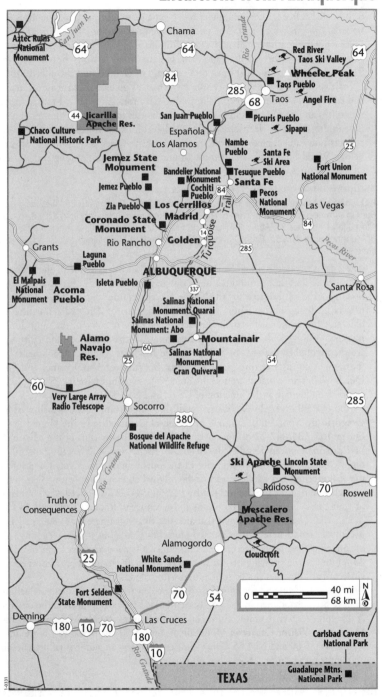

6 through 17, free for children under 6. The charge to take still photographs is $10; *no video cameras are allowed.* The pueblo is open daily in the summer from 8am to 7pm; daily in the spring, fall, and winter from 8am to 4:30pm. One-hour tours begin every 30 minutes; the last tour is scheduled 1 hour before closing. The pueblo is closed the first or second weekend in October and also July 10 to 13.

Start your tour at the **visitor center** at the base of the mesa. There are a **museum** and **cafe** here. A 16-seat **tour bus** climbs through a rock garden of 50-foot sandstone monoliths and past precipitously dangling outhouses to the mesa's summit. There's no running water or electricity in this medieval-looking village; a small reservoir collects rainwater for most purposes, but drinking water is transported up from below. Wood-hole ladders and mica windows are prevalent among the 300-odd adobe structures.

Salinas Pueblo Missions National Monument

P.O. Box 517, Mountainair, NM 87036. ☎ **505/847-2585.** Free admission. Sites, daily 9am–6pm in summer, 9am–5pm the rest of the year. Visitor center in Mountainair, daily 8am–5pm. Closed Jan 1 and Dec 25. Abo is 9 miles west of Mountainair on US 60. Quarai is 8 miles north of Mountainair on NM 55. Gran Quivira is 25 miles south of Mountainair on NM 55. All roads are paved.

The Spanish conquistadors' Salinas Jurisdiction, on the east side of the Manzano Mountains (southeast of Albuquerque), was an important 17th-century trade center because of the salt extracted by the Native Americans from the salt lakes. Franciscan priests, utilizing native labor, constructed missions of Abo red sandstone and blue-gray limestone for the native converts. The ruins of some of the most durable missions—along with evidence of preexisting Anasazi and Mogollon cultures— are the highlights of a visit to Salinas Pueblo Missions National Monument. The monument consists of three separate units: the ruins of Abo, Quarai, and Gran Quivira. They are situated around the quiet town of **Mountainair,** 75 miles southeast of Albuquerque at the junction of US 60 and NM 55.

Abo (☎ **505/847-2400**) boasts the 40-foot-high ruins of the **Mission of San Gregorio de Abo,** a rare example of medieval architecture in the United States. Quarai (☎ **505/847-2290**) preserves the largely intact remains of the **Mission of La Purísima Concepción de Cuarac** (1630). Its vast size, 100 feet long and 40 feet high, contrasts with the modest size of the pueblo mounds. A small museum in the visitor center has a scale model of the original church, along with a selection of artifacts found at the site. **Gran Quivira** (☎ **505/847-2770**) once had a population of 1,500. Las Humanes has 300 rooms and seven kivas. Rooms dating back to 1300 can be seen. There are indications that an older village, dating back to 800, may have previously stood here. Ruins of two churches (one almost 140 feet long) and a Convento have been preserved. A museum with many artifacts from the site, a 40-minute movie showing the excavation of some 200 rooms, plus a short history video of Las Humanes can be seen at the visitor center.

All three pueblos and the churches that were constructed above them are believed to have been abandoned in the 1670s. Self-guided tour pamphlets can be obtained at the units' respective visitor centers and at the **Salinas Pueblo Missions National Monument Visitor Center** in Mountainair, on US 60 one block west of the intersection of US 60 and NM 55. The visitor center offers an audiovisual presentation on the region's history, a bookstore, and an art exhibit.

Coronado State Monument

NM 44 (P.O. Box 95), Bernalillo, NM 87004. ☎ **505/867-5351.** Admission $3 adults, free for children 16 and under. Daily 8am–5pm. Closed major holidays. To get to the site (20 miles north of Albuquerque), take I-25 to Bernalillo and NM 44 west.

When the Spanish explorer Coronado traveled through this region in 1540–41 while searching for the Seven Cities of Cíbola, he wintered at a village on the west bank of the Rio Grande—probably one located on the ruins of the ancient Anasazi Pueblo known as Kuaua. Those excavated ruins have been preserved in this state monument.

Hundreds of rooms can be seen, and a kiva has been restored so that visitors can descend a ladder into the enclosed space, once the site of sacred rites. Unique multicolored murals, depicting human and animal forms, were found on successive layers of wall plaster in this and other kivas here; some examples are displayed in the monument's small archaeological museum.

Jemez State Monument

NM 4 (P.O. Box 143), Jemez Springs, NM 87025. ☎ **505/829-3530.** Admission $3 adults, free for children 16 and under. Daily 8am–5pm. Closed Jan 1, Easter Sunday, Thanksgiving, Dec 25. From Albuquerque, take NM 44 to NM 4, and then continue on NM 4 for about 18 miles.

All that's left of the Mission of San José de los Jemez, founded by Franciscan missionaries in 1621, is preserved at this site. Visitors will find massive walls standing alone; sparse small door and window openings underscore the need for security and permanence in those times. The mission was excavated between 1921 and 1937, along with portions of a prehistoric Jemez Pueblo. The pueblo, near the Jemez Hot Springs, was called Giusewa—"place of the boiling waters."

A small **museum** at the site displays artifacts found during the excavation, describes traditional crafts and foods, and weaves the thread of history of the Jemez peoples to the 21st century in a series of exhibits. An instructional trail winds through the ruins.

✪ THE TURQUOISE TRAIL

Best known as "The Turquoise Trail," NM 14 begins about 16 miles east of downtown Albuquerque, at I-40's Cedar Crest exit, and winds some 46 miles to Santa Fe along the east side of the Sandia Mountains. This state-designated scenic and historic route traverses the revived "ghost towns" of Golden, Madrid, and Cerrillos, where gold, silver, coal, and turquoise were once mined in great quantities. Modern-day settlers, mostly artists and craftspeople, have brought a renewed frontier spirit to the old mining towns.

GOLDEN   Golden is approximately 10 miles north of the Sandia Park junction on NM 14. Its sagging houses, with their missing boards and the wind whistling through the broken eaves, make it a purist's ghost town. There's a general store widely known for its large selection of well-priced jewelry, as well as a bottle seller's "glass garden." Nearby are the ruins of a pueblo called **Paako,** abandoned around 1670. Such communities of mud huts were all that the Spaniards ever found during their avid quest for the gold of Cíbola.

MADRID   Madrid (pronounced with the accent on the first syllable) is about 12 miles north of Golden. Madrid and neighboring Cerrillos were in a fabled turquoise-mining area dating back to prehistory. Gold and silver mines followed, and when they faltered, there was coal. The Turquoise Trail towns supplied fuel for the locomotives of the Santa Fe Railroad until the 1950s, when the railroad converted to diesel fuel. Madrid used to produce 100,000 tons of coal a year, but the mine closed in 1956. Today this is a village of artists and craftspeople seemingly stuck in the 1960s: Its funky, ramshackle houses have many counterculture residents who operate several crafts stores and import shops.

The **Old Coal Mine Museum** (☎ 505/473-0743) invites visitors to go down into a real mine, saved when the town was abandoned. You can see the old mine's offices, steam engines, machines, and tools. It's called a "living" museum because blacksmiths, metalworkers, and leatherworkers ply their trades here in restoring parts and tools found in the mine. It's open daily; admission is $3 for adults and seniors, and $1 for children 6 to 12; children under 6 are free.

Next door, the **Mine Shaft Tavern** continues its colorful career by offering a variety of burgers on the menu and presenting live music Saturday and Sunday afternoons and some Friday and Saturday nights; it attracts folks from Santa Fe and Albuquerque. Next door is the **Madrid Opera House,** possibly the only such establishment on earth with a built-in steam locomotive on its stage. (The structure had been an engine repair shed; the balcony is made of railroad track.)

CERRILLOS   Cerrillos, about 3 miles north of Madrid, is a village of dirt roads that sprawls along Galisteo Creek. It appears to have changed very little since it was founded during a lead strike in 1879; the old hotel, the saloon, and even the sheriff's office look very much like an Old West movie set. It's another 15 miles to Santa Fe and I-25.

Appendix: Useful Toll-Free Numbers & Web Sites

AIRLINES

Air Canada
☎ 800/776-3000
www.aircanada.ca

Alaska Airlines
☎ 800/426-0333
www.alaskaair.com

America West Airlines
☎ 800/235-9292
www.americawest.com

American Airlines
☎ 800/433-7300
www.americanair.com

British Airways
☎ 800/247-9297
☎ 0345/222-111 in Britain
www.british-airways.com

Canadian Airlines International
☎ 800/426-7000
www.cdair.ca

Carnival Airlines
☎ 800/824-7386
www.carnivalair.com

Continental Airlines
☎ 800/525-0280
www.flycontinental.com

Delta Air Lines
☎ 800/221-1212
www.delta-air.com

Hawaiian Airlines
☎ 800/367-5320
www.hawaiianair.com

Kiwi International Air Lines
☎ 800/538-5494
www.jetkiwi.com

Midway Airlines
☎ 800/446-4392

Northwest Airlines
☎ 800/225-2525
www.nwa.com

Southwest Airlines
☎ 800/435-9792
iflyswa.com

Tower Air
☎ 800/34-TOWER (800/348-6937) outside New York
(☎ 718/553-8500 in New York)
www.towerair.com

Trans World Airlines (TWA)
☎ 800/221-2000
www2.twa.com

United Airlines
☎ 800/241-6522
www.ual.com

US Airways
☎ 800/428-4322
www.usairways.com

Virgin Atlantic Airways
☎ 800/862-8621 in Continental U.S.
☎ 0293/747-747 in Britain
www.fly.virgin.com

CAR RENTAL AGENCIES

Advantage
☎ 800/777-5500
www.arac.com

Alamo
☎ 800/327-9633
www.goalamo.com

Avis
☎ 800/331-1212 in the Continental U.S.
☎ 800/TRY-AVIS in Canada
www.avis.com

Budget
☎ 800/527-0700
www.budgetrentacar.com

Dollar
☎ 800/800-4000

Enterprise
☎ 800/325-8007

Hertz
☎ 800/654-3131
www.hertz.com

National
☎ 800/CAR-RENT
www.nationalcar.com

Payless
☎ 800/PAYLESS
www.paylesscar.com

Rent-A-Wreck
☎ 800/535-1391
rent-a-wreck.com

Thrifty
☎ 800/367-2277
www.thrifty.com

Value
☎ 800/327-2501
www.go-value.com

MAJOR HOTEL & MOTEL CHAINS

Best Western International
☎ 800/528-1234
www.bestwestern.com

Clarion Hotels
☎ 800/CLARION
www.hotelchoice.com/cgi-bin/res/webres?clarion.html

Comfort Inns
☎ 800/228-5150
www.hotelchoice.com/cgi-bin/res/webres?comfort.html

Courtyard by Marriott
☎ 800/321-2211
www.courtyard.com

Days Inn
☎ 800/325-2525
www.daysinn.com

Doubletree Hotels
☎ 800/222-TREE
www.doubletreehotels.com

Econo Lodges
☎ 800/55-ECONO
www.hotelchoice.com/cgi-bin/res/webres?econo.html

Fairfield Inn by Marriott
☎ 800/228-2800
www.fairfieldinn.com

Hampton Inn
☎ 800/HAMPTON
www.hampton-inn.com

Hilton Hotels
☎ 800/HILTONS
www.hilton.com

Holiday Inn
☎ 800/HOLIDAY
www.holiday-inn.com

Howard Johnson
☎ 800/654-2000
www.hojo.com/hojo.html

Hyatt Hotels & Resorts
☎ 800/228-9000
www.hyatt.com

ITT Sheraton
☎ 800/325-3535
www.sheraton.com

La Quinta Motor Inns
☎ 800/531-5900
www.laquinta.com

Marriott Hotels
☎ 800/228-9290
www.marriott.com

Motel 6
☎ 800/4-MOTEL6 (800/466-8536)

Quality Inns
☎ 800/228-5151
www.hotelchoice.com/cgi-bin/res/webres?quality.html

Radisson Hotels International
☎ 800/333-3333
www.radisson.com

Ramada Inns
☎ 800/2-RAMADA
www.ramada.com

Red Carpet Inns
☎ 800/251-1962

Red Lion Hotels & Inns
☎ 800/547-8010
www.travelweb.com

Red Roof Inns
☎ 800/843-7663
www.redroof.com

Residence Inn by Marriott
☎ 800/331-3131
www.residenceinn.com

Rodeway Inns
☎ 800/228-2000
www.hotelchoice.com/cgi-bin/res/webres?rodeway.html

Super 8 Motels
☎ 800/800-8000
www.super8motels.com

Travelodge
☎ 800/255-3050

Vagabond Hotels
☎ 800/255-3050
www.vagabondinns.com

Index

See also separate Accommodations and Restaurant indexes, below.

SANTA FE ACCOMMODATIONS

SANTA FE RESTAURANTS

TAOS ACCOMMODATIONS

TAOS RESTAURANTS

ALBUQUERQUE ACCOMMODATIONS

FROMMER'S COMPLETE TRAVEL GUIDES
(Comprehensive guides to destinations around the world, with selections in all price ranges—from deluxe to budget)

Acapulco, Ixtapa & Zihuatenejo
Alaska
Amsterdam
Arizona
Atlanta
Australia
Austria
Bahamas
Barcelona, Madrid & Seville
Belgium, Holland & Luxembourg
Bermuda
Boston
Budapest & the Best of Hungary
California
Canada
Cancún, Cozumel & the Yucatán
Cape Cod, Nantucket & Martha's Vineyard
Caribbean
Caribbean Cruises & Ports of Call
Caribbean Ports of Call
Carolinas & Georgia
Chicago
Colorado
Costa Rica
Denver, Boulder & Colorado Springs
England

Europe
Florida
France
Germany
Greece
Hawaii
Hong Kong
Honolulu, Waikiki & Oahu
Ireland
Israel
Italy
Jamaica & Barbados
Japan
Las Vegas
London
Los Angeles
Maryland & Delaware
Maui
Mexico
Miami & the Keys
Montana & Wyoming
Montréal & Québec City
Munich & the Bavarian Alps
Nashville & Memphis
Nepal
New England
New Mexico
New Orleans
New York City
Northern New England
Nova Scotia, New Brunswick & Prince Edward Island
Paris

Philadelphia & the Amish Country
Portugal
Prague & the Best of the Czech Republic
Provence & the Riviera
Puerto Rico
Rome
San Antonio & Austin
San Diego
San Francisco
Santa Fe, Taos & Albuquerque
Scandinavia
Scotland
Seattle & Portland
South Pacific
Spain
Switzerland
Thailand
Tokyo
Toronto
Tuscany & Umbria
U.S.A.
Utah
Vancouver & Victoria
Vienna & the Danube Valley
Virgin Islands
Virginia
Walt Disney World & Orlando
Washington, D.C.
Washington & Oregon

FROMMER'S DOLLAR-A-DAY BUDGET GUIDES
(The ultimate guides to low-cost travel)

Australia from $50 a Day
Berlin from $50 a Day
California from $60 a Day
Caribbean from $60 a Day
Costa Rica & Belize from $35 a Day
England from $60 a Day
Europe from $50 a Day
Florida from $50 a Day
Greece from $50 a Day
Hawaii from $60 a Day

India from $40 a Day
Ireland from $45 a Day
Israel from $45 a Day
Italy from $50 a Day
London from $60 a Day
Mexico from $35 a Day
New York from $75 a Day
New Zealand from $50 a Day
Paris from $70 a Day
San Francisco from $60 a Day
Washington, D.C., from $50 a Day

FROMMER'S PORTABLE GUIDES

(Pocket-size guides for travelers who want everything in a nutshell)

Charleston & Savannah
Dublin
Las Vegas
Maine Coast

New Orleans
Puerto Vallarta,
　Manzanillo &
　Guadalajara

San Francisco
Venice
Washington, D.C.

FROMMER'S FAMILY GUIDES

(The complete guides for successful family vacations)

California with Kids
Los Angeles with Kids

New York City with Kids
San Francisco with Kids

Washington, D.C.,
with Kids

FROMMER'S AMERICA ON WHEELS

*(Everything you need for a successful road trip, including full-color
road maps and ratings for every hotel)*

California & Nevada
Florida
Great Lake States &
　Midwest

Mid-Atlantic
New England & New York
Northwest & Great Plains

South-Central States
　& Texas
Southeast
Southwest

FROMMER'S WALKING TOURS

(Memorable neighborhood strolls through the world's great cities)

London
New York
Paris

San Francisco
Spain's Favorite Cities
Tokyo

Venice
Washington, D.C.

SPECIAL-INTEREST TITLES

Arthur Frommer's Branson!
Arthur Frommer's New World of Travel
The Civil War Trust's Official Guide to
　the Civil War Discovery Trail
Frommer's America's 100 Best-Loved
　State Parks
Frommer's Caribbean Hideaways
Frommer's Complete Hostel Vacation Guide
　to England, Scotland & Wales
Frommer's Europe's Greatest
　Driving Tours
Frommer's Food Lover's Companion
　to France
Frommer's Food Lover's Companion to Italy

New York Times Weekends
Outside Magazine's Adventure Guide
　to New England
Outside Magazine's Adventure Guide
　to Northern California
Outside Magazine's Adventure Guide
　to the Pacific Northwest
Outside Magazine's Guide
　to Family Vacations
Places Rated Almanac
Retirement Places Rated
Wonderful Weekends from NYC
Wonderful Weekends from San Francisco

FROMMER'S IRREVERENT GUIDES

(Wickedly honest guides for sophisticated travelers)

Amsterdam
Chicago
London

Manhattan
Miami
New Orleans

Paris
San Francisco
Santa Fe

U.S. Virgin Islands
Walt Disney World
Washington, D.C.

UNOFFICIAL GUIDES

(Get the unbiased truth from these candid, value-conscious guides)

Atlanta
Branson, Missouri
Chicago
Cruises
Disneyland

The Great Smoky
　& Blue Ridge
　Mountains
Las Vegas

Miami & the Keys
Mini-Mickey
New Orleans
Skiing in the West

Walt Disney World
Walt Disney World
　Companion
Washington, D.C.

BAEDEKER

(With four-color photographs and a free pull-out map)

Amsterdam	Crete	Lisbon	Scandinavia
Athens	Florence	London	Scotland
Austria	Florida	Mexico	Singapore
Bali	Germany	New York	South Africa
Belgium	Great Britain	New Zealand	Spain
Berlin	Greece	Paris	Switzerland
Brazil	Greek Islands	Portugal	Thailand
Budapest	Hawaii	Prague	Tokyo
California	Hong Kong	Provence	Turkish Coast
Canada	Ireland	Rome	Tuscany
Caribbean	Israel	San Francisco	Venice
China	Italy	St. Petersburg	Vienna
Copenhagen			

FROMMER'S BY NIGHT GUIDES

(The series for those who know that life begins after dark)

Amsterdam	Los Angeles	Manhattan	Paris
Chicago	Madrid	Miami	Prague
Las Vegas	& Barcelona	New Orleans	San Francisco
London			Washington, D.C.

FROMMER'S BEST BEACH VACATIONS

(The top places to sun, stroll, shop, stay, play, party, and swim, with ratings for each beach)

California	Florida	Mid-Atlantic
Carolinas & Georgia	Hawaii	New England

FROMMER'S BED & BREAKFAST GUIDES

(Selective guides with four-color photos and full descriptions of the best inns in each region)

California	Mid-Atlantic	The Rockies
Caribbean	New England	Southeast
Great American Cities	Pacific Northwest	Southwest
Hawaii		

FROMMER'S DRIVING TOURS

(Four-color photos and detailed maps outlining spectacular scenic driving routes)

America	California	Ireland	Scotland
Australia	Florida	Italy	Spain
Austria	France	New England	Switzerland
Britain	Germany	Scandinavia	Western Europe

FROMMER'S BORN TO SHOP

(The ultimate guides for travelers who love to shop)

Caribbean Ports	Great Britain	London	New York
of Call	Hong Kong	Mexico	Paris
France	Italy	New Egnland	

TRAVEL & LEISURE GUIDES

(Sophisticated pocket-size guides for discriminating travelers)

Amsterdam	Hong Kong	New York	San Francisco
Boston	London	Paris	Washington, D.C.

WHEREVER YOU TRAVEL, *H*ELP IS NEVER FAR AWAY.

From planning your trip to providing travel assistance along the way, American Express® Travel Service Offices are always there to help you do more.

Santa Fe

Pajarito Travel (R)
2801 Rodeo Road
Suite B
Santa Fe
505/474-7177